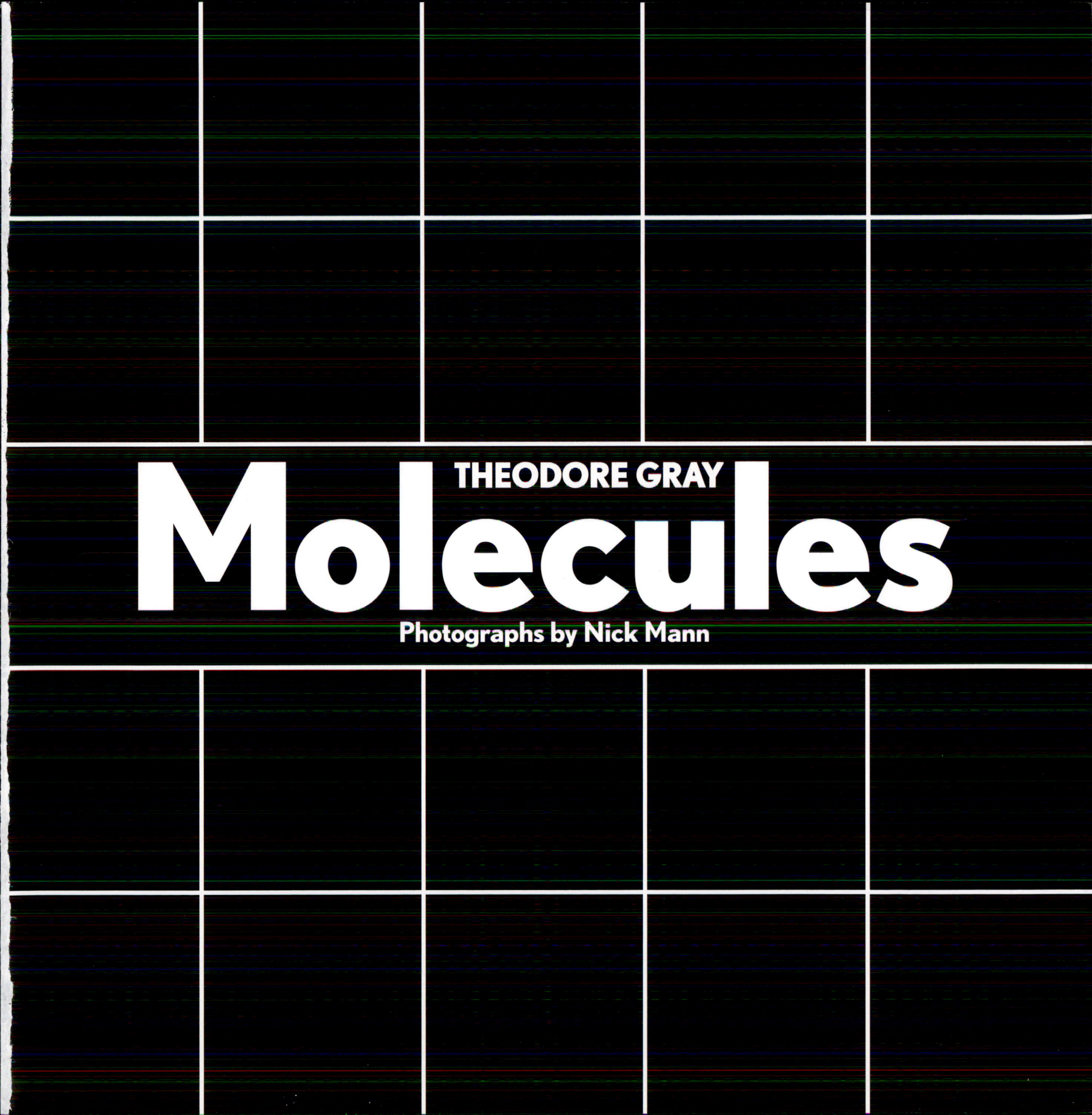

THEODORE GRAY

Molecules

Photographs by Nick Mann

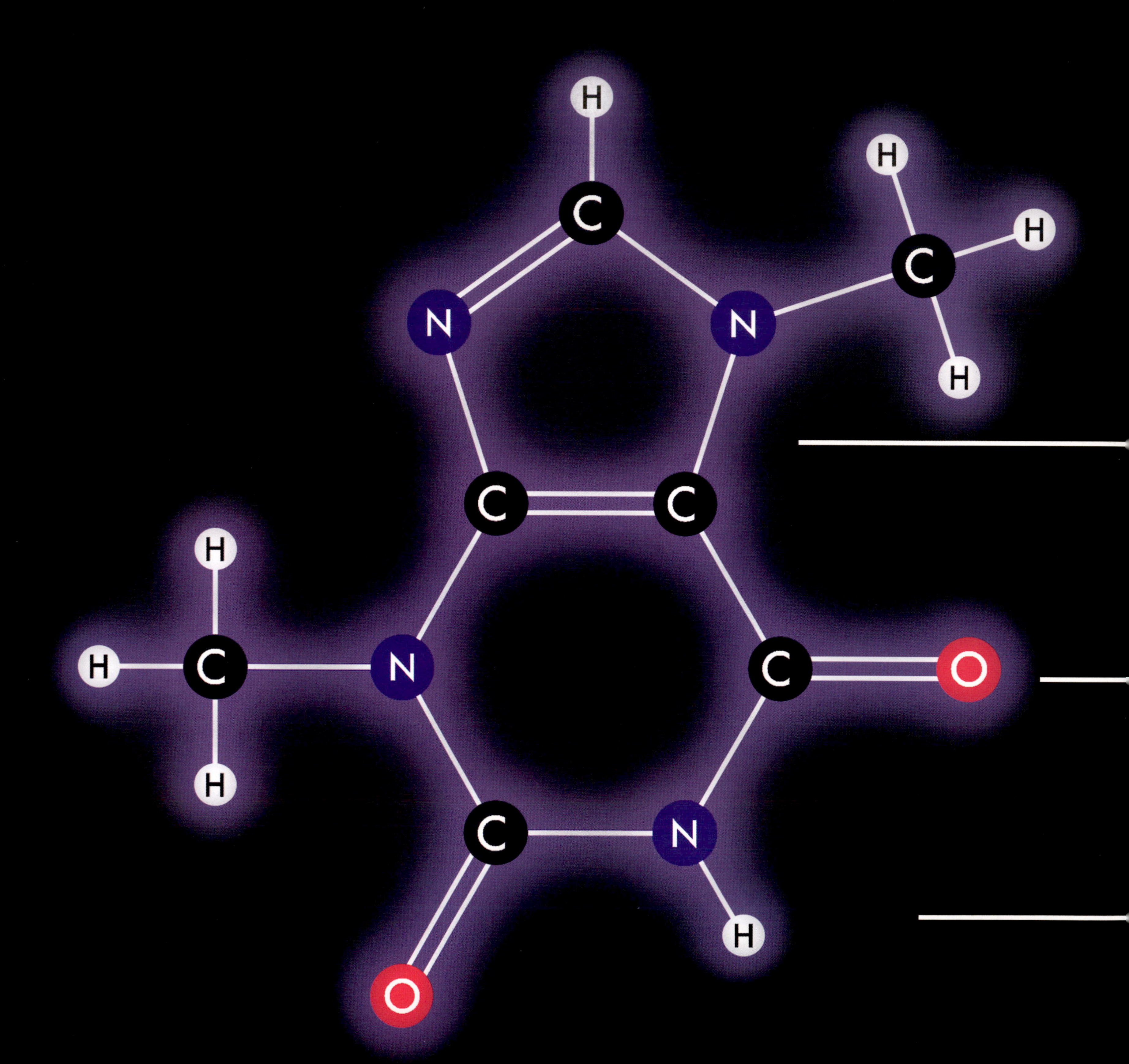
H
H
C
H
C
N
N
H
C
C
H
H
C
N
C
O
H
C
N
H
O

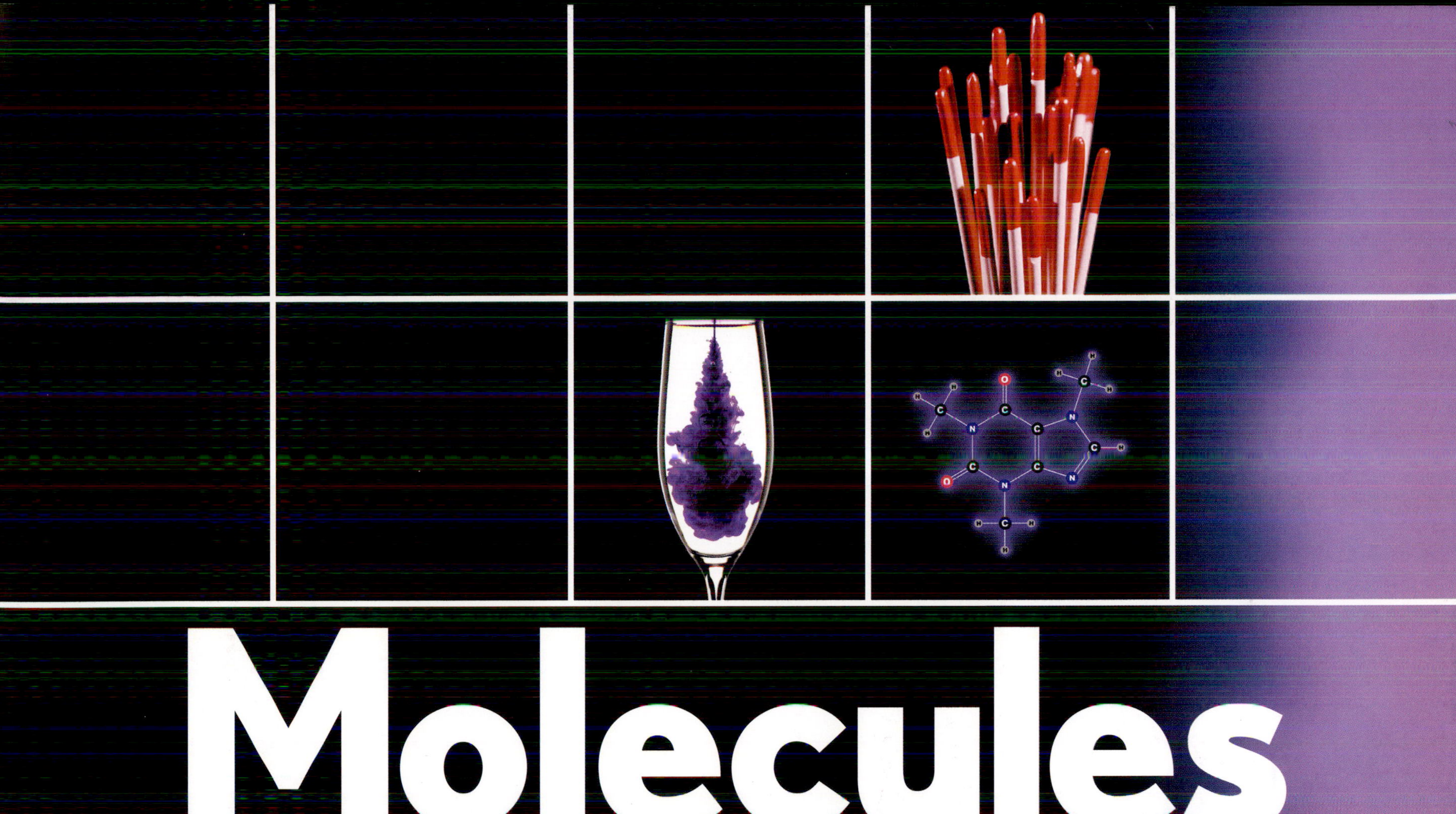

Molecules

The Elements and the Architecture of Everything

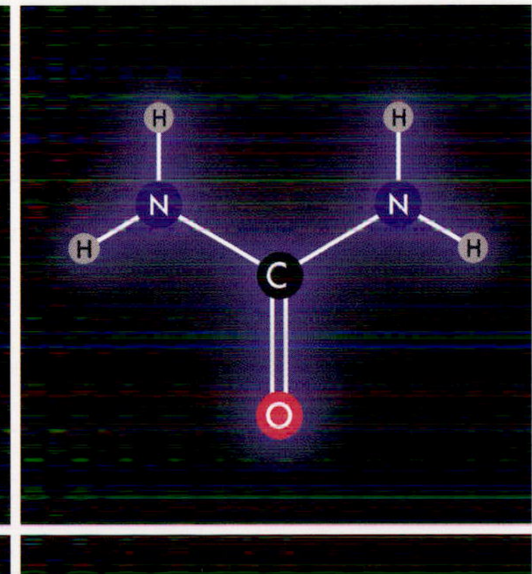

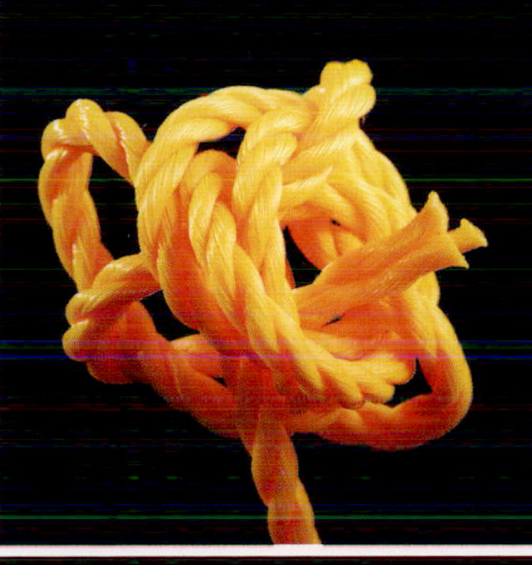

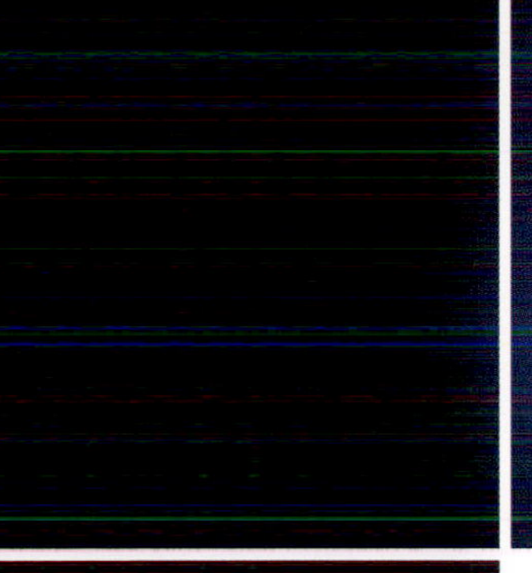

THEODORE GRAY

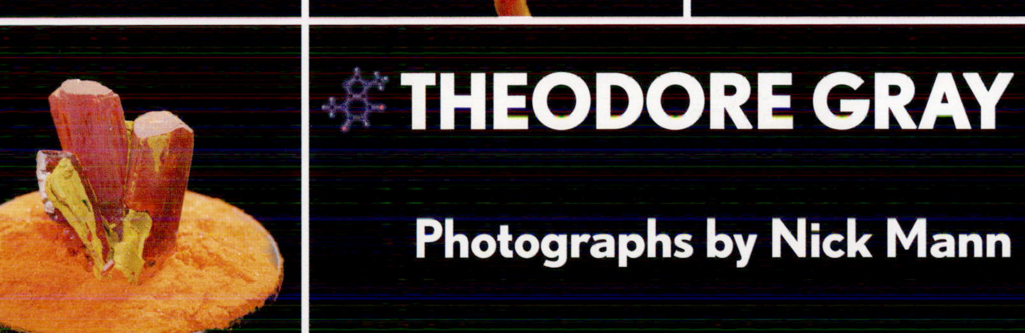

Photographs by Nick Mann

BLACK DOG
& LEVENTHAL
PUBLISHERS
NEW YORK

Published by
Black Dog & Leventhal Publishers, Inc.
151 West 19th Street
New York, NY 10011

Distributed by
Workman Publishing Company
225 Varick Street
New York, NY 10014

Manufactured in China

Cover and interior design by Matthew Riley Cokeley

ISBN-13: 978-1-57912-971-2

h g f e d c b a

Library of Congress Cataloging-in-Publication Data available on file.

Contents

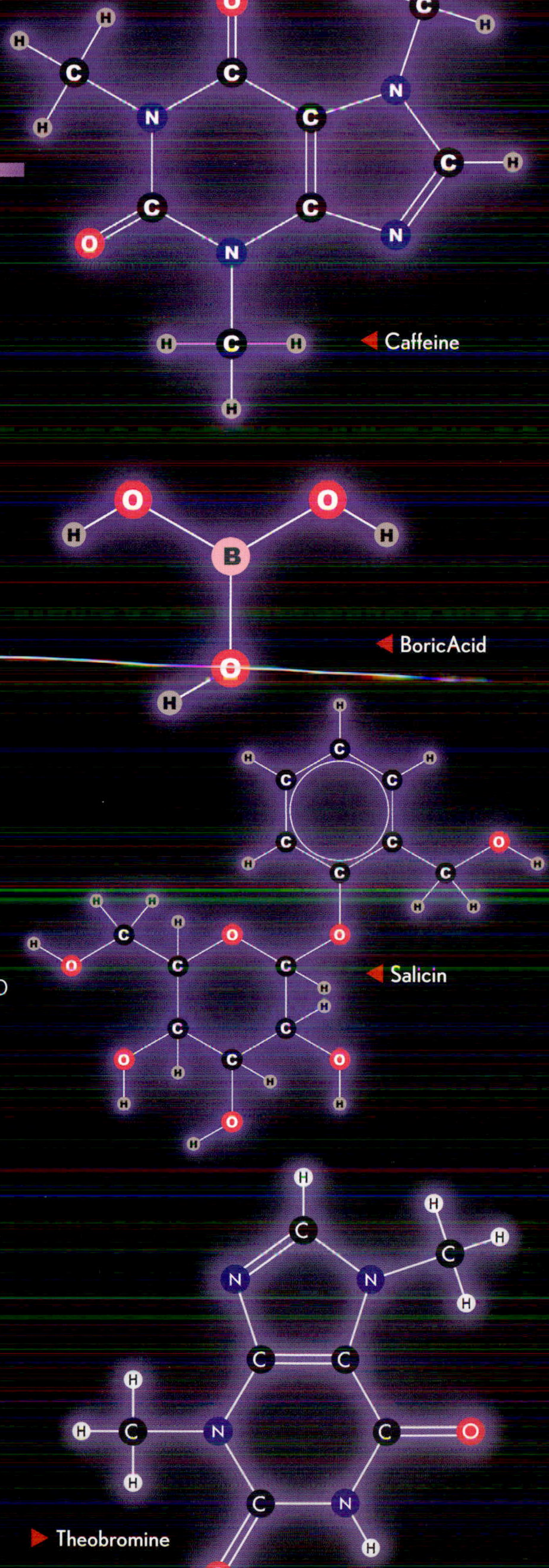

Caffeine
BoricAcid
Salicin
Theobromine

Introduction

THE PERIODIC TABLE IS COMPLETE: We know those hundred or so *elements* are all we ever need to worry about. But there is no catalog of all the *molecules* in the universe, and there can't be. There may be only six different chess pieces, but it's out of the question to list all the ways of arranging them on a chess board.

Even putting molecules into logical groups (in order to write a book that at least covers all the categories) is a losing battle. There are almost as many categories of molecules as there are molecules. I take that to mean that I have the freedom to write about only the interesting ones, and the ones that illustrate the deeper connections and broader concepts that unify them all.

If you're looking for a standard presentation of compounds, such as you might find in a chemistry textbook, you'll be disappointed. There is no chapter on acids and bases in this book. I do talk about acids, of course, but in connection with other things that I personally find more fascinating, like soap (which is made by using a strong base to turn a weak acid into a soluble salt that makes oil and water mix).

In that sense this book is more like the one collection of compounds every kid should have: a chemistry set. It's a little of everything, put together not to be complete, but to be *interesting*. It will teach you something about how the world of chemistry works, and give you a sense of the scope of the subject.

I hope you enjoy reading this book as much as I enjoyed writing it.

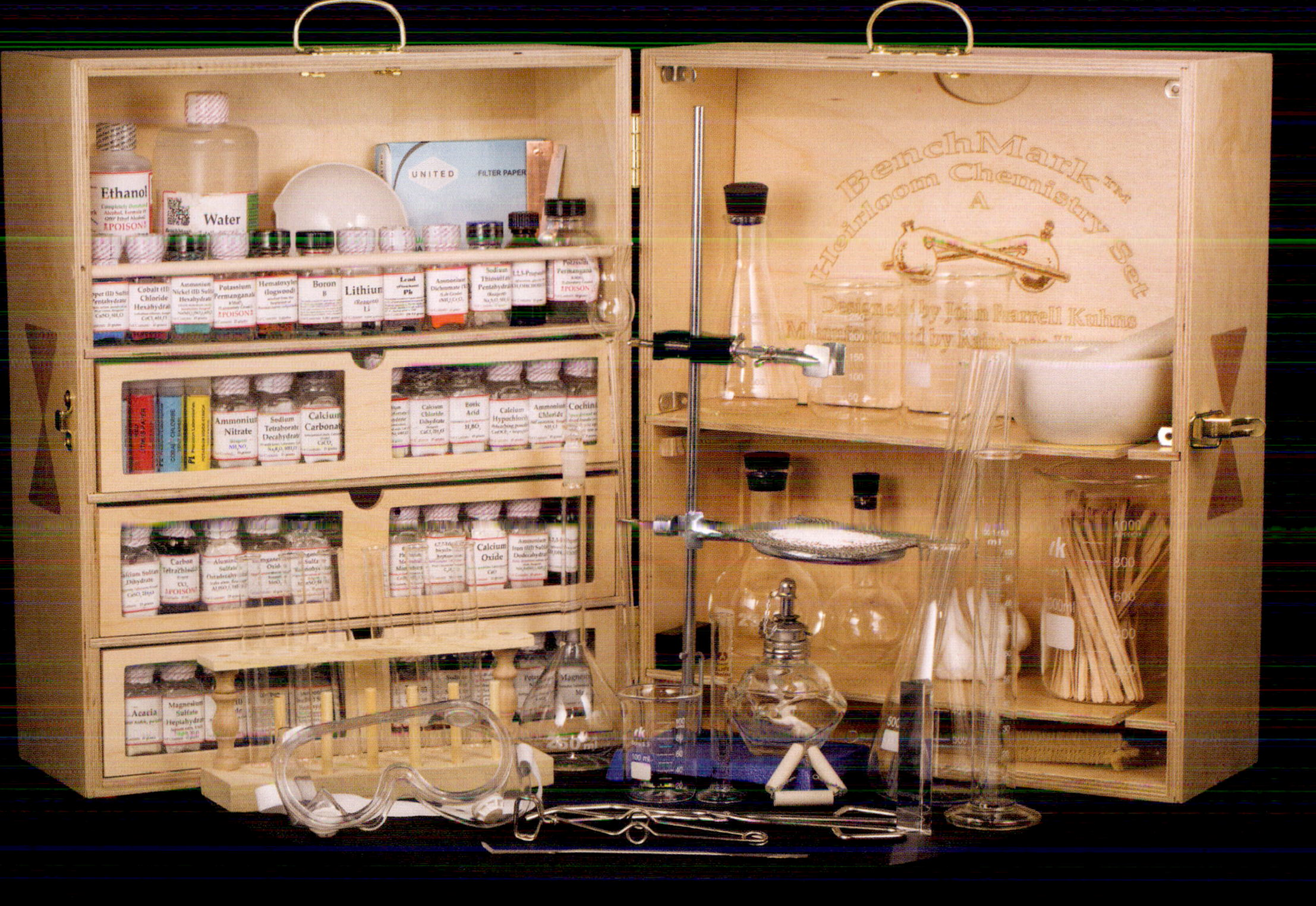

Chemistry sets were more popular a few generations ago. It's a common lament of older scientists that kids today just don't have access to the proper tools for discovery and learning. Just *try* to make an explosive using the typical chemistry set you get today. It's almost as if they were trying to make it hard. But as with much lamenting about the world today and all the great things you can't get anymore, a little digging often shows that what you're missing still exists, it's just moved to the internet. This set, a Kickstarter project, is every bit as complete and filled with opportunities for mischief as any set from the past hundred years. Like this book, it doesn't shy away from the more interesting compounds just because they might have a bit of an edge to them. And like this book, it comes with clear warnings about the very real dangers that chemicals pose when handled carelessly or without an understanding of their power.

The world of compounds is so wide and diverse that you could make up a large chemistry set focused on even a tiny fraction of it. This lovely antique set, for example, contains only simple inorganic compounds of interest to someone wanting to learn about the operation of foundries and metal refineries. So it has ores, alloys, clays, fire-resistant brick materials, and other such things. (See Chapter 6 for more about ores.)

Chapter 1

A House Built of Elements

Just two elements, carbon and hydrogen, create an astonishing number of compounds called hydrocarbons. Add oxygen to the mix and you can make a carbohydrate, like this light brown sugar.

ALL PHYSICAL THINGS in the world are made of the elements of the periodic table. I wrote a whole other book about that, and about all the places you can find each of the elements. Sometimes they exist on their own, as in aluminum pans or copper wires. But usually they are found combined with each other in compounds like table salt (which is made of vast arrays of sodium and chlorine atoms in a crystalline grid) or in molecules like sugar (which is made of tightly connected groups of twelve carbon, twenty-two hydrogen, and eleven oxygen atoms).

Molecules and compounds are what this book is all about.

In daily life, we encounter vastly more molecules and compounds than elements (countless thousands vs. dozens) because atoms can connect to each other in so many different ways. Using just hydrogen and carbon, you can make the entire class of compounds called hydrocarbons, which includes oils, greases, solvents, fuels, paraffins, and plastics. Add oxygen to the mix, and you can make carbohydrates including sugars, starches, waxes, fats, painkillers, pigments, more plastics, and a great many other compounds. Add just a few more elements, and you have all the compounds needed to make a living creature, including proteins, enzymes, and the mother of all molecules, DNA.

But what holds these atoms together with such great diversity? And why do I keep saying compounds *and* molecules: is there a difference?

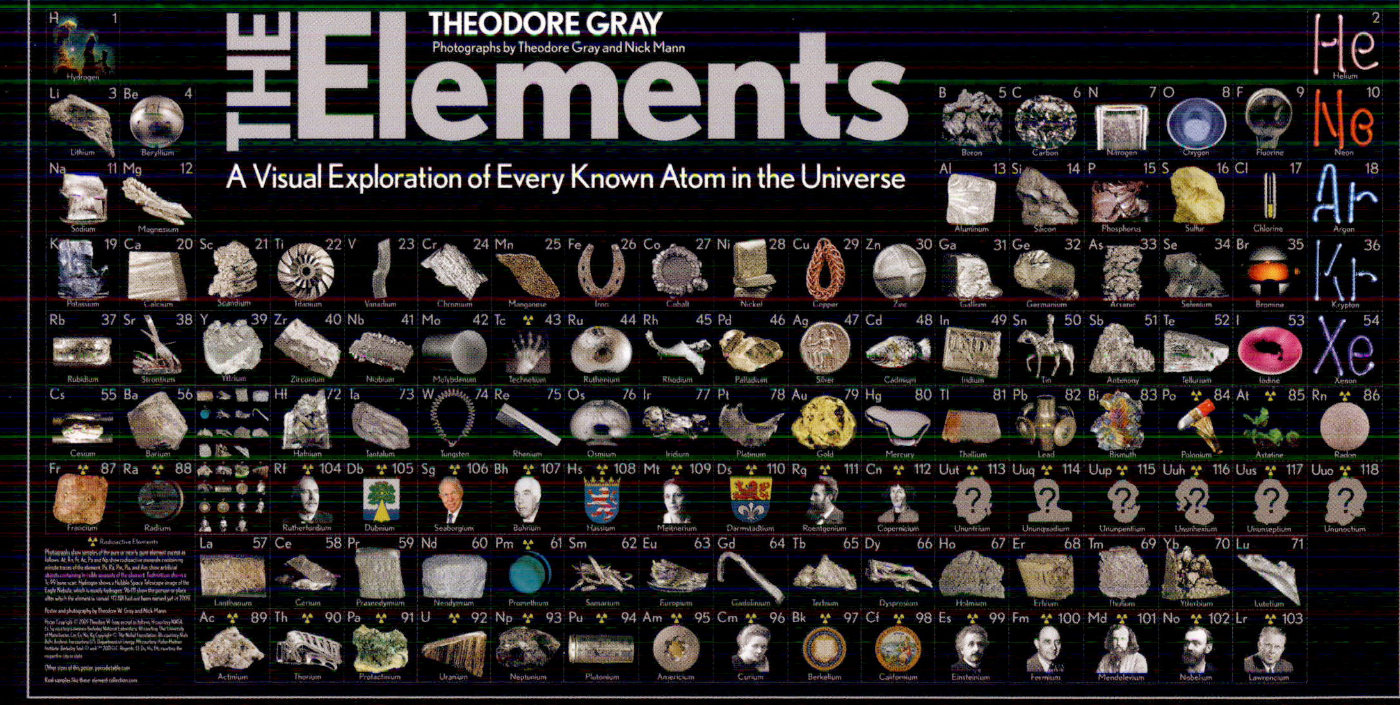

The periodic table is a catalog of all the kinds of atoms that exist or can exist in the universe. Everything is made of these few kinds of atoms, but they can be combined in a huge number of different ways. To learn more about the elements, you can read the whole book I wrote about them called *The Elements: A Visual Exploration of Every Known Atom in the Universe*.

▼ Elemental chlorine is normally a gas but can be liquefied under high pressure, as it is in this quartz ampule. Chlorine kills rapidly and painfully on contact with the lungs.

▲ The pure element sodium is a bright, silvery metal that explodes on contact with water. This sodium has been made into the shape of a duck for absolutely no good reason.

▲ Sodium chloride is a compound, a combination of equal numbers of atoms of the elements sodium and chlorine. Individually, these elements are alarmingly dangerous, but when combined this way, they are as harmless as table salt (which is another name for sodium chloride). Not only are these two elements harmless when combined, but the combination tastes good as well—both to us and to other animals. This is a "salt lick" given to horses to make sure they get enough salt in their diet.

▶ Just two elements, carbon and hydrogen, create an astonishing number of compounds. Well over a hundred thousand molecules made of nothing but these two elements have been studied and given names. Vastly more exist anonymously.

◀ Hydrocarbons include a wide range of liquids, from solvents lighter than water through all grades of oil to the goopiest crank case grease. The more carbon atoms connected together in each molecule, the more viscous the hydrocarbon becomes until eventually the compound turns waxy and finally to solid plastic.

▶ Polyethylene plastic, used in everything from flimsy grocery bags to fancy, cut-resistant gloves, is also a hydrocarbon, made with nothing but carbon and hydrogen. Its molecules contain tens or even hundreds of thousands of connected atoms.

The Force at the Heart of Chemistry

THE FORCE THAT holds compounds together and drives all of chemistry is the electrostatic force. It's the same force that holds a balloon to the wall after you rub it on your shirt or makes your hair stand on end when you shuffle on the right kind of carpet.

It's easy to start describing this force. Any material can carry an electric charge, which can be either positive or negative. If two things have charges of the same sign, then they repel each other. If they have charges of the opposite sign, then they attract each other. (It's a bit like with magnets, where two north poles or two south poles repel, but a north pole and a south pole attract.)

We know a lot about how this force works—how strong it is, how quickly it weakens with distance, how fast it can be transmitted through space, and so on. These details can be described with great precision and mathematical sophistication. But what the electrostatic force actually *is* remains a complete and utter mystery.

It's quite marvelous that something so fundamental is fundamentally unknown. But that's not a practical problem, because a description of how the force works, not a true understanding of it, is all that's needed to make creative use of all the ways that atoms can combine with each other.

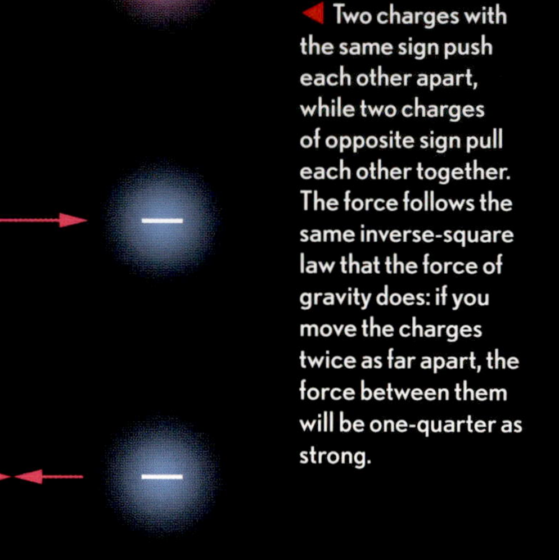

Two charges with the same sign push each other apart, while two charges of opposite sign pull each other together. The force follows the same inverse-square law that the force of gravity does: if you move the charges twice as far apart, the force between them will be one-quarter as strong.

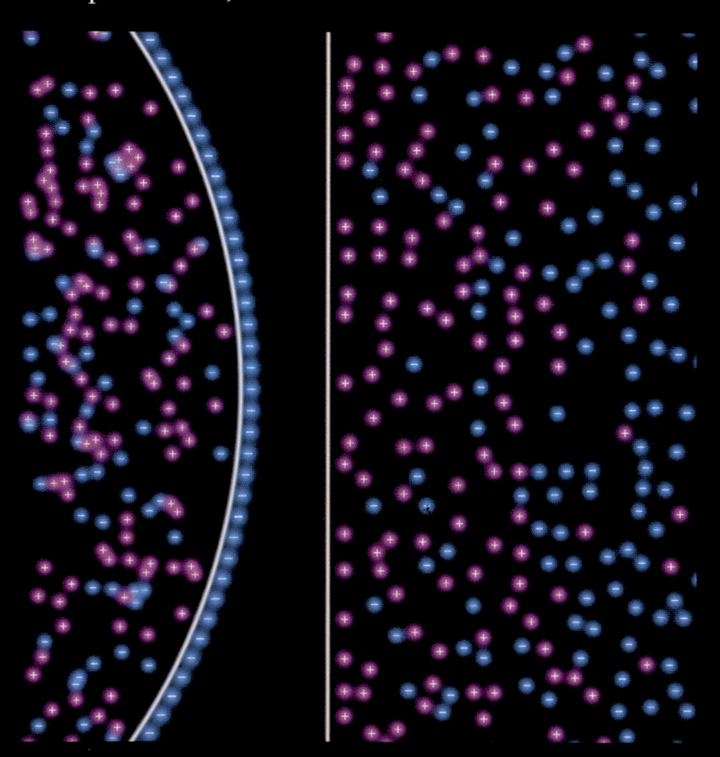

A small amount of electric charge accumulates on the surface of a balloon when it's rubbed against another material, such as a T-shirt. When the balloon is near a wall, this charge pulls on charges of the opposite sign in the wall, moving them closer to the surface and leading to an overall attraction between the wall and the balloon. You may hear the term *Van der Waals force* used in connection with molecules: it's the same idea, just happening on a molecular scale rather than a living-room scale.

A Van de Graaff generator accumulates large amounts of electric charge, leading to delightful results. The charge travels along individual strands of hair, causing them to repel each other because they have the same type of charge.

When a negative electric charge (i.e., a large number of electrons) is deposited on the two parts of this device, the repulsive force between the electrons pushes the needle away from the bar holding it. By measuring how far the needle swings, you can measure, crudely, how many extra electrons have been placed on it. Fancier instruments can count individual electrons and measure the forces from them precisely.

Atoms

ATOMS HAVE a small, dense nucleus containing protons and neutrons. The protons have a positive electric charge, and the neutrons have no charge, so overall each nucleus has a positive charge equal to its number of protons.

Surrounding the nucleus are a number of electrons, which have a negative electric charge. Because negative charges are attracted to positive charges, the electrons are held close to the nucleus, and it takes energy to pull them away. We say the electrons are *bound* to the nucleus by their electric charge.

The negative charge on an electron has exactly the same strength, but opposite sign, as the positive charge on a proton. So when an atom has the same number of electrons and protons, the overall charge on the atom is zero; it is a neutral atom.

There's a name for the number of protons in a nucleus: it's called the *atomic number*, and it defines which element you are looking at. For example, if you've got an atom with six protons in its nucleus, you have carbon, and you can make graphite or diamond out of it. If you've got eleven protons in each nucleus, you've got sodium, which you can combine with chlorine to make salt or throw in a lake to make an explosion when it reacts with water.

An atom's nucleus determines which element you have, but it's the electrons around the outside that control how that element behaves. Chemistry is really all about the behavior of electrons.

▼ You will often see pictures of atoms that show a small nucleus surrounded by electrons drawn as little balls, with lines that imply they are whizzing around the nucleus like planets around the sun. But those pictures lie. The small nucleus part is OK, but electrons are simply not little balls, and they aren't moving around the nucleus in the conventional sense of the word *moving*. They exist as delocalized objects—puffs of probability—that, in a weird, quantum mechanical way, may or may not be at any one place at any particular time. The best you can do in talking about electrons is to mathematically describe the likelihood that they will be in particular places. And it turns out that these probability distributions have beautiful shapes called atomic orbitals. The electrons don't move around these orbitals, and they are not shaped like these orbitals. Instead, the orbitals, drawn this way, show the likelihood of finding an electron in a given location around the nucleus: brighter areas are more likely to contain an electron, if you were to look there. If you don't look, then the electron is everywhere and nowhere at the same time. Yes, it's very weird. Einstein didn't like it any more than you do, but this math works to describe our world with greater precision than any other theory yet devised. The best you can do is to get used to it.

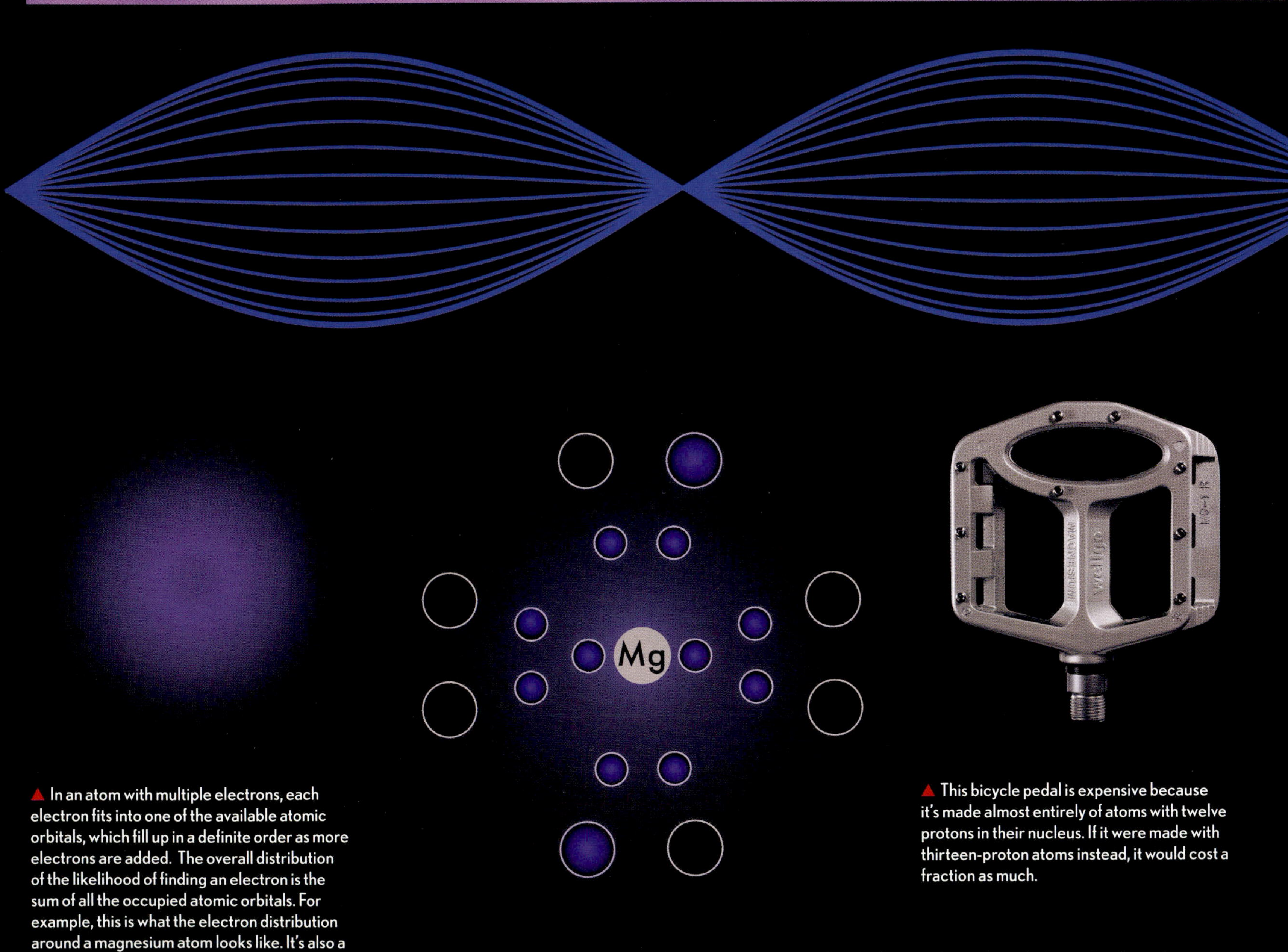

▲ In an atom with multiple electrons, each electron fits into one of the available atomic orbitals, which fill up in a definite order as more electrons are added. The overall distribution of the likelihood of finding an electron is the sum of all the occupied atomic orbitals. For example, this is what the electron distribution around a magnesium atom looks like. It's also a demonstration of why you pretty much never see pictures like this in chemistry books: there are twelve separate electrons shown in this diagram, and you can't make out a single one of them because they blend perfectly into each other, forming a symmetrical, uniform distribution of probability density around the whole nucleus. I'm showing you the picture only to show you why it's pointless to show you the picture.

▲ This bicycle pedal is expensive because it's made almost entirely of atoms with twelve protons in their nucleus. If it were made with thirteen-proton atoms instead, it would cost a fraction as much.

▲ Although I hate to have conventional diagrams of electrons as little balls around a nucleus, the fact is that these diagrams are useful because they let you actually see and count the electrons, and they show that the electrons are arranged in "shells" around the nucleus, each of which can hold a certain number of electrons. As the number of electrons around the nucleus increases, the shells fill up one electron at a time from the inside out. It turns out that for nearly all of the elements of interest to us in this book (except hydrogen) the outermost of these shells can hold up to eight electrons. How many electrons are actually in that outermost shell, called the valence shell, depends on which element you have. For example, magnesium, shown here, has two electrons in its valence shell. These valence electrons are what give magnesium its chemical properties. Diagrams like this do *not* represent anything about the actual physical location of any of the electrons! They are just a handy way of showing how many electrons are in each shell, particularly the valence shell.

▼ How can an electron be everywhere and nowhere at the same time? Electrons, like many quantum mechanical objects, behave sometimes like a wave and sometimes like a particle. Imagine the space around an atom as being a bit like a violin string, and the electron a bit like a vibration, a wave, on that string. Where on the string is that wave located? Well, it isn't anywhere in particular on the string, and it's everywhere on the string at the same time. That is, in a sense, the way in which an electron too can be everywhere and nowhere at the same time. When the electron is probed, its behavior becomes more like that of a particle, and it materializes, or localizes in the language of quantum mechanics, in a particular spot.

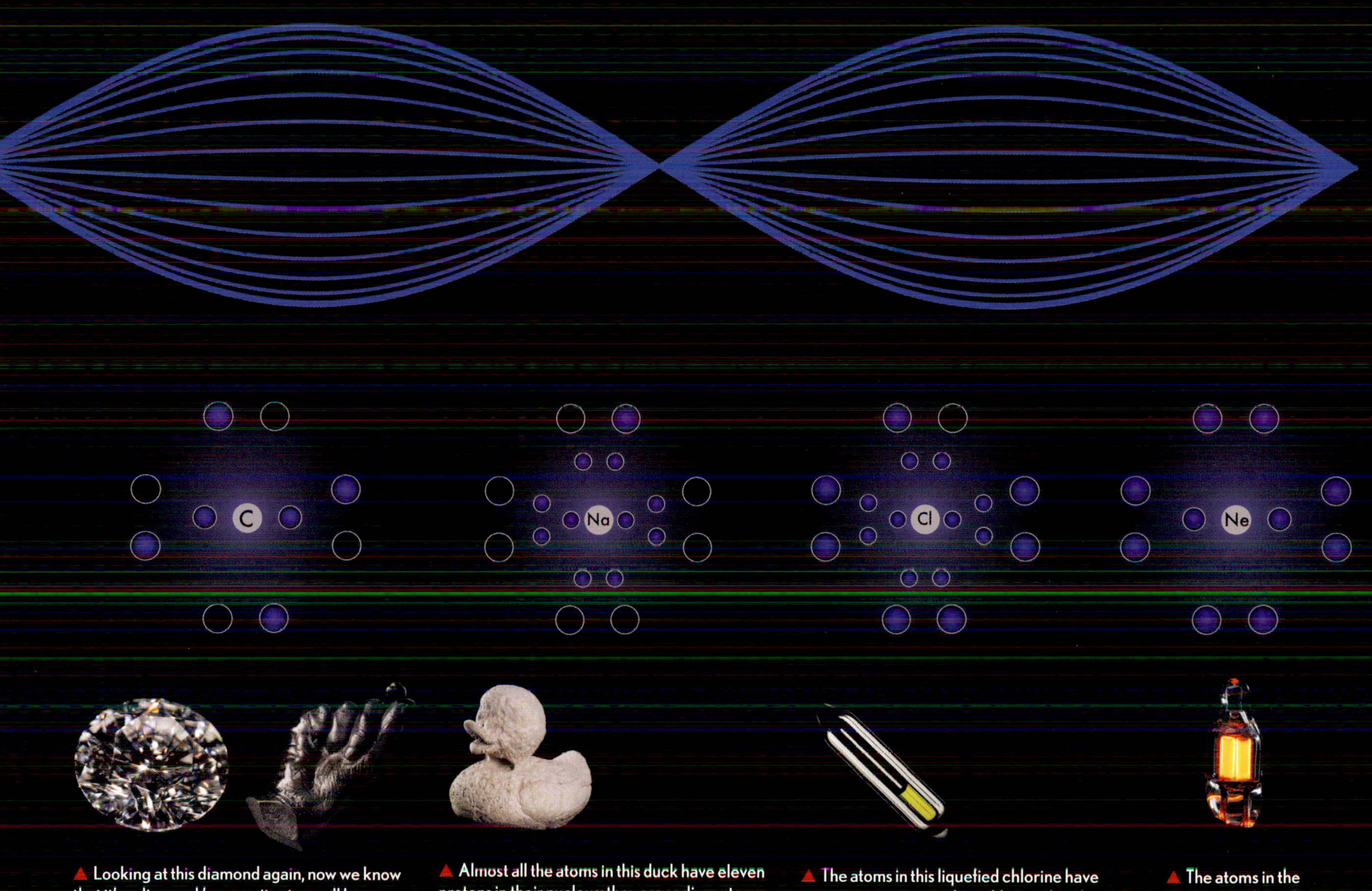

▲ Looking at this diamond again, now we know that it's a diamond *because* its atoms all have six protons in them. Graphite might seem like a completely different substance, but its atoms also have six protons in each nucleus, so it too is made of carbon. Notice that carbon has four valence electrons in its outermost shell, with room for four more. This fact is *crucial* to the existence of life on earth, and to most of the rest of this book.

▲ Almost all the atoms in this duck have eleven protons in their nucleus; they are sodium atoms, so this is a sodium duck. A few on the surface have only eight; those are oxygen atoms from the air that have combined to form the compound sodium oxide, a white powder. Throughout the duck there are a few atoms with various other numbers of protons; those are contaminants, elements other than sodium that have no business being in a sodium duck. Notice that sodium has a single, lone electron in its outermost shell. This fact singlehandedly explains nearly all of sodium's chemical behavior.

▲ The atoms in this liquefied chlorine have seventeen protons in them. Notice that the outermost shell of electrons in chlorine is missing one electron. That tells you just about everything you need to know about chlorine's chemistry.

▲ The atoms in the neon gas in this indicator light have ten protons. Notice that the outermost shell of electrons is completely filled. This makes neon an extremely un-reactive element: When an atom's outermost shell is filled, it is in a happy place.

Compounds

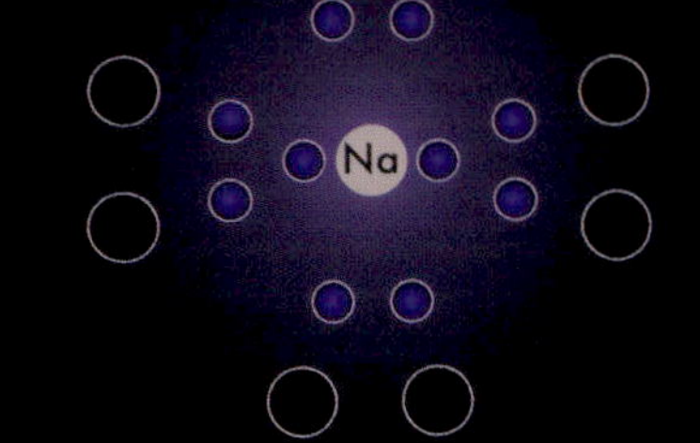

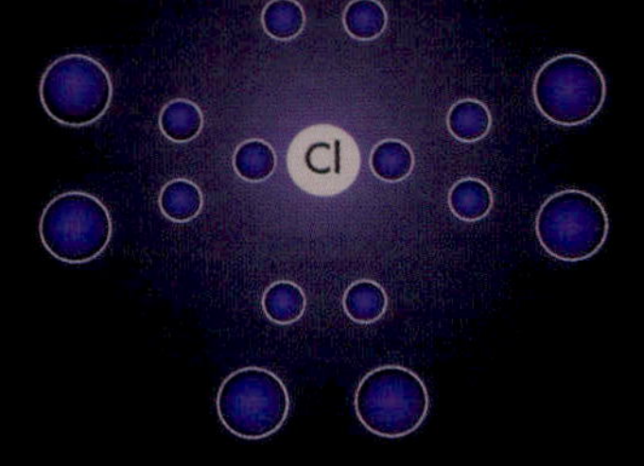

THE ELECTROSTATIC FORCE is what holds electrons and protons together in a single atom, and it's also what holds atoms to each other in compounds and molecules. When an individual atom has exactly the same number of protons and electrons, it has no overall charge, so there's no electrostatic force between it and any other neutral atom. To get them to connect to each other, you have to move the electrons around from one atom to another, creating an electrostatic force between them.

Look again at the atomic diagrams on the previous pages. Notice that some of them (such as neon) have "full" outer shells, while others (such as carbon, sodium, and chlorine) have gaps that indicate missing electrons. Each shell has a fixed number of electrons that it can hold (either two or eight, depending on which layer it is). The inner shells fill up completely, but there may not be enough electrons to completely fill the outermost, or valence, shell. When that shell is not full, you are dealing with an unhappy atom, and you have a golden opportunity to move electrons around.

Atoms are willing to go to great lengths to get a complete shell, even if that means not being electrically neutral anymore. But they do have preferences. Some like to fill in holes with extra electrons, while others choose to shed a few stragglers in their outermost shell. Still others prefer to share electrons with neighbors in a way that allows a single electron to, at least partially, satisfy two atoms at once.

Any time you have two or more atoms connected to each other, it's called a molecule. If there are at least two different kinds of elements in your molecule, then it's also called a compound.

▲ Here are the diagrams for sodium (eleven protons) and chlorine (seventeen protons) again. Notice that in both of them the outermost shell of electrons is incomplete. Sodium has room for eight electrons in its outer shell, but only has one. Chlorine also has room for eight, and is missing only one. Sodium and chlorine are very angry about this; both of them are *very* reactive substances that viciously attack anything in their vicinity. Sodium rips up any water it comes near, while chlorine contents itself with ripping up your lungs if you breathe it.

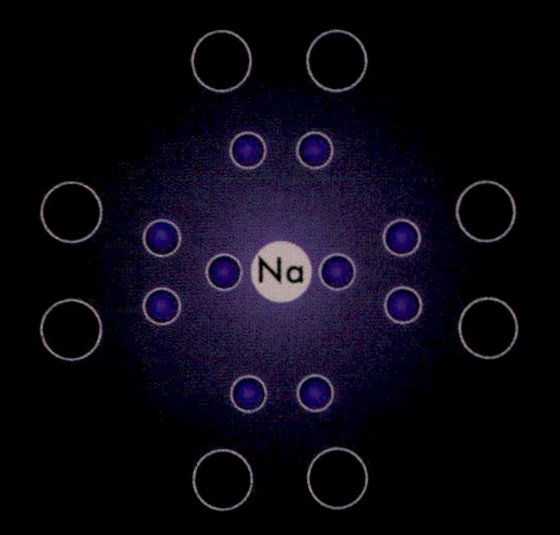

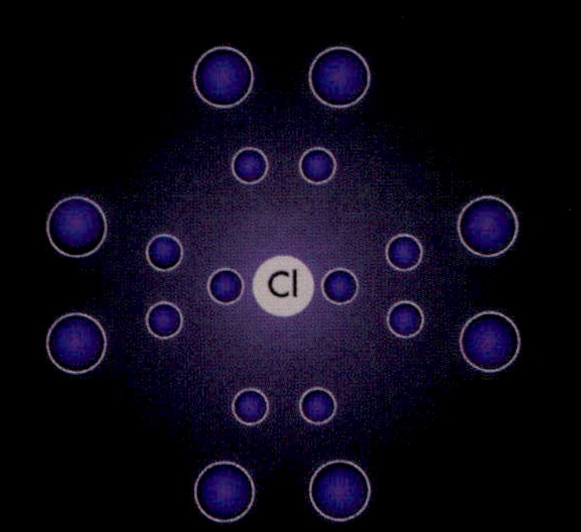

▲ Moving a single electron from a sodium atom to a chlorine atom solves both their problems, because each now has a filled outer shell. (The completely empty shell drawn around sodium is only to show where the electron used to be: what counts is the filled shell just inside it.) Once the electron has moved, the sodium atom has a positive charge, while the chlorine atom has a negative charge. Because the atoms now have opposite charges, they are attracted to each other. They stick together to form a compound known as sodium chloride, or, more commonly, table salt.

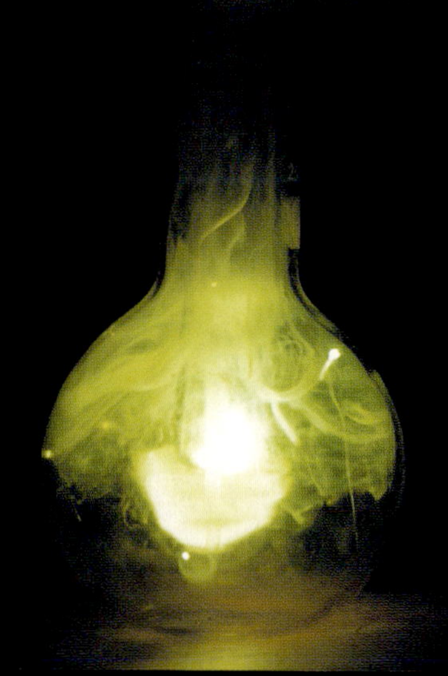

▶ Atoms of sodium and chlorine are very happy to exchange electrons to form sodium chloride. In this context, "happy" means that, as the electrons move into their favorable arrangement, the process releases a lot of energy. Chemical reactions that release energy do so in the form of heat, light, and sound. The more happy elements are to combine (i.e., the more energy they release when doing so), the less likely you are to find them in isolation in nature. Highly reactive elements like sodium and chlorine are absolutely never found that way: if you see pure sodium or pure chlorine, you know someone has gone to a lot of trouble to tear them out of their happy union with other elements.

▶ When atoms have an electric charge, as in salt, they are called "ions." The sodium ion has a +1 charge (i.e., it's missing one negatively charged electron), while the chloride ion has a -1 charge. The bond formed between two ions is called an ionic bond, and compounds formed out of ionic bonds are called ionic compounds. Sodium chloride, thus, is an example of an ionic compound. Many compounds have this kind of bond, and many of them are known generically as salts. Because there are only two kinds of charge, positive or negative, compounds held together exclusively by ionic bonds are always pretty simple. Each negative charge attracts all the nearby positive charges indiscriminately, and vice versa. So the elements pack together as tightly as they can in a simple repeating arrangement called a crystal. Shown here is a sodium chloride crystal. If you strictly follow the definition of a molecule as any set of atoms bonded to each other, an entire grain of salt is a single molecule. But usually people shy away from this idea and say the salt grain is an ionic crystal, not really a molecule.

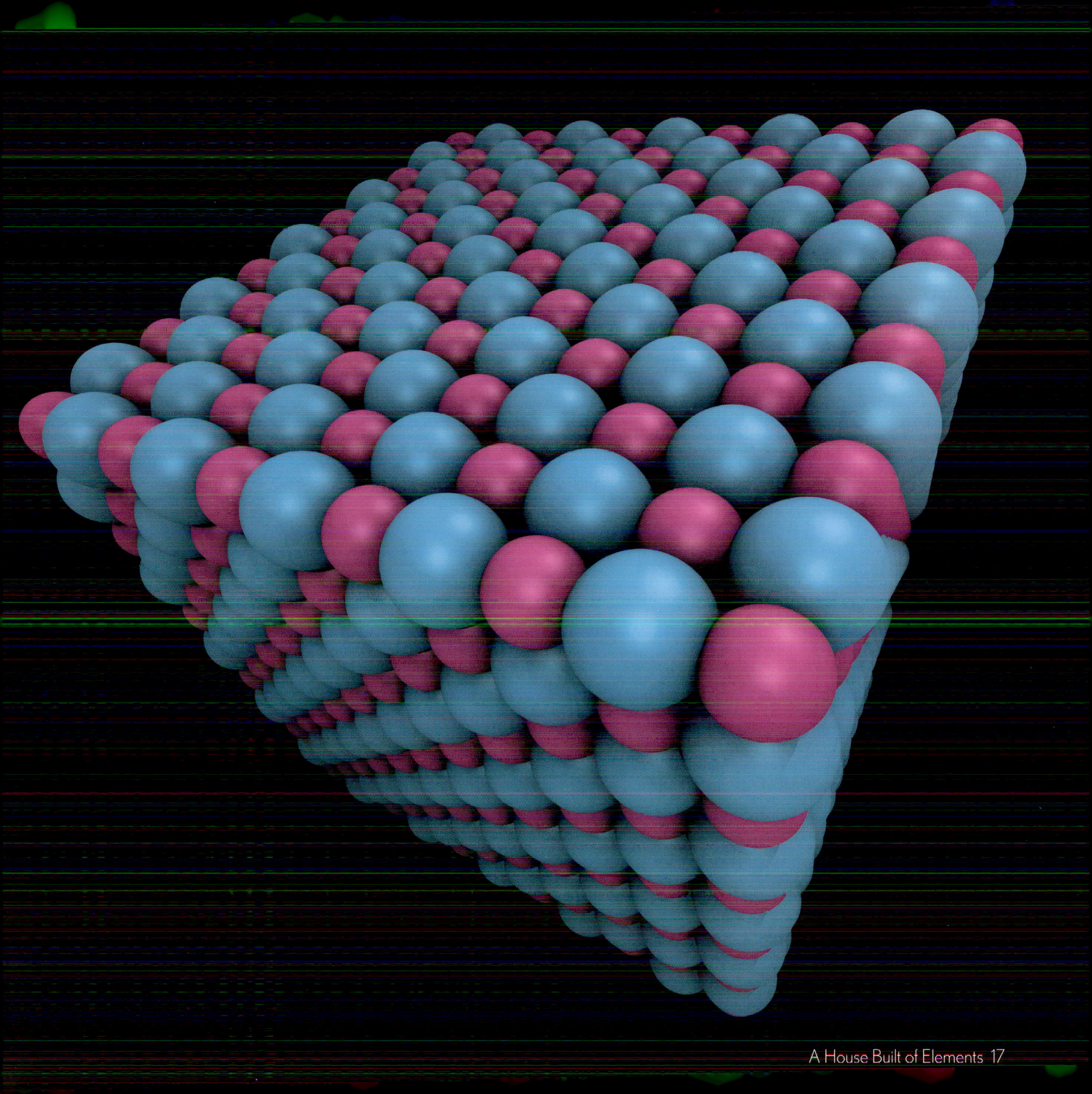

Molecules

ODIUM AND CHLORINE form *ionic* onds because chlorine really wants o acquire an extra electron, and odium is very happy to be rid of what t considers an excess electron. Other toms are less strong-willed: rather han gain or lose electrons altogether, hey prefer to share electrons with ach other. When atoms share one or nore electrons, they form *covalent* onds.

Covalent bonds allow for omplicated structures because, nlike i c bonds, they are personal; hey ex between specific pairs of toms.

Each kind of atom has a characteristic number of electrons that it likes to share with neighboring atoms. For example, carbon, which is missing four electrons from its outer shell, likes to take a share of four electrons from other atoms so that it can pretend it has a full outer shell of eight. Oxygen likes to take a share of two. Hydrogen is incredibly generous: it has only one electron but is happy to share it with other atoms.

These rules allow atoms to work like LEGO® blocks that snap together in particular ways. And when they do, the result is called a molecule.

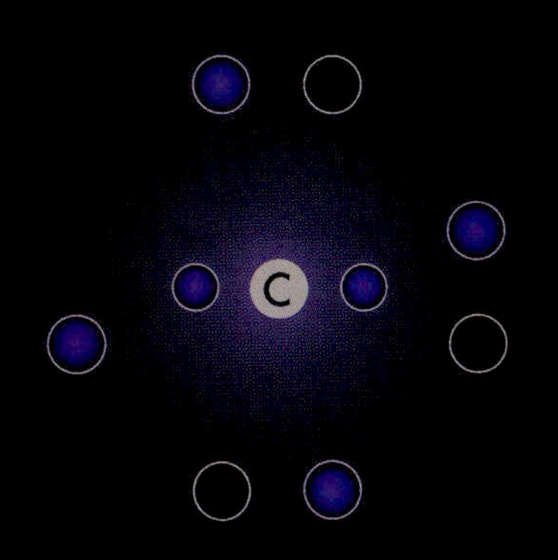

Car has four electrons in a shell meant or eigh is means carbon is often found onded our other atoms, sharing electrons vith each of them to complete its shell.

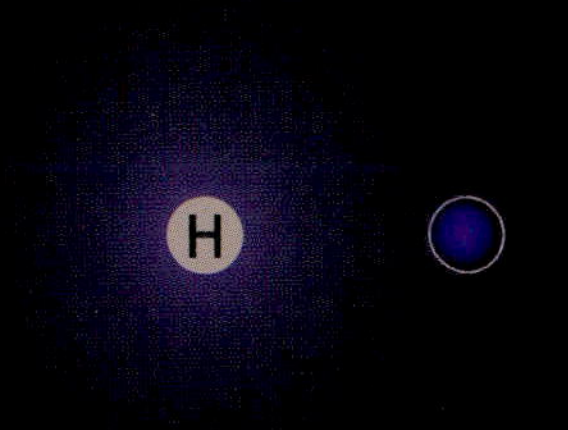

Hydrogen has one electron in a shell meant or two. That means it likes to bond with a single ther atom.

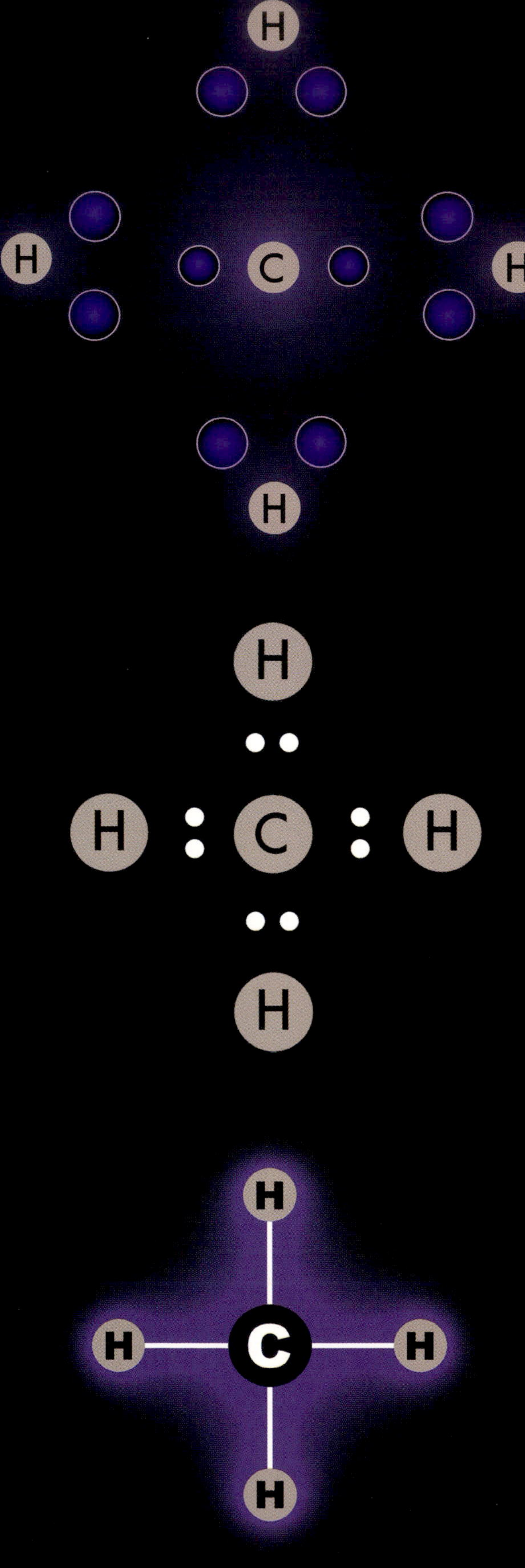

(top) When four hydrogen atoms combine with one carbon atom, the result makes all of the atoms fairly happy. The outer shell of the carbon atom is populated with a total of eight electrons, four of them from the carbon and one from each of the four hydrogens. The carbon pretends all eight electrons belong to it, creating a full shell, while the hydrogen atoms each pretend they have two electrons to fill their own shells.

A group of atoms arranged this way is called a methane molecule.

(middle) The fuzzy diagram above doesn't represent the real locations of electrons in a methane molecule, but it does conveniently let you count them and see how they fill up the atoms' outermost shells. A more schematic form of this diagram is called a "Lewis dot" structure. Each dot represents one electron in the valence shell. You will find Lewis dot structures in chemistry textbooks explaining why particular kinds of atoms bond in particular ways.

(bottom) Showing all the individual electrons in the atoms that make up a molecule, either fuzzy or as Lewis dots, quickly becomes impractical. Therefore we are going to draw molecules the way you usually see them in chemistry books, with lines that show where electrons are being shared. Each line represents one pair of shared electrons. I've left a soft glow around the lines as a reminder that they are symbolic and do not reflect what the atoms actually look like. There are no strings or rods in a real atom, only fuzzy, diffuse electrons swimming around and between the atomic nuclei, gluing them together with electrostatic force.

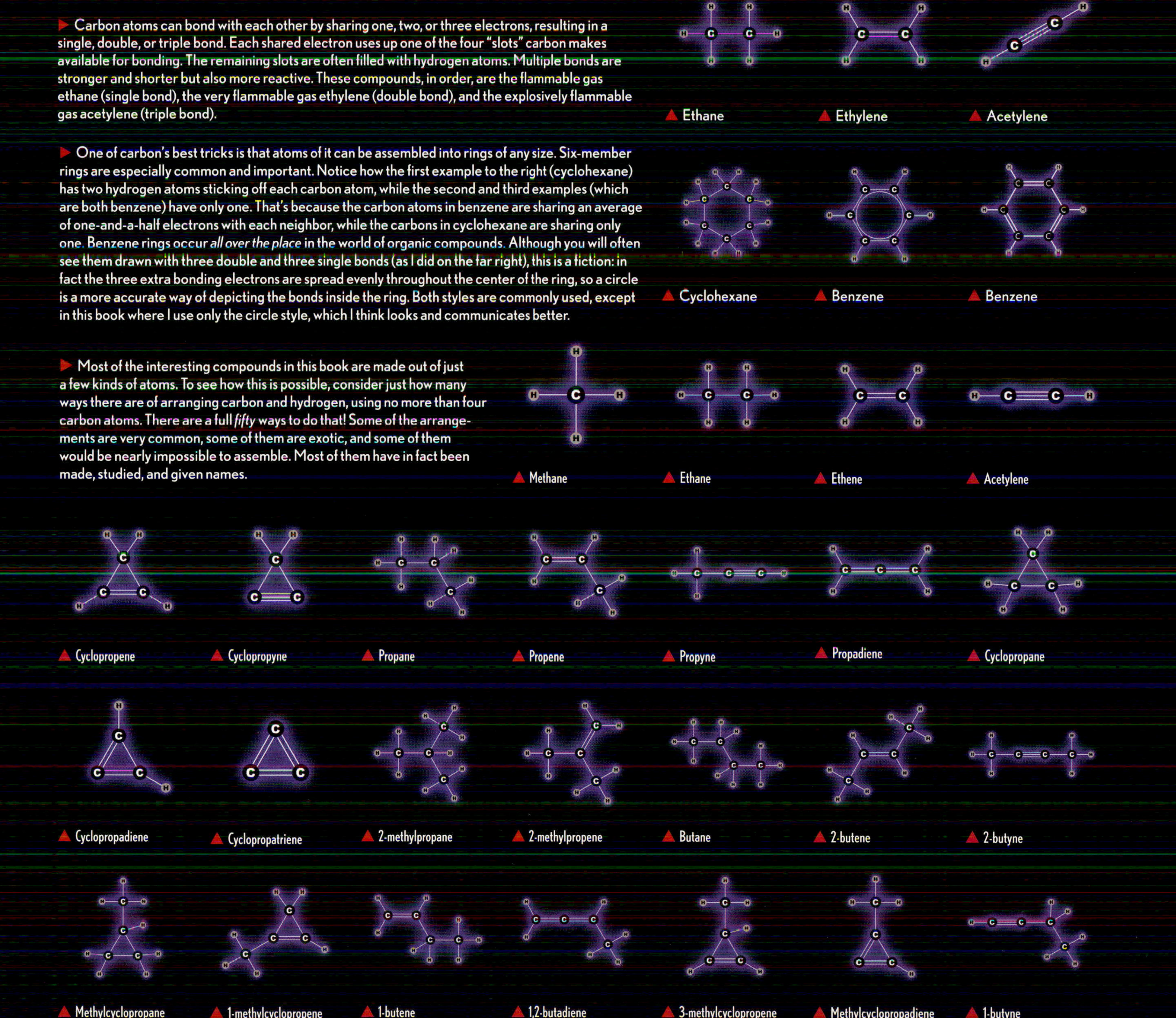

Carbon atoms can bond with each other by sharing one, two, or three electrons, resulting in a single, double, or triple bond. Each shared electron uses up one of the four "slots" carbon makes available for bonding. The remaining slots are often filled with hydrogen atoms. Multiple bonds are stronger and shorter but also more reactive. These compounds, in order, are the flammable gas ethane (single bond), the very flammable gas ethylene (double bond), and the explosively flammable gas acetylene (triple bond).

One of carbon's best tricks is that atoms of it can be assembled into rings of any size. Six-member rings are especially common and important. Notice how the first example to the right (cyclohexane) has two hydrogen atoms sticking off each carbon atom, while the second and third examples (which are both benzene) have only one. That's because the carbon atoms in benzene are sharing an average of one-and-a-half electrons with each neighbor, while the carbons in cyclohexane are sharing only one. Benzene rings occur *all over the place* in the world of organic compounds. Although you will often see them drawn with three double and three single bonds (as I did on the far right), this is a fiction: in fact the three extra bonding electrons are spread evenly throughout the center of the ring, so a circle is a more accurate way of depicting the bonds inside the ring. Both styles are commonly used, except in this book where I use only the circle style, which I think looks and communicates better.

Most of the interesting compounds in this book are made out of just a few kinds of atoms. To see how this is possible, consider just how many ways there are of arranging carbon and hydrogen, using no more than four carbon atoms. There are a full *fifty* ways to do that! Some of the arrangements are very common, some of them are exotic, and some of them would be nearly impossible to assemble. Most of them have in fact been made, studied, and given names.

Molecules

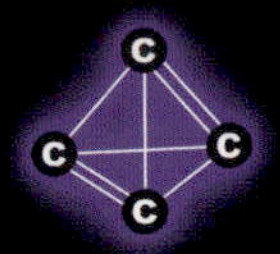

Tetrahedradiene

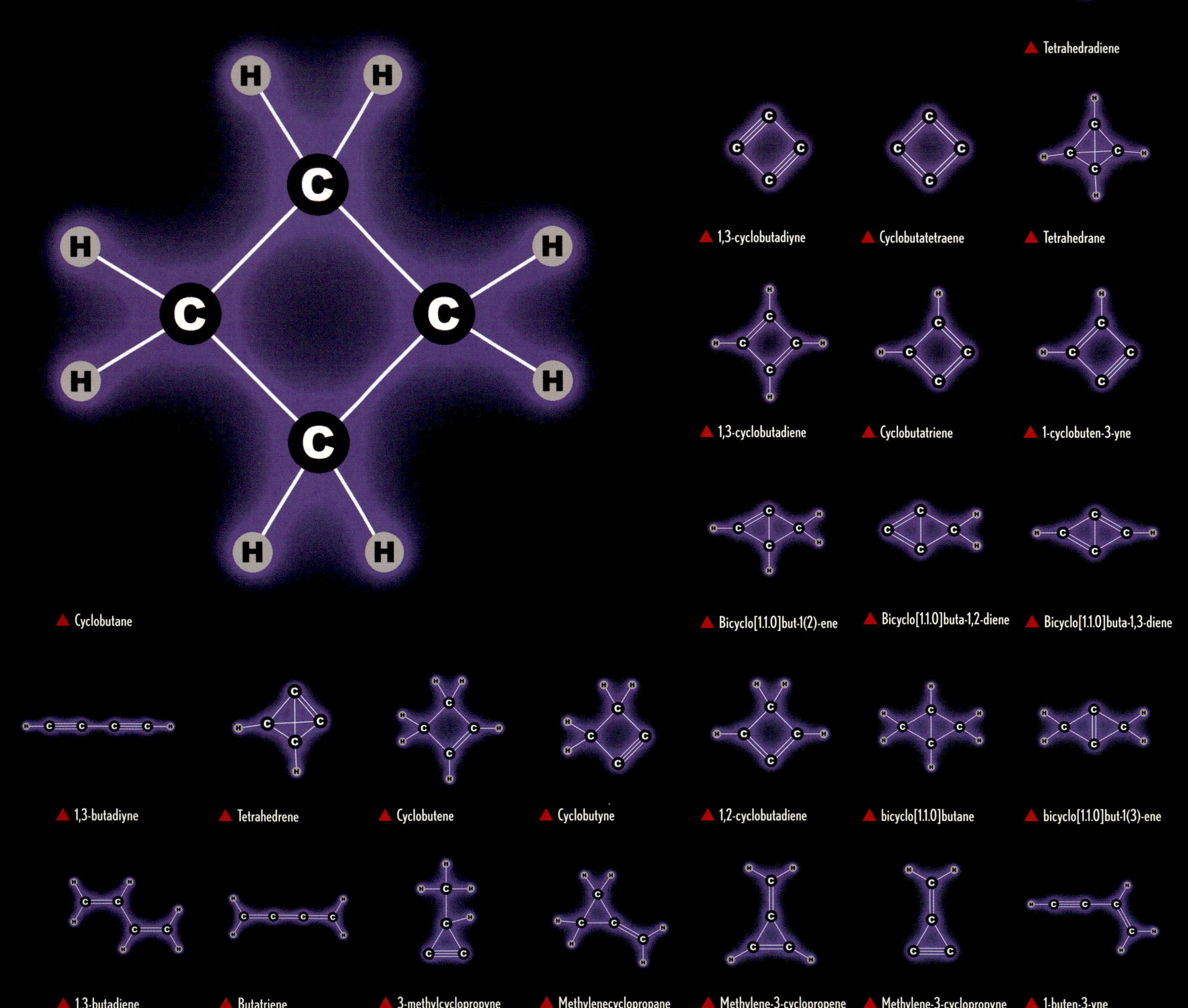

1,3-cyclobutadiyne
Cyclobutatetraene
Tetrahedrane
Cyclobutane
1,3-cyclobutadiene
Cyclobutatriene
1-cyclobuten-3-yne
Bicyclo[1.1.0]but-1(2)-ene
Bicyclo[1.1.0]buta-1,2-diene
Bicyclo[1.1.0]buta-1,3-diene
1,3-butadiyne
Tetrahedrene
Cyclobutene
Cyclobutyne
1,2-cyclobutadiene
bicyclo[1.1.0]butane
bicyclo[1.1.0]but-1(3)-ene
1,3-butadiene
Butatriene
3-methylcyclopropyne
Methylenecyclopropane
Methylene-3-cyclopropene
Methylene-3-cyclopropyne
1-buten-3-yne

An Architecture of Atoms

CHEMICAL DIAGRAMS like the ones we've been looking at show how atoms are connected to each other. They make molecules look flat, which is not the case at all: molecules are very three-dimensional objects. However, drawing the structures to appear flat makes it easier to see how each atom is bonded to its neighbors, so that's how people usually draw them.

Physical models can show the real, three-dimensional shapes of the molecules. Computer renderings can do the same, especially when displayed live on a computer screen so they can be rotated and zoomed.

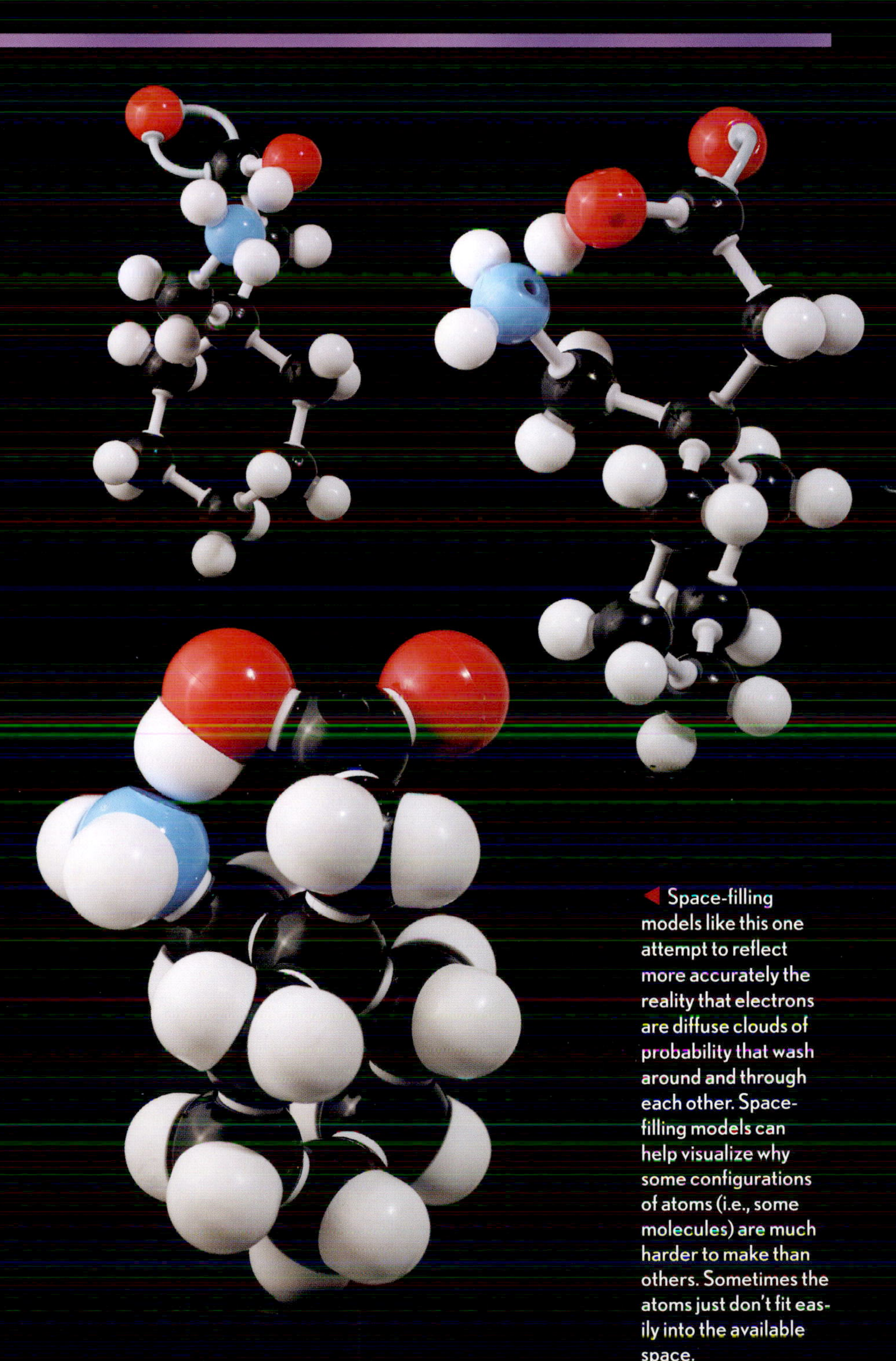

▶ This plastic ball-and-stick model of gabapentin shows its three-dimensional structure reasonably well, but only if you can turn it around; from any one viewpoint, some parts are hard to make out. As with flat diagrams of molecules, the lines are a lie: there are no rods and no hard spheres in a real molecule.

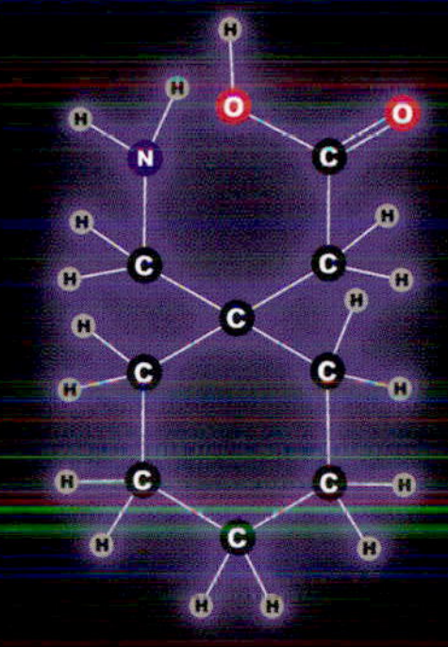

▲ This diagram of the molecule known as gabapentin, a drug used to treat nerve pain, shows how its atoms are connected together. It's a logical mapping of what kinds of atoms are in the molecule and how they are bonded to each other. But it really doesn't show the three-dimensional structure well at all.

◀ Space-filling models like this one attempt to reflect more accurately the reality that electrons are diffuse clouds of probability that wash around and through each other. Space-filling models can help visualize why some configurations of atoms (i.e., some molecules) are much harder to make than others. Sometimes the atoms just don't fit easily into the available space.

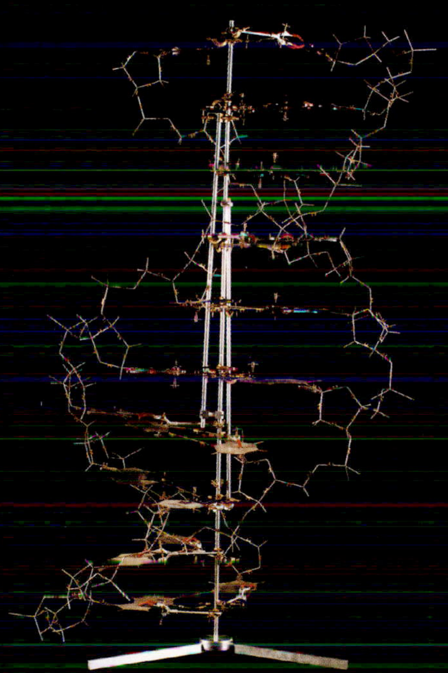

▲ Chemists continue to use physical models for relatively small molecules. But before computers took over the job of displaying giant molecules, they too were made into physical models. This model of a tiny fragment of DNA was built by Francis Crick and James Watson to explore how its atoms might fit together. When they eventually got it just right, they used this model to explain to the world that DNA was a double helix.

An Explosion of Possibilities

In Chapter 3, we'll learn how the synthesis of this compound forced everyone who understood it to rethink the deepest questions of life.

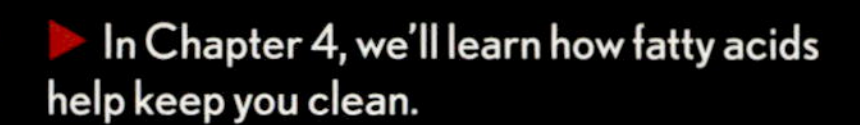

In Chapter 4, we'll learn how fatty acids help keep you clean.

In Chapter 5, we'll learn why this stuff is so icky.

IT'S ASTONISHING how much of chemistry involves only about half a dozen elements. Pretty much the entire fields of organic chemistry and biochemistry are to do with carbon, hydrogen, oxygen, nitrogen, sulfur, sodium, potassium, and phosphorus, with a very few other elements showing up from time to time and in relatively small quantities.

A greater diversity of elements is found in the world of inorganic compounds, but quite frankly the entire range of interesting inorganic compounds would fit in a small corner of one room in the house of chemistry (sorry, inorganic chemists). The real action in modern chemistry is centered on carbon, because carbon is the element of life—the basic building block of the great majority of molecules that are significant to living things.

In the remainder of this book, we will visit the rooms of the house of chemistry, the house built of elements. It is lovingly decorated with molecules, organic and inorganic, safe and unsafe, beloved and despised. Just as every living creature has a place and a role (even mosquitoes), so too every compound wants to be known and appreciated for what it contributes to the richness of the natural world (even thimerosal).

In Chapter 2, we'll learn about sweet oil of vitriol and how chemical compounds have three names.

In Chapter 6, we'll learn where compounds come from.

▲ In Chapter 7, we'll learn about a molecule shaped like a shoe.

▲ In Chapter 8, we'll learn what these darts are used to inject and about the power of poppies.

▲ In Chapter 9, we'll learn why one of these bowls is so much smaller than the other.

▲ In Chapter 10 we'll learn why natural vanilla extract is radioactive, but synthetic vanilla is not.

▲ In Chapter 11 we'll learn what this device is used for.

▶ In Chapter 12, we'll learn why being colorful is a rare thing for a molecule.

▶ In Chapter 13, we'll learn why this molecule caused a dangerous movement.

▶ In Chapter 14, we'll learn about molecules that are more like computers than molecules.

MARTKQTARK
STGGKAPRKQ
LATKAARKSA
PATGGVKKPH
RYRPGTVALR
EIRRYQKSTE
LLIRKLPFQR
LVREIAQDFK
TDLRFQSSAV
MALQEASEAY
LVGLFEDTNL
CAIHAKRVTI
MPKDIQLARR
IRGERA

Chapter 2

The Power of Names

I DECIDED TO take a class in organic chemistry for what was quite possibly the silliest of reasons: I liked the names of the compounds. It wasn't so much the sound of them, but the fact that assembled together they form a system that connects to a deep and beautiful body of knowledge. As I considered what these names mean and how each one gives meaning to the others, for the first time I truly appreciated the power that comes from giving a thing a name.

Just as T. S. Eliot said of cats, many chemical compounds have three names.

If they have been known for a long time, they have an ancient, alchemical name. These poetic names usually describe where the stuff comes from rather than what it is, because back then no one really had any idea what they were working with.

For example, in the alchemical language, sweet oil of vitriol is obtained by distilling oil of vitriol with the spirit of wine. (And oil of vitriol, in case you didn't know, is the liquid obtained by roasting green vitriol. The spirit of wine is the first thing that evaporates when you heat wine.)

I love these names for the images they form of wizards and potions, but they don't tell you the first thing about the true nature of the substances they refer to.

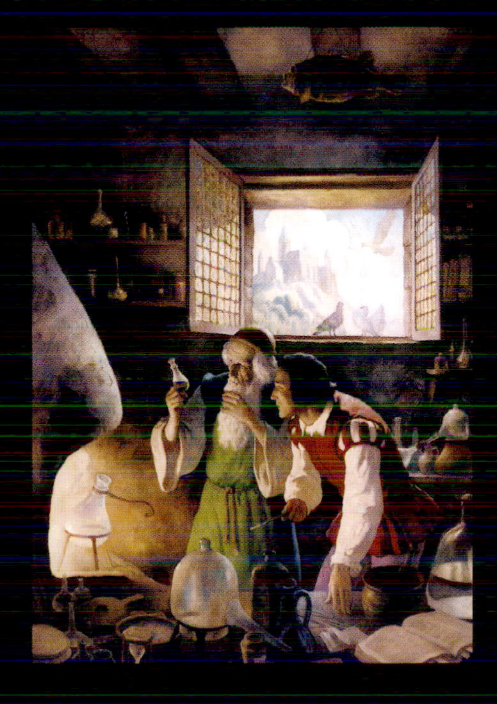

The alchemists, though commonly looked down on today as superstitious quacks trying to turn lead into gold, were actually serious students of nature who made many early discoveries. They set the stage for the emergence in the 1700s of the modern science of chemistry.

A vile and fuming substance, but what is its name?

Alchemical Names

HERE ARE TWO chemical reactions written out using the alchemical names for each substance. This language sounds beautiful, but what does it mean? Notice that oil of vitriol appears on both sides of the second reaction: it isn't actually consumed or altered in this process but must be present in order for the spirit of wine to be transformed. Why?

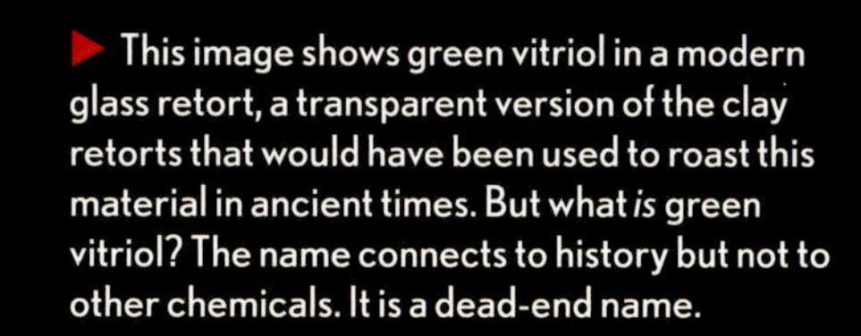

▶ This image shows green vitriol in a modern glass retort, a transparent version of the clay retorts that would have been used to roast this material in ancient times. But what *is* green vitriol? The name connects to history but not to other chemicals. It is a dead-end name.

Green vitriol

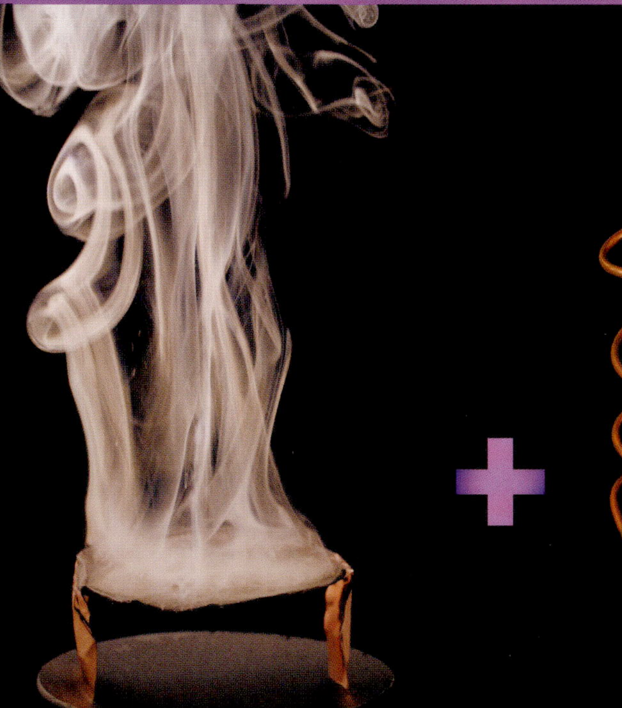

◀ Distillation of wine into spirits is one of the oldest of all chemical processes. It is the physical separation principally of two compounds, water and the spirit of wine. You can probably guess the modern name for the spirit of wine.

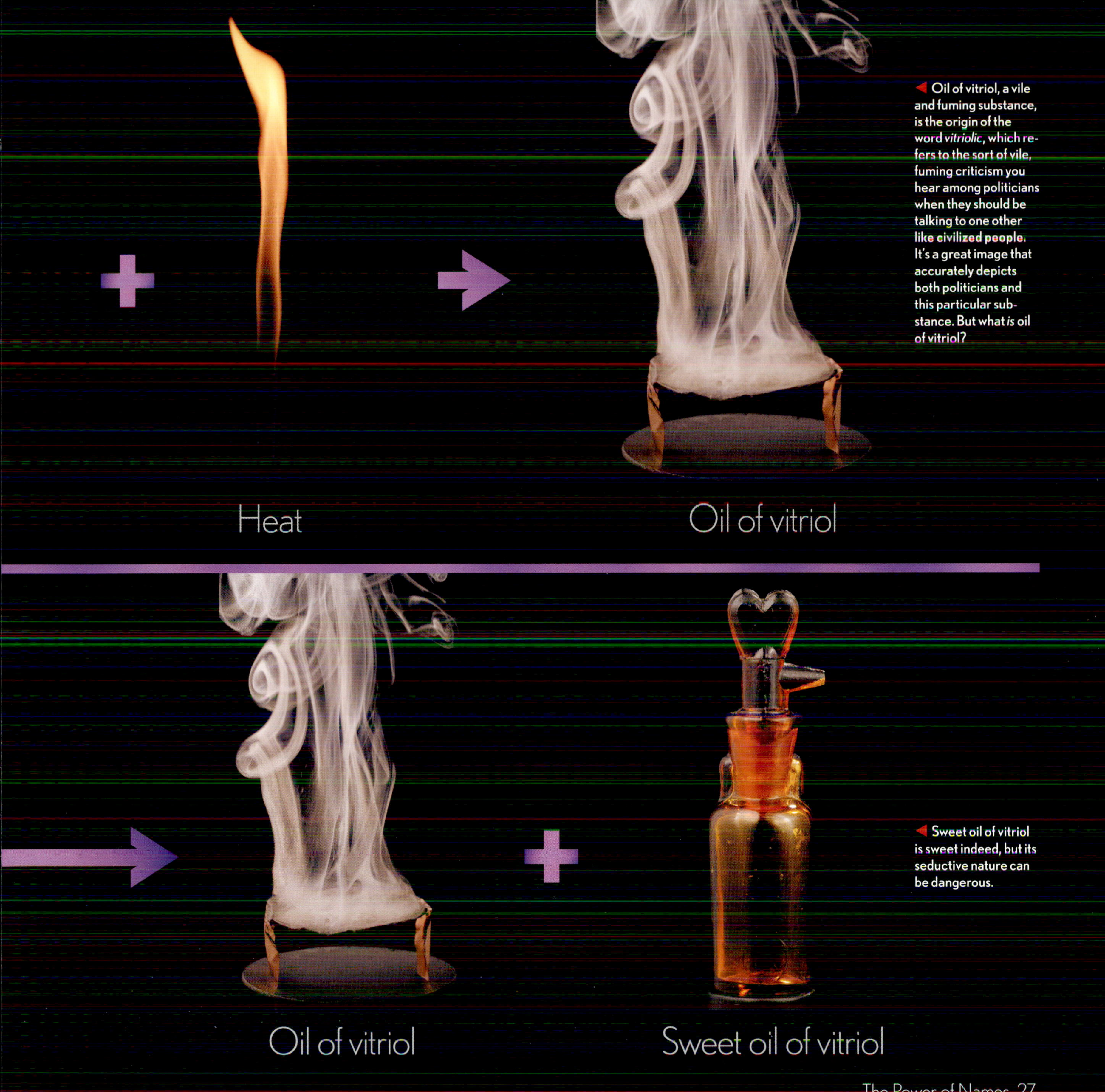

Oil of vitriol, a vile and fuming substance, is the origin of the word *vitriolic*, which refers to the sort of vile, fuming criticism you hear among politicians when they should be talking to one other like civilized people. It's a great image that accurately depicts both politicians and this particular substance. But what *is* oil of vitriol?

Sweet oil of vitriol is sweet indeed, but its seductive nature can be dangerous.

Common Names

COMPOUNDS IN WIDE use today all have common names by which they are known and traded. Today we know oil of vitriol, for example, as battery acid, chamber acid, or Glover acid, depending on its concentration. You have probably heard of battery acid. Although the name tells you *how* it is used, does it really tell you anything about *what* the substance is?

Sweet oil of vitriol is common ether, which has been used as a surgical anesthetic in the past. Green vitriol doesn't have a modern common name, but it is sometimes known by its mineral form, rozenite.

Spirit of wine is, of course, grain alcohol. Again, this name is familiar. You may know that there's an important difference between grain alcohol and wood alcohol, but what is that difference?

To really understand these substances, you need to know their third name—the kind of name that gives you power over a thing.

Rozenite

▲ Rozenite is a mineral form of green vitriol. I still haven't told you what it is.

▶ Written this way, the reactions appear more familiar, but they still don't make a whole lot of sense. Why on earth would battery acid plus booze give you a gas that knocks you out? Well OK, maybe that does make a certain kind of sense, but why, *chemically*, would it do that?

Battery acid

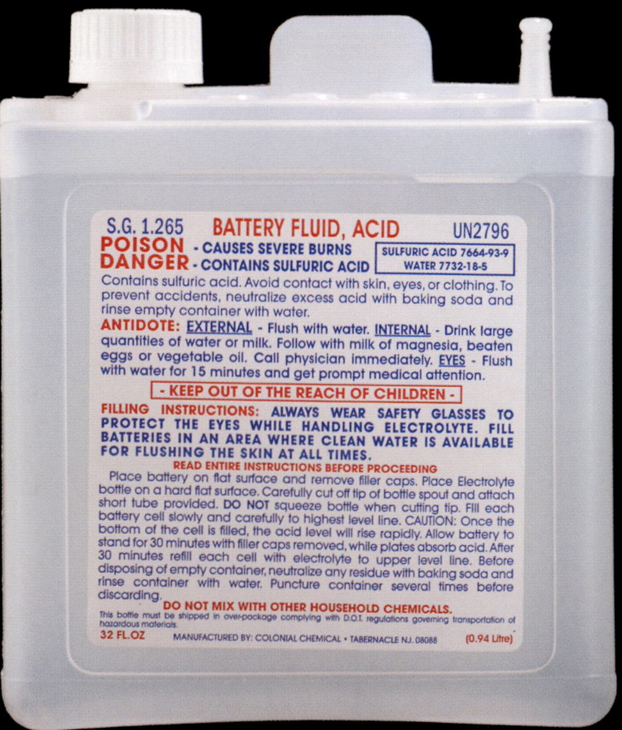
S.G. 1.265
BATTERY FLUID, ACID
UN2796
POISON - CAUSES SEVERE BURNS
DANGER - CONTAINS SULFURIC ACID
SULFURIC ACID 7664-93-9
WATER 7732-18-5
Contains sulfuric acid. Avoid contact with skin, eyes, or clothing. To prevent accidents, neutralize excess acid with baking soda and rinse empty container with water.
ANTIDOTE: EXTERNAL - Flush with water. INTERNAL - Drink large quantities of water or milk. Follow with milk of magnesia, beaten eggs or vegetable oil. Call physician immediately. EYES - Flush with water for 15 minutes and get prompt medical attention.
- KEEP OUT OF THE REACH OF CHILDREN -
FILLING INSTRUCTIONS: ALWAYS WEAR SAFETY GLASSES TO PROTECT THE EYES WHILE HANDLING ELECTROLYTE. FILL BATTERIES IN AN AREA WHERE CLEAN WATER IS AVAILABLE FOR FLUSHING THE SKIN AT ALL TIMES.
READ ENTIRE INSTRUCTIONS BEFORE PROCEEDING
Place battery on flat surface and remove filler caps. Place Electrolyte bottle on a hard flat surface. Carefully cut off tip of bottle spout and attach short tube provided. DO NOT squeeze bottle when cutting tip. Fill each battery cell slowly and carefully to highest level line. CAUTION: Once the bottom of the cell is filled, the acid level will rise rapidly. Allow battery to stand for 30 minutes with filler caps removed, while plates absorb acid. After 30 minutes refill each cell with electrolyte to upper level line. Before disposing of empty container, neutralize any residue with baking soda and rinse container with water. Puncture container several times before discarding.
DO NOT MIX WITH OTHER HOUSEHOLD CHEMICALS.
This bottle must be shipped in over-package complying with D.O.T. regulations governing transportation of hazardous materials.
32 FL.OZ
MANUFACTURED BY: COLONIAL CHEMICAL • TABERNACLE NJ. 08088
(0.94 Litre)

+

Grain alcohol

190
GEM CLEAR
Grain Alcohol
190
GEM CLEAR
Grain Alcohol
U.S.P.
FLAMMABLE HANDLE WITH CARE

+

Heat

◀ The purest form of alcohol you can buy for the purposes of human consumption is 95 percent grain alcohol and 5 percent water. (The "proof" of an alcoholic beverage is just the percentage of alcohol multiplied by two, so this stuff is sold as 190 proof alcohol.)

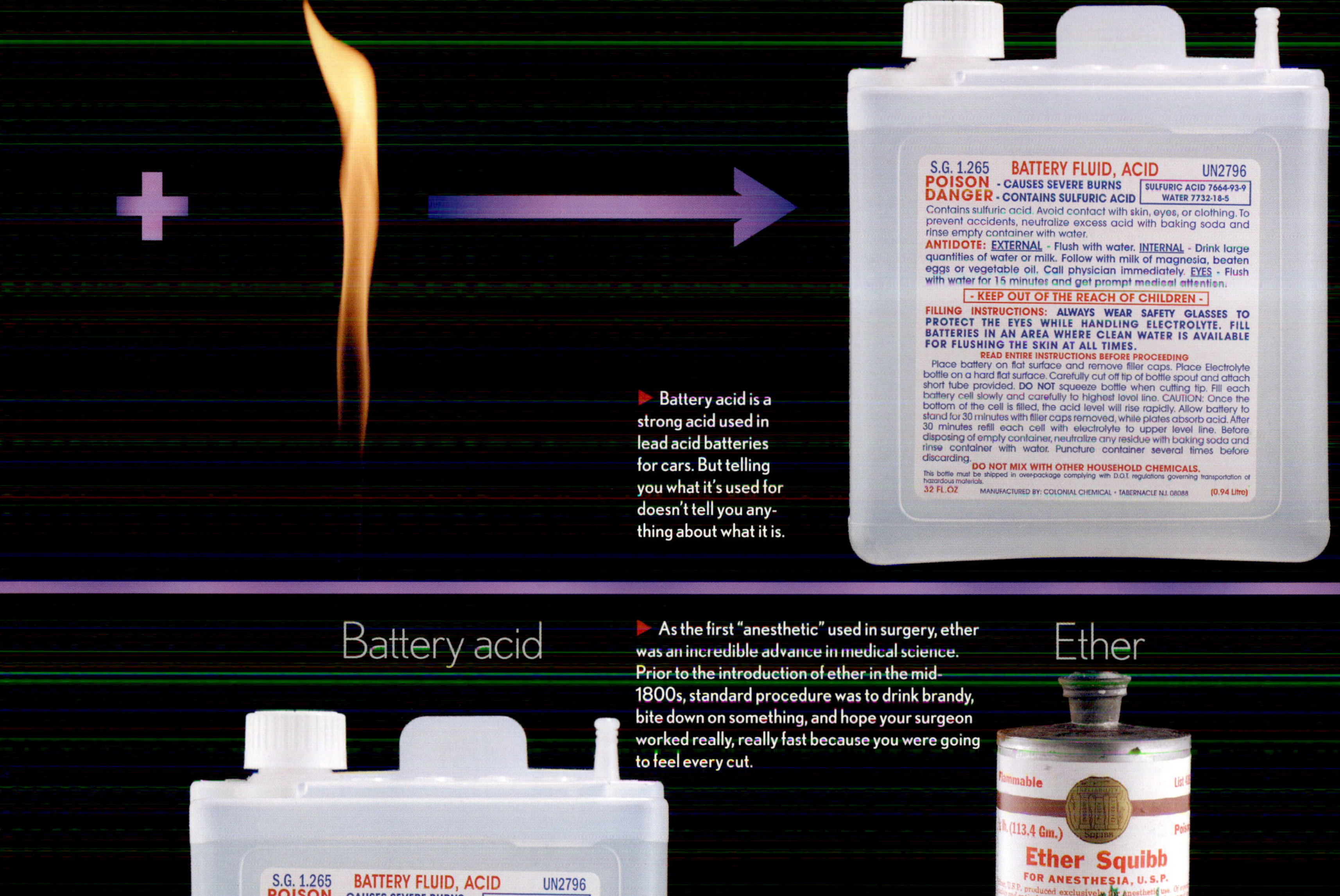
S.G. 1.265 BATTERY FLUID, ACID UN2796
POISON DANGER - CAUSES SEVERE BURNS - CONTAINS SULFURIC ACID
SULFURIC ACID 7664-93-9 WATER 7732-18-5
Contains sulfuric acid. Avoid contact with skin, eyes, or clothing. To prevent accidents, neutralize excess acid with baking soda and rinse empty container with water.
ANTIDOTE: EXTERNAL - Flush with water. INTERNAL - Drink large quantities of water or milk. Follow with milk of magnesia, beaten eggs or vegetable oil. Call physician immediately. EYES - Flush with water for 15 minutes and get prompt medical attention.
- KEEP OUT OF THE REACH OF CHILDREN -
FILLING INSTRUCTIONS: ALWAYS WEAR SAFETY GLASSES TO PROTECT THE EYES WHILE HANDLING ELECTROLYTE. FILL BATTERIES IN AN AREA WHERE CLEAN WATER IS AVAILABLE FOR FLUSHING THE SKIN AT ALL TIMES.
READ ENTIRE INSTRUCTIONS BEFORE PROCEEDING
DO NOT MIX WITH OTHER HOUSEHOLD CHEMICALS.
32 FL.OZ MANUFACTURED BY: COLONIAL CHEMICAL • TABERNACLE N.J. 08088 (0.94 Litre)

Battery acid is a strong acid used in lead acid batteries for cars. But telling you what it's used for doesn't tell you anything about what it is.

Battery acid

Ether

As the first "anesthetic" used in surgery, ether was an incredible advance in medical science. Prior to the introduction of ether in the mid-1800s, standard procedure was to drink brandy, bite down on something, and hope your surgeon worked really, really fast because you were going to feel every cut.

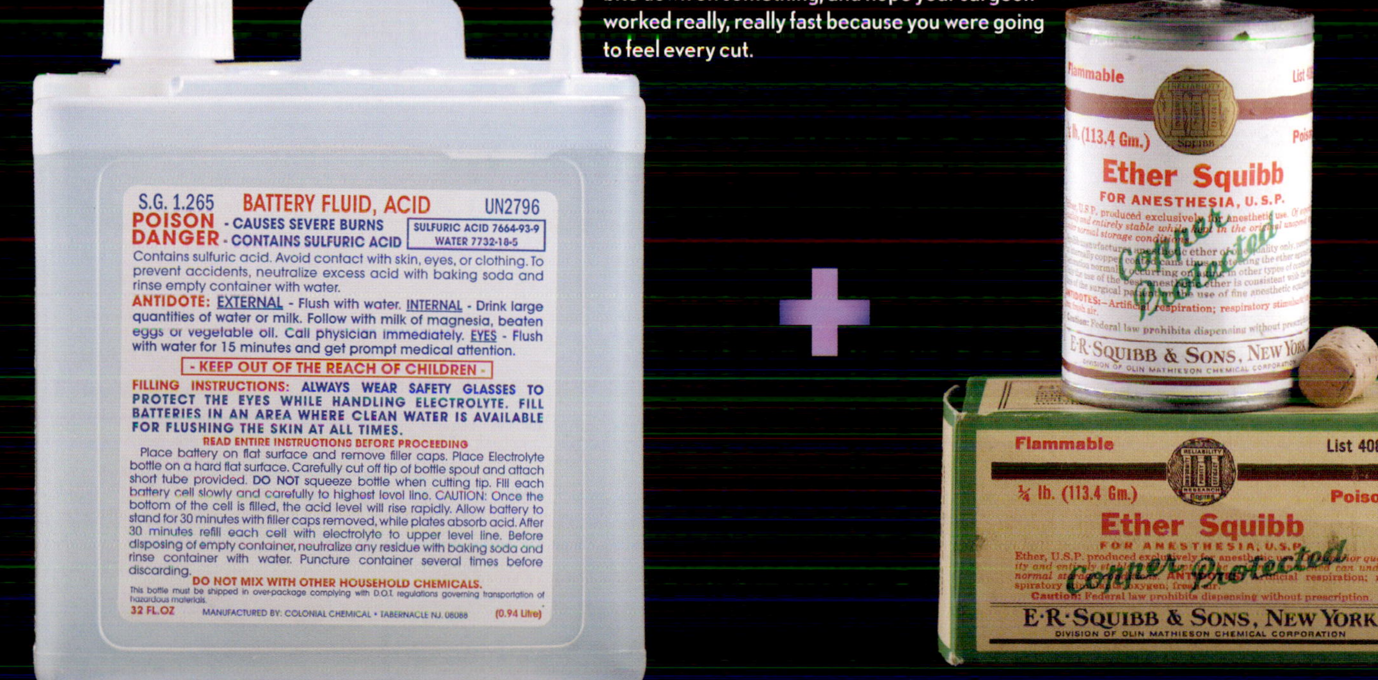
S.G. 1.265 BATTERY FLUID, ACID UN2796
POISON DANGER - CAUSES SEVERE BURNS - CONTAINS SULFURIC ACID
- KEEP OUT OF THE REACH OF CHILDREN -
DO NOT MIX WITH OTHER HOUSEHOLD CHEMICALS.
Flammable
List 4082
¼ lb. (113.4 Gm.)
Poison
Ether Squibb
FOR ANESTHESIA, U.S.P.
Copper Protected
E·R·SQUIBB & SONS, NEW YORK
DIVISION OF OLIN MATHIESON CHEMICAL CORPORATION

Systematic Names

IN THE EARLY 1800S, it became clear that chemical compounds were different arrangements of specific types of atoms in definite proportions. For example, today we know that green vitriol is composed of molecules that contain exactly one iron atom, one sulfur atom, and four oxygen atoms. Further, we know that the four oxygen atoms are strongly bonded to the sulfur atom and that this group of five atoms is bonded in a different way to the iron atom.

All this information is encoded in the separate parts of the modern systematic name for green vitriol, iron(II) sulfate, and its chemical formula, $FeSO_4$. Let's pick apart this particular name.

"Sulfate" always refers to a group of four oxygen atoms around a single sulfur atom; that's the SO_4 part. You will find this name embedded in the names of countless compounds (we'll see examples later on). "Iron," of course, refers to the element iron, whose symbol, for historical reasons, is Fe. The (II) refers to the electric charge the iron atom has in this particular compound, positive two, meaning it gave up two electrons when the compound was formed.

Each of the compounds in these reactions has a modern name that encodes deep knowledge about its true nature. We'll look at each one in more detail over the next few pages. Systematic names help us understand the substances and, more importantly, understand why they transform themselves in the particular and repeatable ways they do.

The power of these names is at the very heart of chemistry.

▶ Using modern systematic names and chemical formulas makes clear what is happening in these reactions: the exact same elements appear on both sides in the same numbers (i.e., the equation is balanced). The elements are simply being rearranged into new groupings that represent new compounds. See how, in the reaction below, the two smaller molecules of ethanol are merged into one large molecule of ether, with a water molecule left over where they joined? *That explains so many things!* See how H_2SO_4 (sulfuric acid) appears unchanged on both sides of the reaction? That means it's a *catalyst* that causes the reaction to happen but is not consumed in the course of the reaction (though it does become more and more diluted by the water that is created by the reaction).

▶ Roasting green vitriol (iron(II) sulfate) with water to create oil of vitriol (sulfuric acid) *means* this reaction.

▶ Heating oil of vitriol (sulfuric acid) with the spirit of wine (ethanol) *means* this:

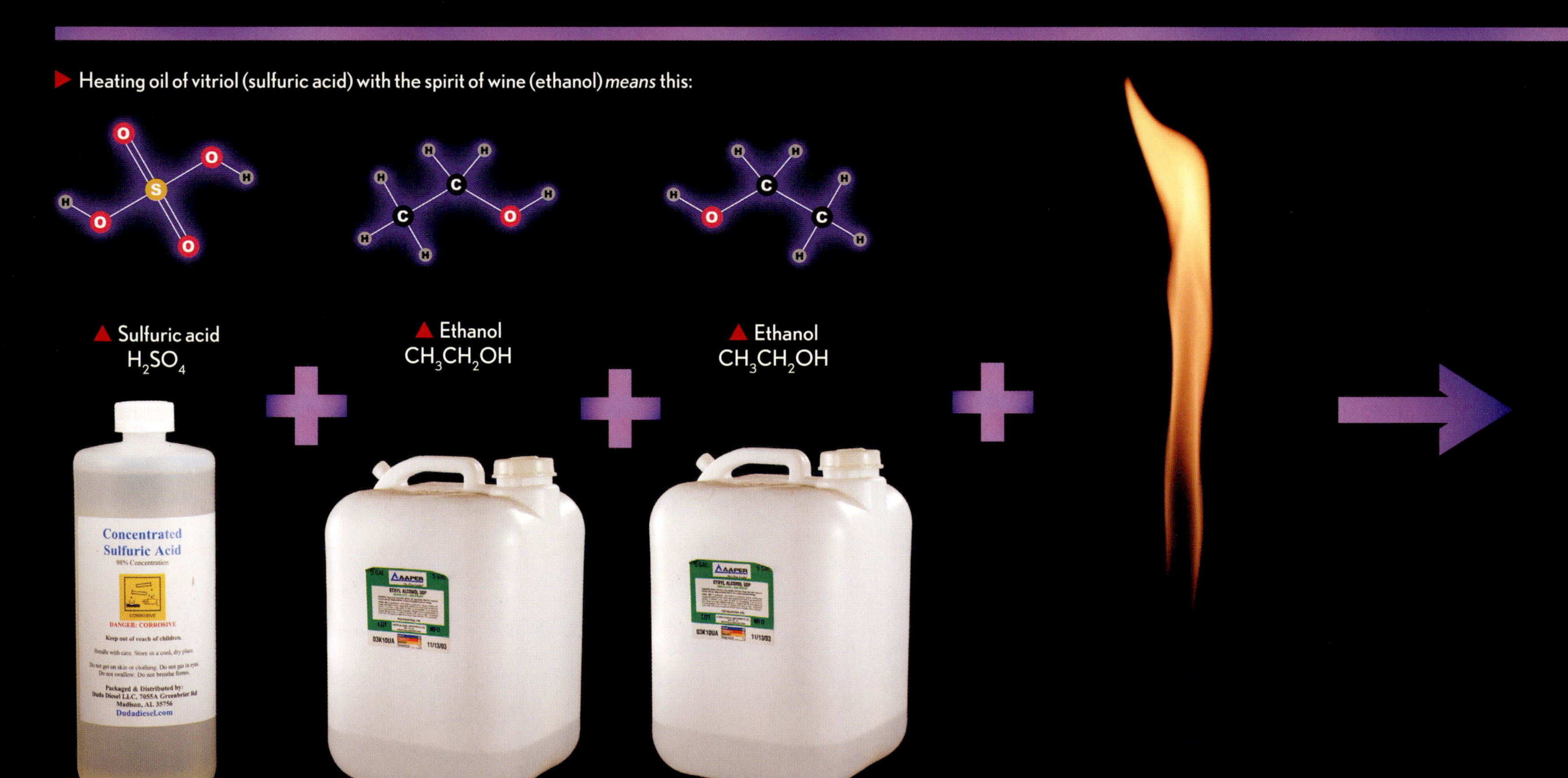

▲ Sulfuric acid H_2SO_4

▲ Ethanol CH_3CH_2OH

▲ Ethanol CH_3CH_2OH

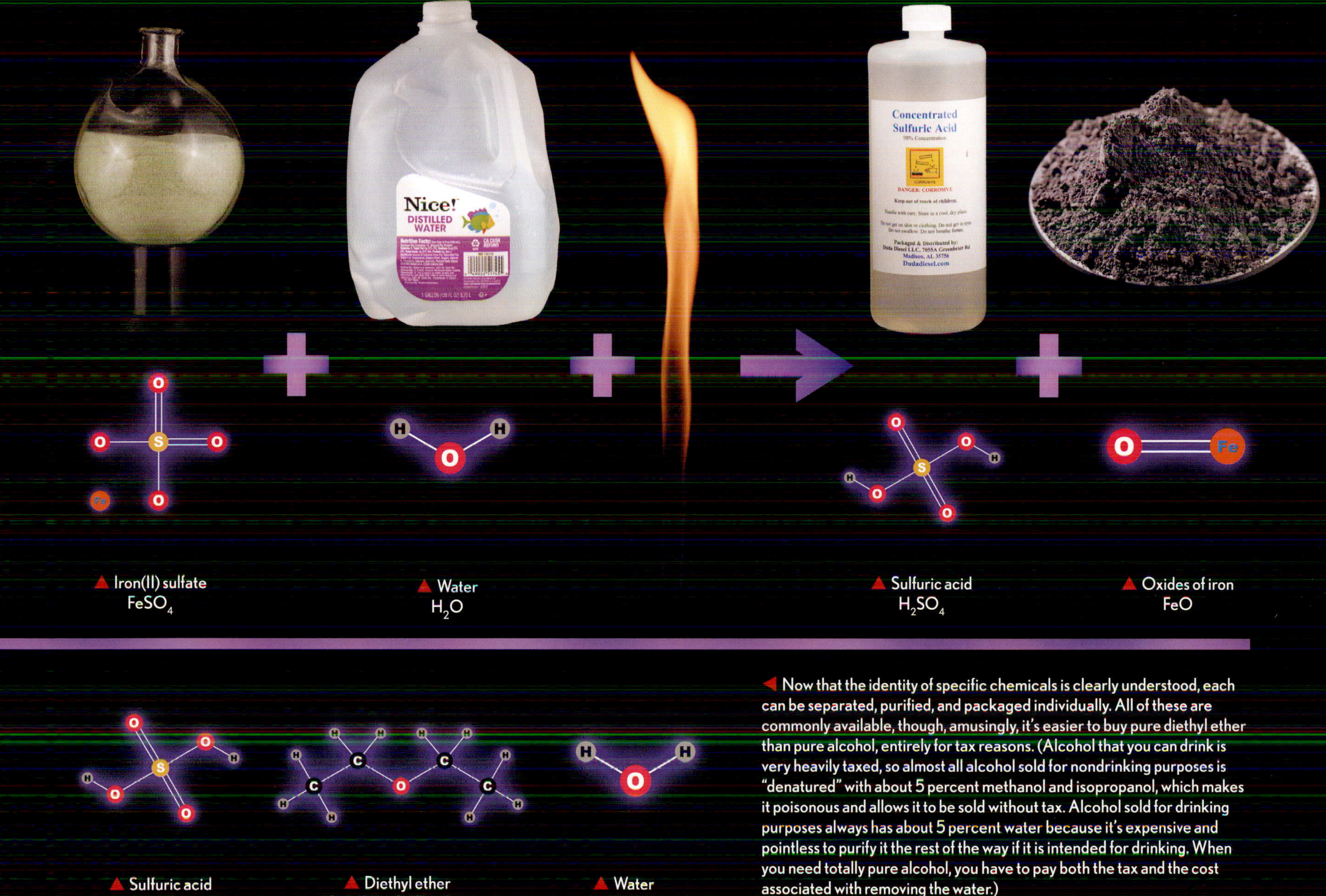

Now that the identity of specific chemicals is clearly understood, each can be separated, purified, and packaged individually. All of these are commonly available, though, amusingly, it's easier to buy pure diethyl ether than pure alcohol, entirely for tax reasons. (Alcohol that you can drink is very heavily taxed, so almost all alcohol sold for nondrinking purposes is "denatured" with about 5 percent methanol and isopropanol, which makes it poisonous and allows it to be sold without tax. Alcohol sold for drinking purposes always has about 5 percent water because it's expensive and pointless to purify it the rest of the way if it is intended for drinking. When you need totally pure alcohol, you have to pay both the tax and the cost associated with removing the water.)

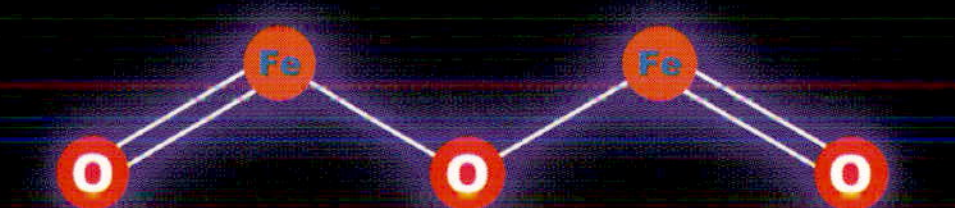

You may have noticed that we added an extra compound in the reaction above, FeO or Iron(II) oxide. This is an oversimplification. The reaction is more likely to produce some combination of different iron oxides such as Fe_2O_3 (iron(III) oxide) or Fe_3O_4 (Iron(II,III) oxide), but that's not important. The point is that knowing the chemical formula for the reactants and products made it obvious that the previous representations of the reaction were incomplete. The old names did not capture the essential truth that all substances are made of elements and that those elements are eternal. What you put in must balance perfectly with what comes out, because chemistry is a game of rearranging atoms, not creating or destroying them.

Where Names Take You: **Salts**

THE WONDERFUL THING about systematic names is that they invite variation. Start with green iron(II) sulfate ($FeSO_4$), change the "II" to a "III", and you get a yellow powder, $Fe_2(SO_4)_3$. Each SO_4 group still has a +2 charge, but each iron atom has a +3 charge, so in order to get the charges to add up to zero, you need two irons for every three sulfates.

Replace the iron with copper, and you get copper(II) sulfate, which grows into lovely, big, blue crystals. Stick with iron but substitute carbonate (CO_3) for the sulfate, and you get iron(II) carbonate, which makes pale, lustrous crystals. Substitute both and you get copper(II) carbonate, the green of weathered copper.

All of these compounds are examples of mineral salts. From their systematic names, you can tell exactly what elements they are made of and in what proportions.

▶ $Fe_2(SO_4)_3$

▲ Iron and sulfate in a two-to-three ratio yields the yellowish power known as ferric sulfate, iron(III) sulfate, and various mineral names—none of them particularly common and all of them of mixed composition.

▼ A sulfate group plus iron in equal numbers gives the green substance known variously as green vitriol, ferrous sulfate, iron(II) sulfate, and the mineral rozenite.

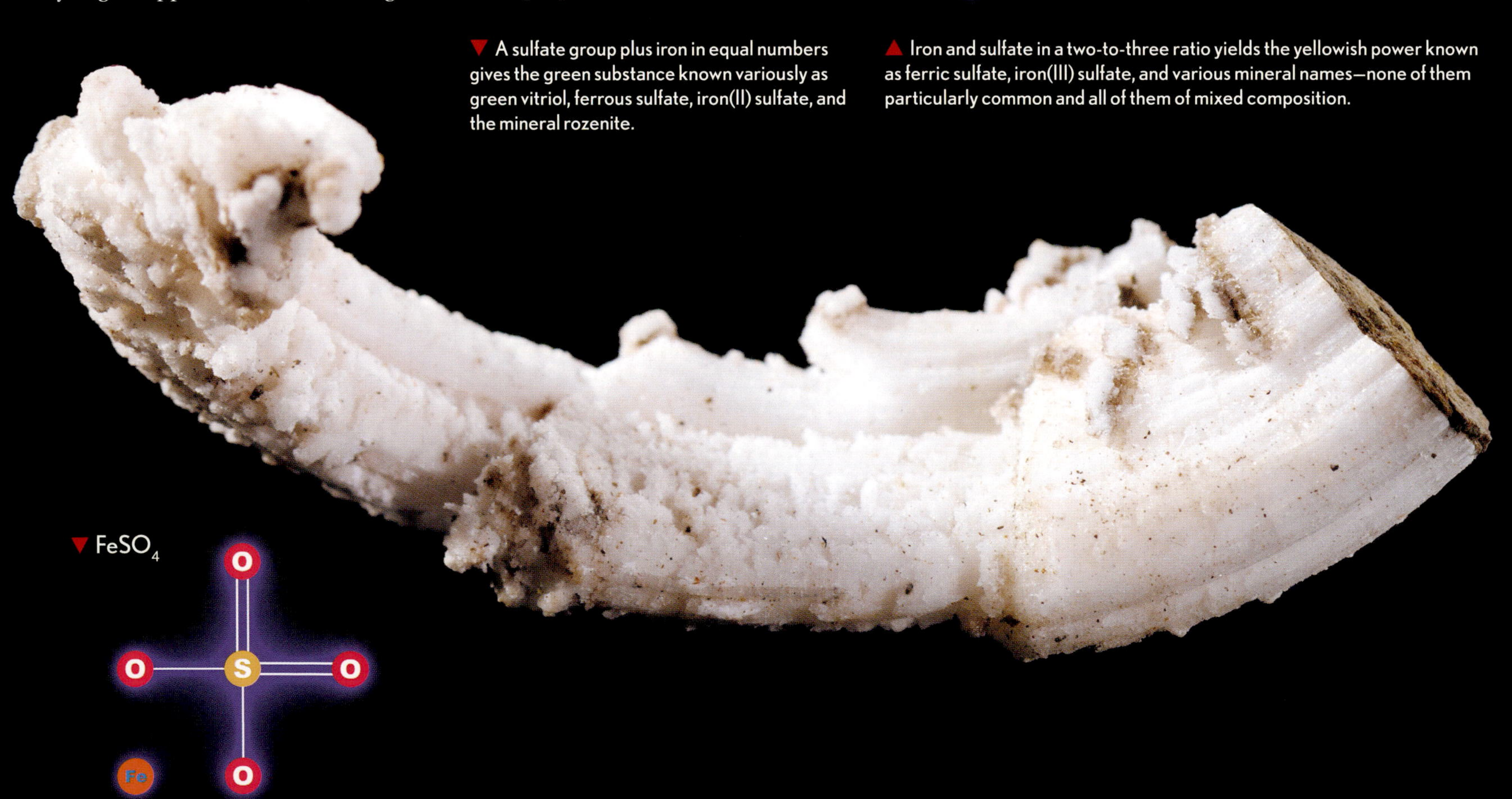

▼ $FeSO_4$

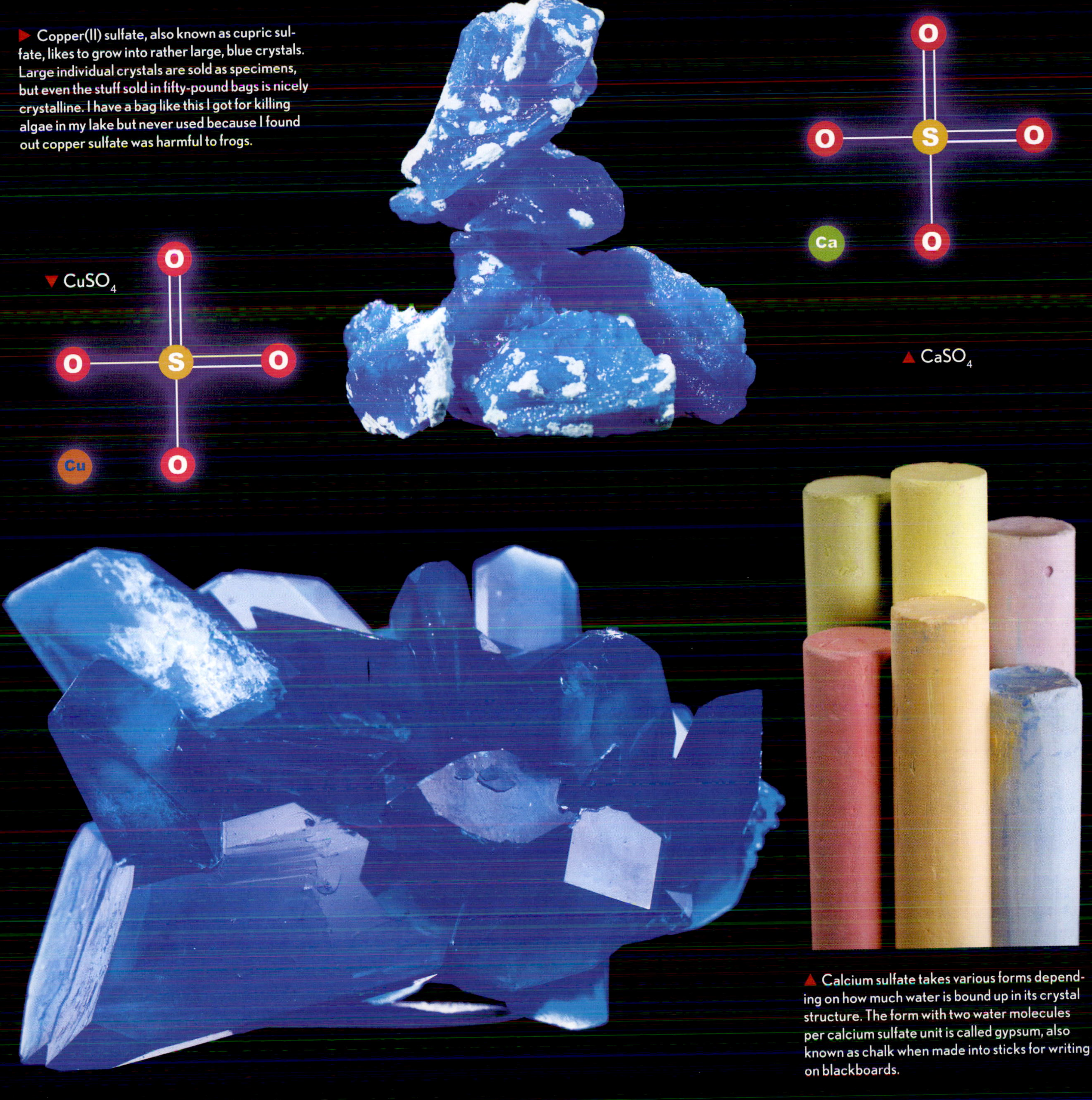

Copper(II) sulfate, also known as cupric sulfate, likes to grow into rather large, blue crystals. Large individual crystals are sold as specimens, but even the stuff sold in fifty-pound bags is nicely crystalline. I have a bag like this I got for killing algae in my lake but never used because I found out copper sulfate was harmful to frogs.

$CuSO_4$

$CaSO_4$

Calcium sulfate takes various forms depending on how much water is bound up in its crystal structure. The form with two water molecules per calcium sulfate unit is called gypsum, also known as chalk when made into sticks for writing on blackboards.

Where Names Take You: **Salts**

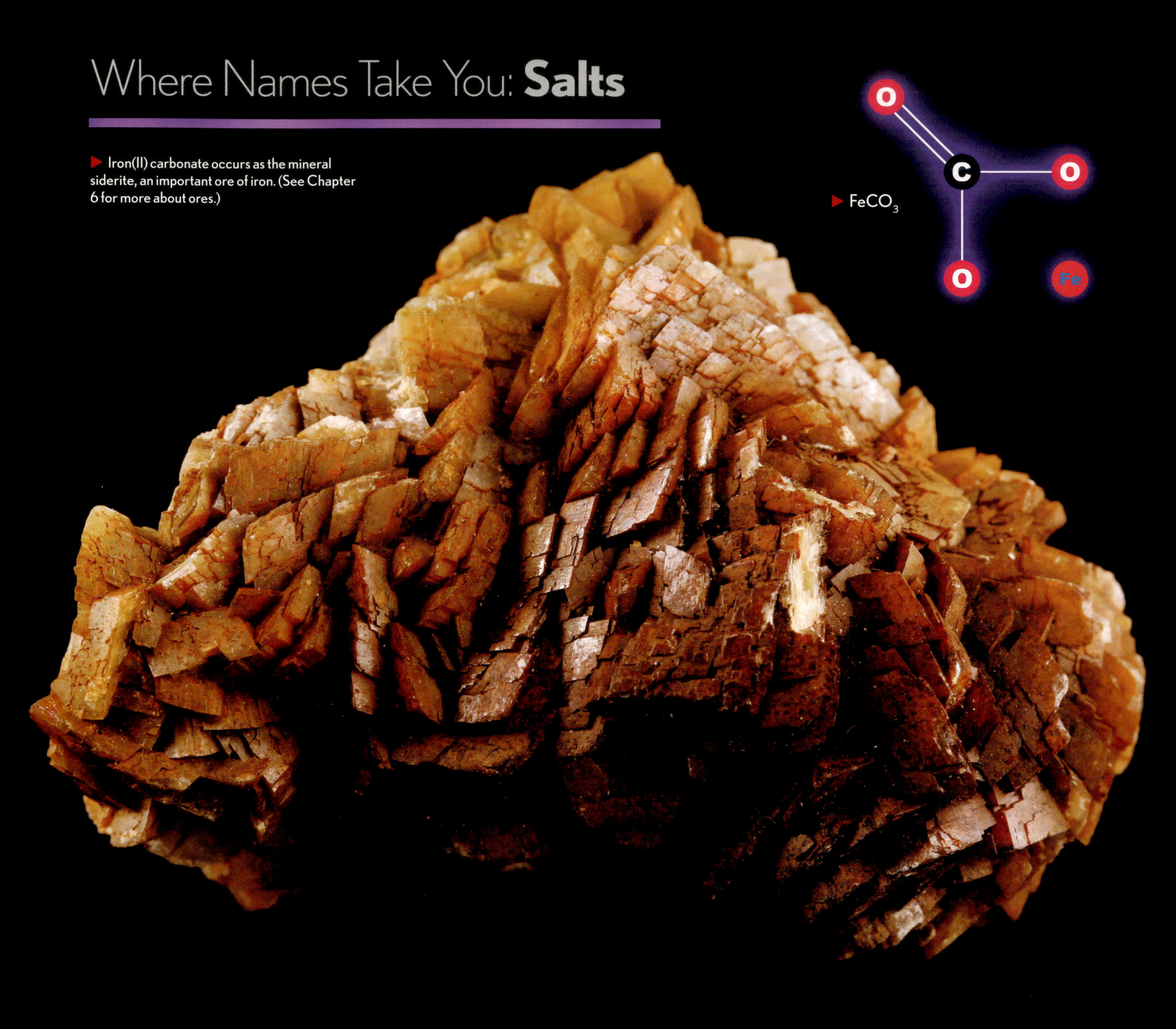

Iron(II) carbonate occurs as the mineral siderite, an important ore of iron. (See Chapter 6 for more about ores.)

Copper carbonate, together with copper hydroxide, form the green patina on copper roofs known as verdigris.

Sea shells are made of calcium carbonate, and so is limestone. This isn't a coincidence; a significant fraction of all the limestone in the world is composed of the broken-up remnants of sea creatures of one sort or another, which can be seen on microscopic examination. Imagine the great depths of time required for countless generations of corals, clams, and microorganisms to live out their lives, pass on their legacy, then die and sink to the ocean floor—just so you can have a gravel driveway. Our own lives, of course, are far less meaningful. We leave nothing, decaying into plant food and fertilizer in just a few years. These creatures built mountains. Our cities rest on their bones.

Where Names Take You: **Acids**

SULFURIC ACID (H_2SO_4) contains the same group of one sulfur and four oxygen atoms that iron(II) sulfate ($FeSO_4$) does, but it's loosely paired with two hydrogen atoms instead of one iron atom. It's these loosely bound hydrogen atoms that make it an acid.

The term *acid* refers very particularly to substances that release free hydrogen ions (H^+) when dissolved in water. It's these hydrogen ions that do the work of eating your face off—or whatever else the acid is called upon to do. The rest of the molecule that hangs out with the hydrogen is important mainly in determining how much hydrogen will be liberated (how strong the acid is).

Acids vary widely in how easily they release hydrogen ions, from very strong ones that dissociate completely to weak ones that release only a small fraction of their hydrogen payload. They can be harsh inorganic compounds or tame, even delicate, organic compounds, depending on what sort of molecule is delivering the hydrogen.

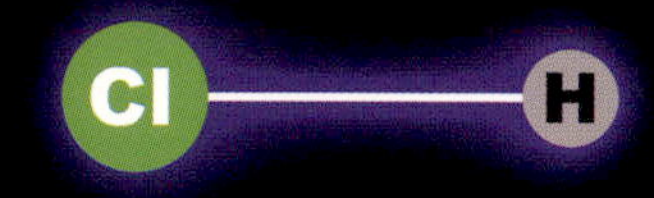

▲ Hydrochloric acid

▲ If you replace the sulfate (SO_4) group in H_2SO_4 with chlorine, you get HCl, hydrochloric acid, which is also a vile, fuming, powerfully corrosive acid that would just as soon eat you as give you the time of day. Here we see concentrated HCl, sold in hardware stores under its common name muriatic acid, being poured onto crushed limestone such as one finds commonly in gravel drives and paths. Acid attacks limestone.

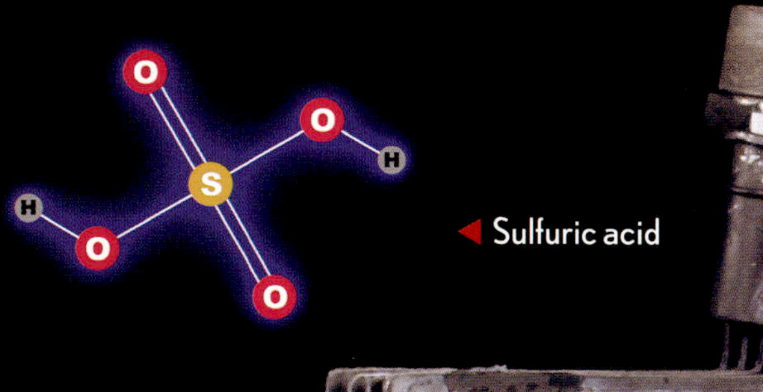

◀ Sulfuric acid

▶ Battery acid, which is about 30 percent sulfuric acid dissolved in water, gets its name because it is very widely used in lead-acid storage batteries used to start cars, motorcycles, and other motorized vehicles. This type of battery can deliver very high currents to run powerful starter motors, but lead acid batteries are very heavy due to the lead metal plates immersed in the acid.

▶ Lysergic acid diethylamide

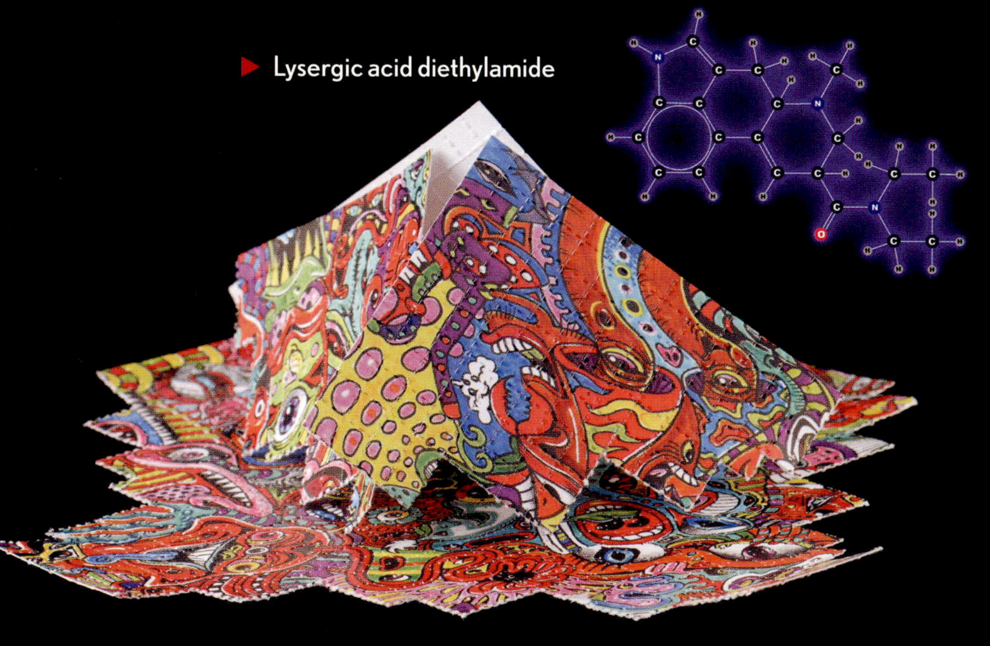

▲ Some acids are dangerous because they can eat your skin. This one is dangerous because it might make you eat your own skin. LSD, the acid being referred to in such phrases as, "Hey man, let's drop some acid," is known chemically as lysergic acid diethylamide. It is indeed a weak acid, but that is hardly the point with a substance like this. Shown here are "blotters," which are small squares of paper that are traditionally soaked in LSD and meant to be placed on the tongue to deliver the drug. (These particular ones are made for art only—no drug content—and are thus perfectly legal for nostalgic hippies to own.)

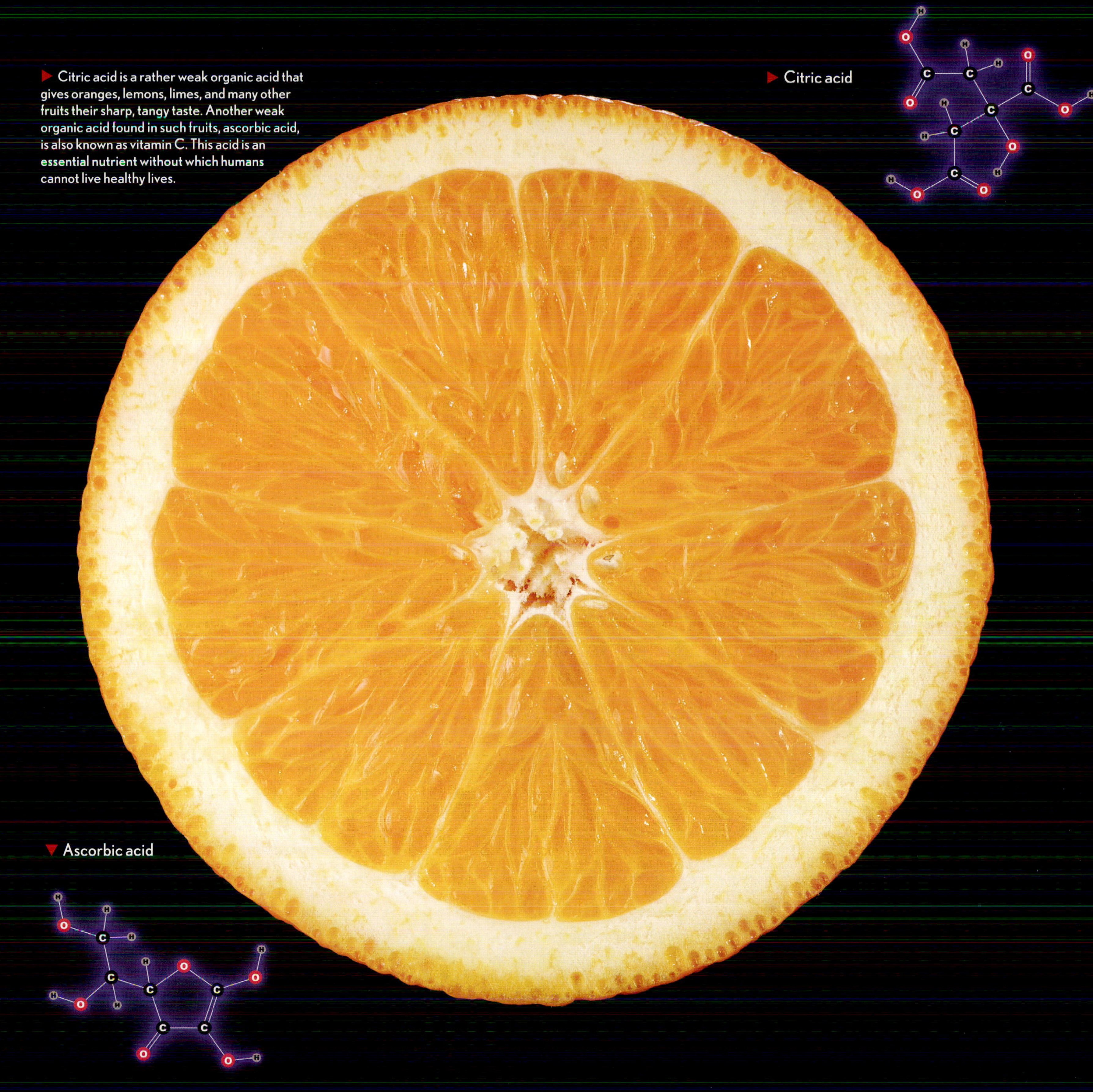

Citric acid is a rather weak organic acid that gives oranges, lemons, limes, and many other fruits their sharp, tangy taste. Another weak organic acid found in such fruits, ascorbic acid, is also known as vitamin C. This acid is an essential nutrient without which humans cannot live healthy lives.

Where Names Take You: **The Spirit of Wine**

OF THE COMPOUNDS we've been talking about, it's the spirit of wine, ethanol, that has the most potential to lead us to interesting places (and not just because it gets humans and other mammals drunk). It's an example of an organic compound, as are the large majority of all chemical compounds that have been studied and named. (In Chapter 3, we'll go into more detail about what exactly makes a compound organic.)

On pages 19 and 20, we saw the great variety of compounds you can make using just carbon and hydrogen. Ethanol is an example of what you can create if you add oxygen to the mix—a nice little constellation of compounds that includes alcohols, aldehydes, ketones, acids, and esters.

It's this specific collection of names that first got me interested in organic chemistry, and I'd like to show you how they all relate to each other by building up a series of compounds of increasing complexity. This is how chemists think about molecules: as building blocks that can be snapped together in many different ways, using techniques that have been developed over centuries.

▶ Methanol

H—H

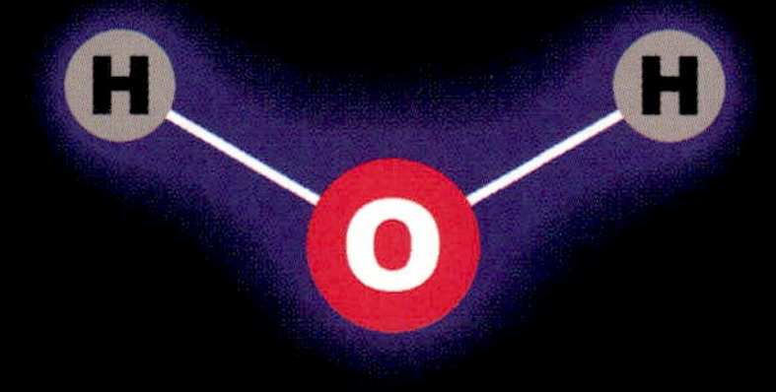

▲ This is the simplest possible thing made out of more than one atom: a molecule of hydrogen gas, H_2. Hydrogen is an element, but in pure form at room temperature, it is always paired up like this, so even though it's an element, it's also a molecule. (But it's not a compound because it's only got one kind of element in it.)

▲ If we insert an oxygen atom between the two hydrogen atoms, the result is H_2O, better known as water. (This insertion is very easy to do; water is what you get when you burn hydrogen in air.)

▶ If you replace one of the two hydrogen atoms in water with a simple carbon "side chain," you get an alcohol. Adding one carbon yields methanol (wood alcohol), and adding two carbons yields ethanol (grain alcohol). Countless other alcohols are known, all of them defined by the fact that they have an oxygen + hydrogen group stuck on them. This -OH group is what defines the word *alcohol*. In the modern naming system, the names of alcohols all end in "ol".

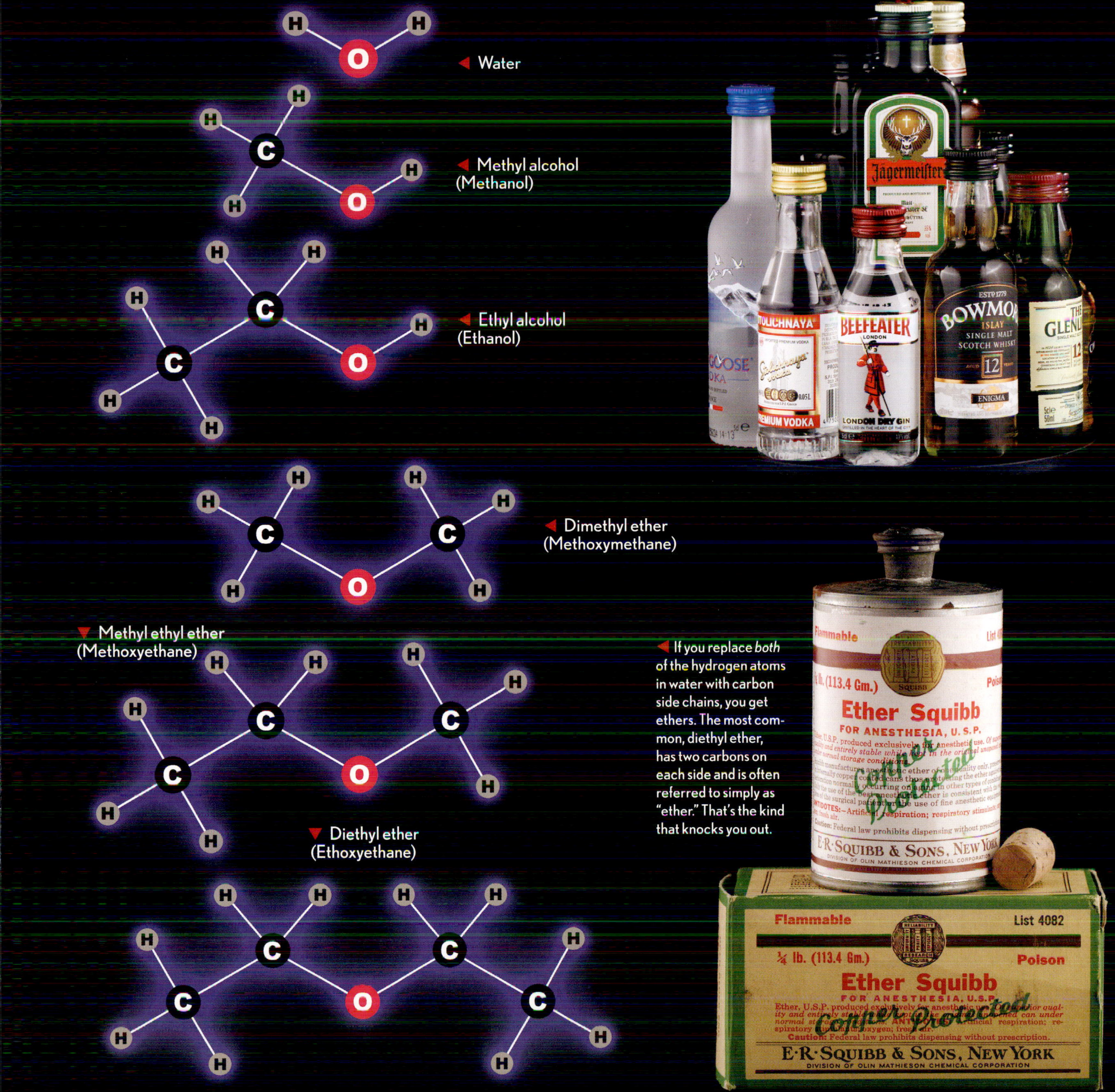

If you replace *both* of the hydrogen atoms in water with carbon side chains, you get ethers. The most common, diethyl ether, has two carbons on each side and is often referred to simply as "ether." That's the kind that knocks you out.

We can go in a slightly different direction and replace the oxygen atom in water with a carbon double-bonded to an oxygen (which is called a carbonyl group). The simplest case, with a hydrogen atom on both sides, is formaldehyde, the slightly scary liquid of preserved-dead-animals fame. (It is called methanal in the formal system.)

If you replace one hydrogen in methanal with a single carbon side chain, you get ethanal. As with alcohols, thousands of aldehydes can be constructed, all containing a -COH group and all with names ending in "al".

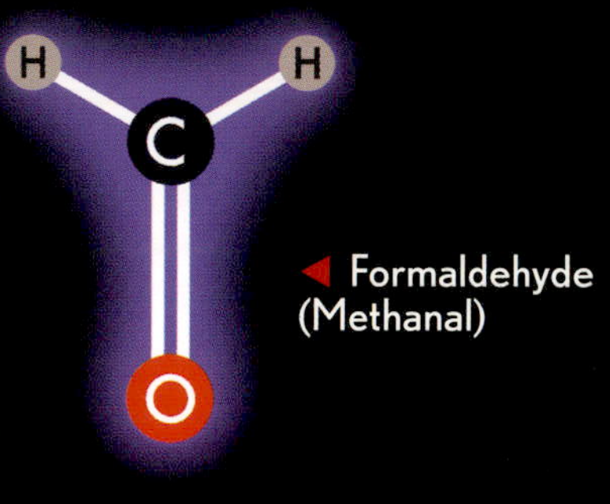

Formaldehyde (Methanal)

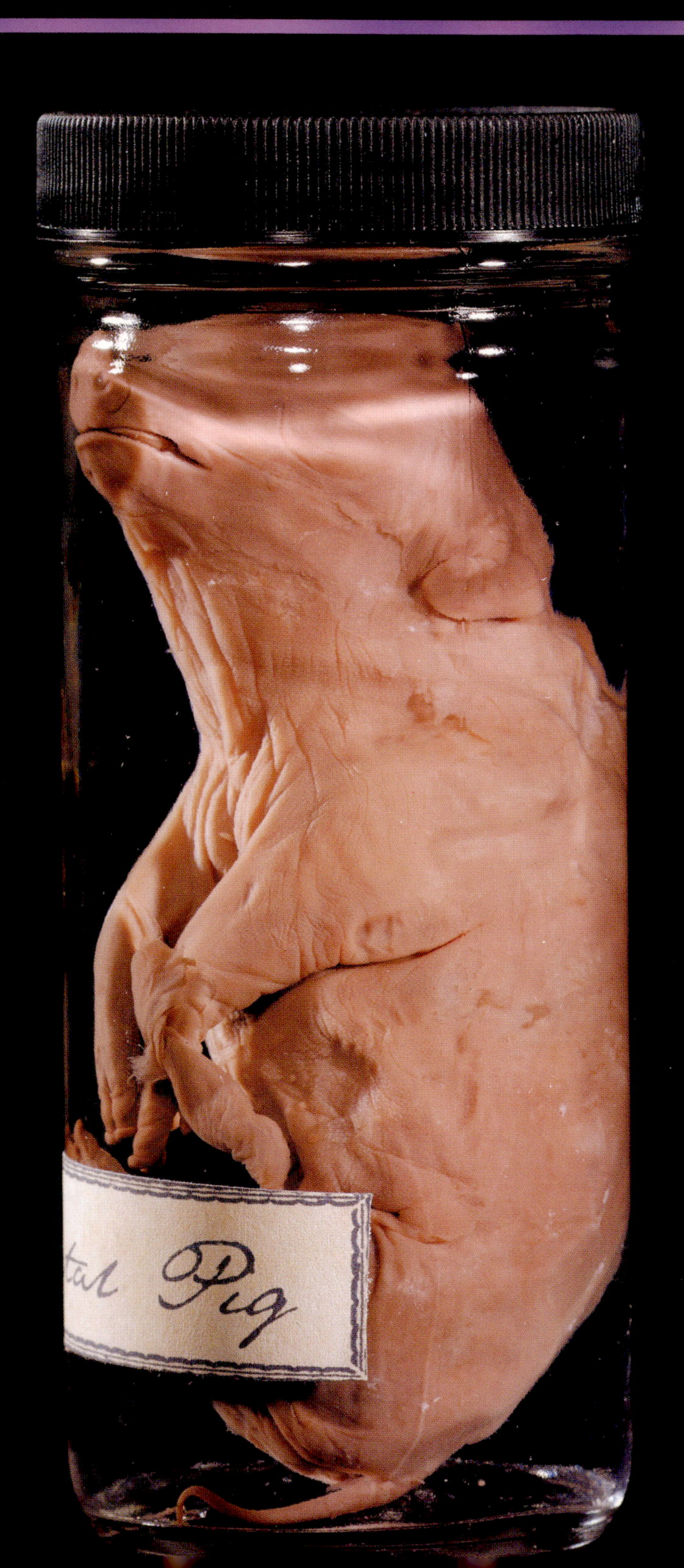

Acetaldehyde (Ethanal)

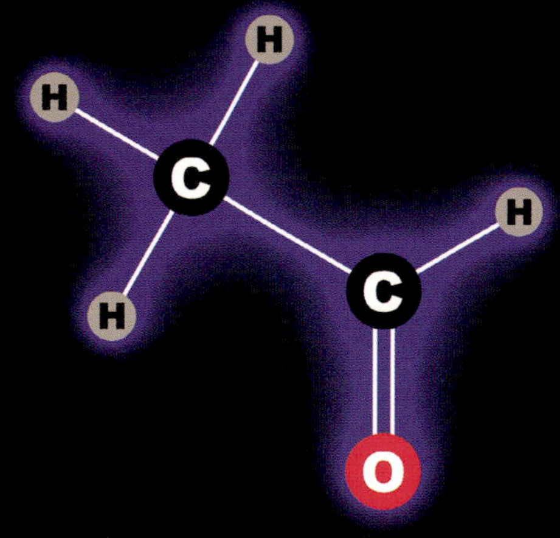

Propionaldehyde (Propanal)

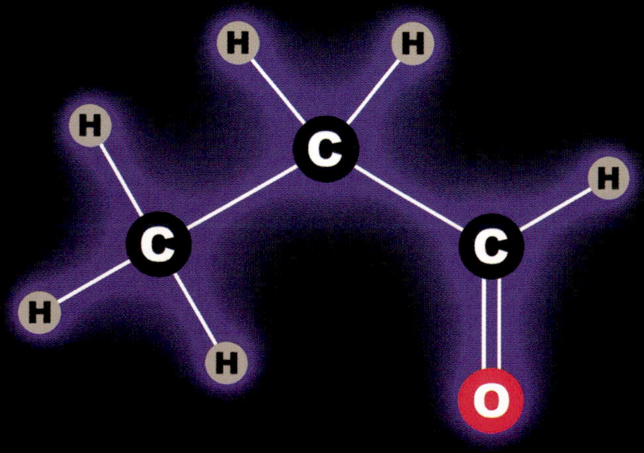

Where Names Take You: **Ketones**

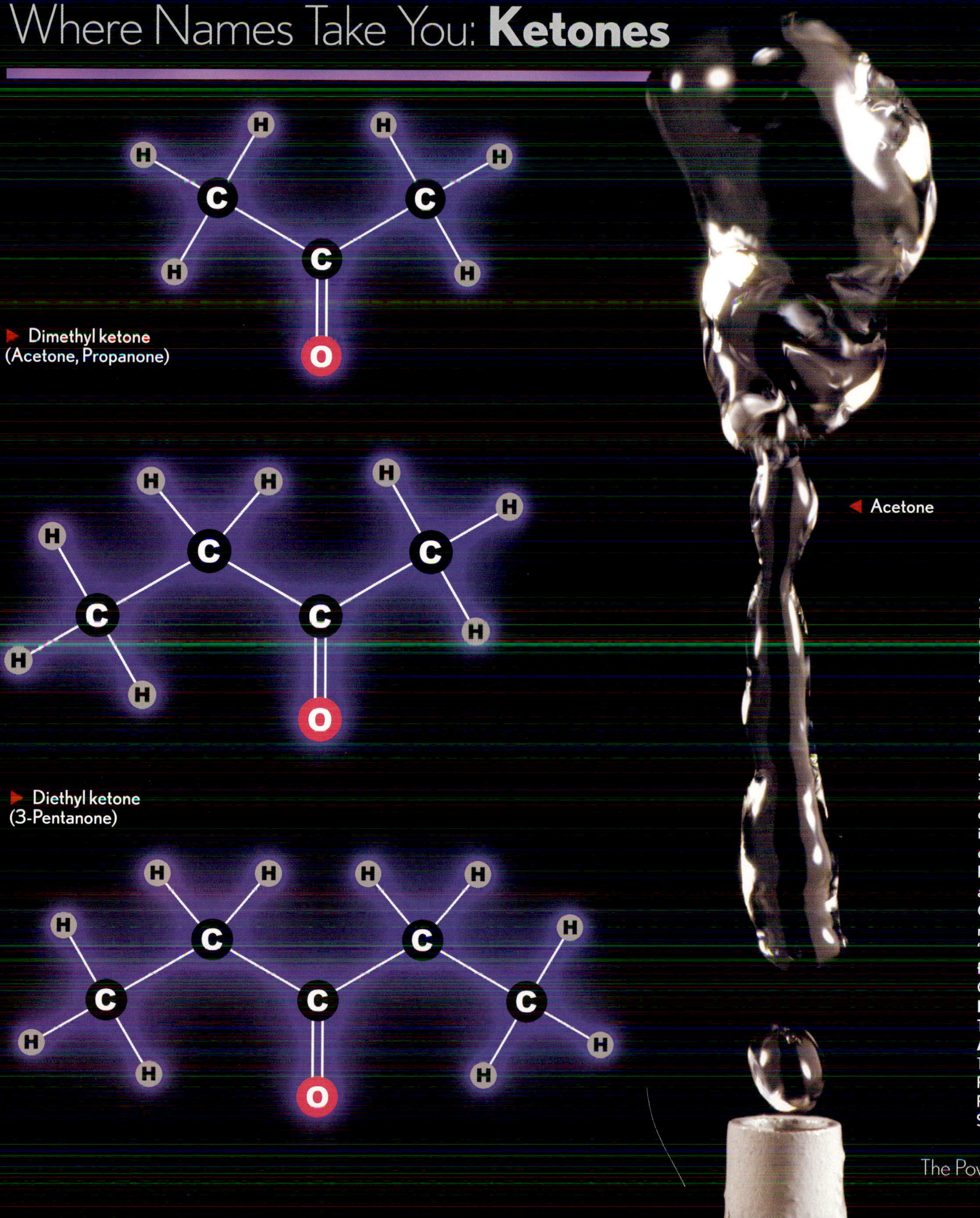

Dimethyl ketone (Acetone, Propanone)

Diethyl ketone (3-Pentanone)

If you replace both hydrogens in methanal with carbon side chains, you get ketones. The simplest, acetone, is a volatile, super-flammable solvent that is created inside the body as one of the three "ketone bodies" that result from eating a high-fat, low-carbohydrate diet. (This diet is used to treat epilepsy and, by some people, to achieve weight loss.) Acetone is largely a waste product in this diet, but the other ketones created along with it are valuable sources of energy for the body.

Acetone

All the names you're seeing on these pages are built from a set of roots that say how many carbon atoms are in each part of the molecule. For example, the prefixes "meth-" and "form-" mean one carbon, so "methanol" is an alcohol with one carbon, formaldehyde is an aldehyde with one carbon. The first four have special names, after that they become Greek and Latin number names, as in "penta" for five, "hexa" for six, and so on.

One carbon: **Meth-, Form-**
Two carbons: **Eth-, Acet-**
Three carbons: **Prop-**
Four carbons: **But-**
Five carbons: **Pent-**
Six carbons: **Hex-**

Where Names Take You: **Organic Acids**

▶ We can get more elaborate by combining the carbonyl (-CO-) and alcohol (-OH) groups we've talked about, resulting in a -COOH group. Any organic molecule with one of these groups on it is called an organic acid. The simplest, formic acid, has just a single carbon atom. Add one extra carbon, and you get acetic acid, which creates the sour taste in vinegar.

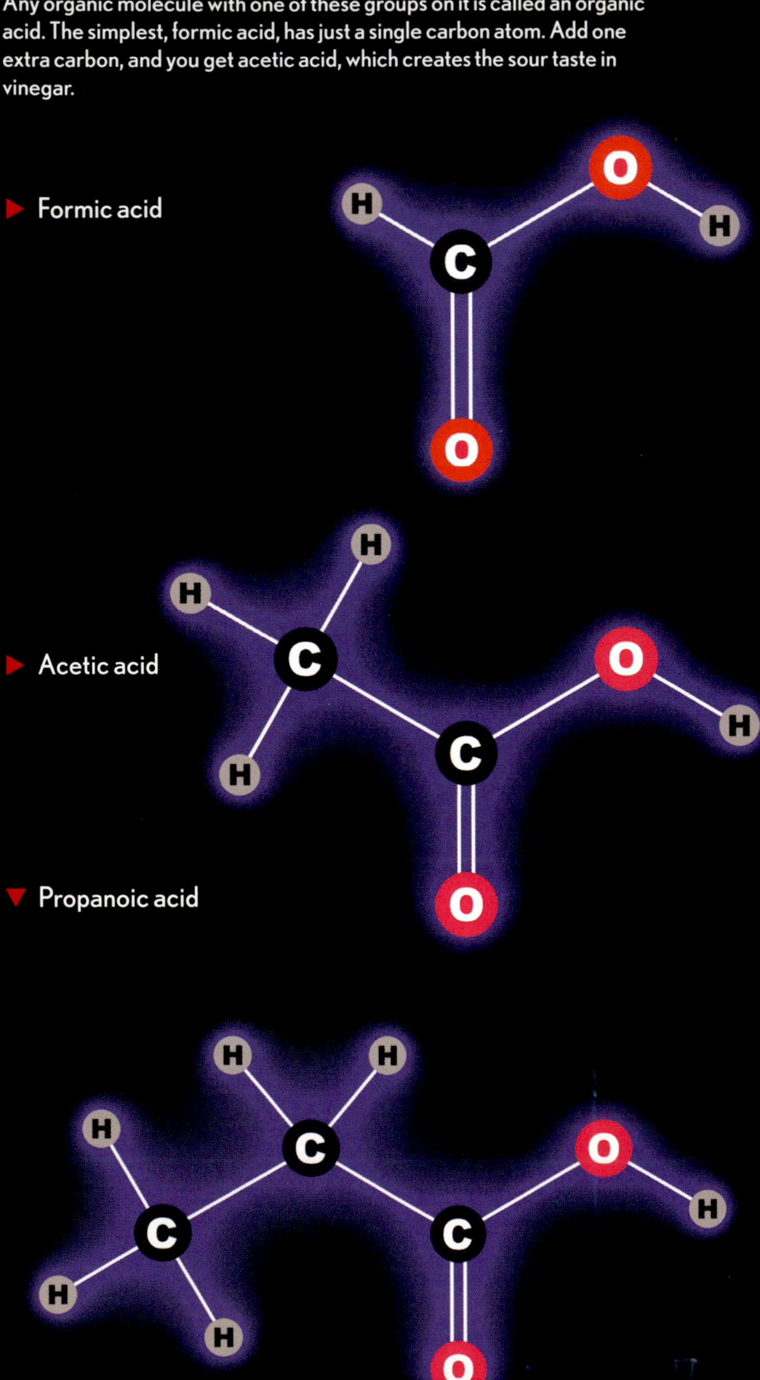

▶ Formic acid

▶ Acetic acid

▼ Propanoic acid

Where Names Take You: **Esters**

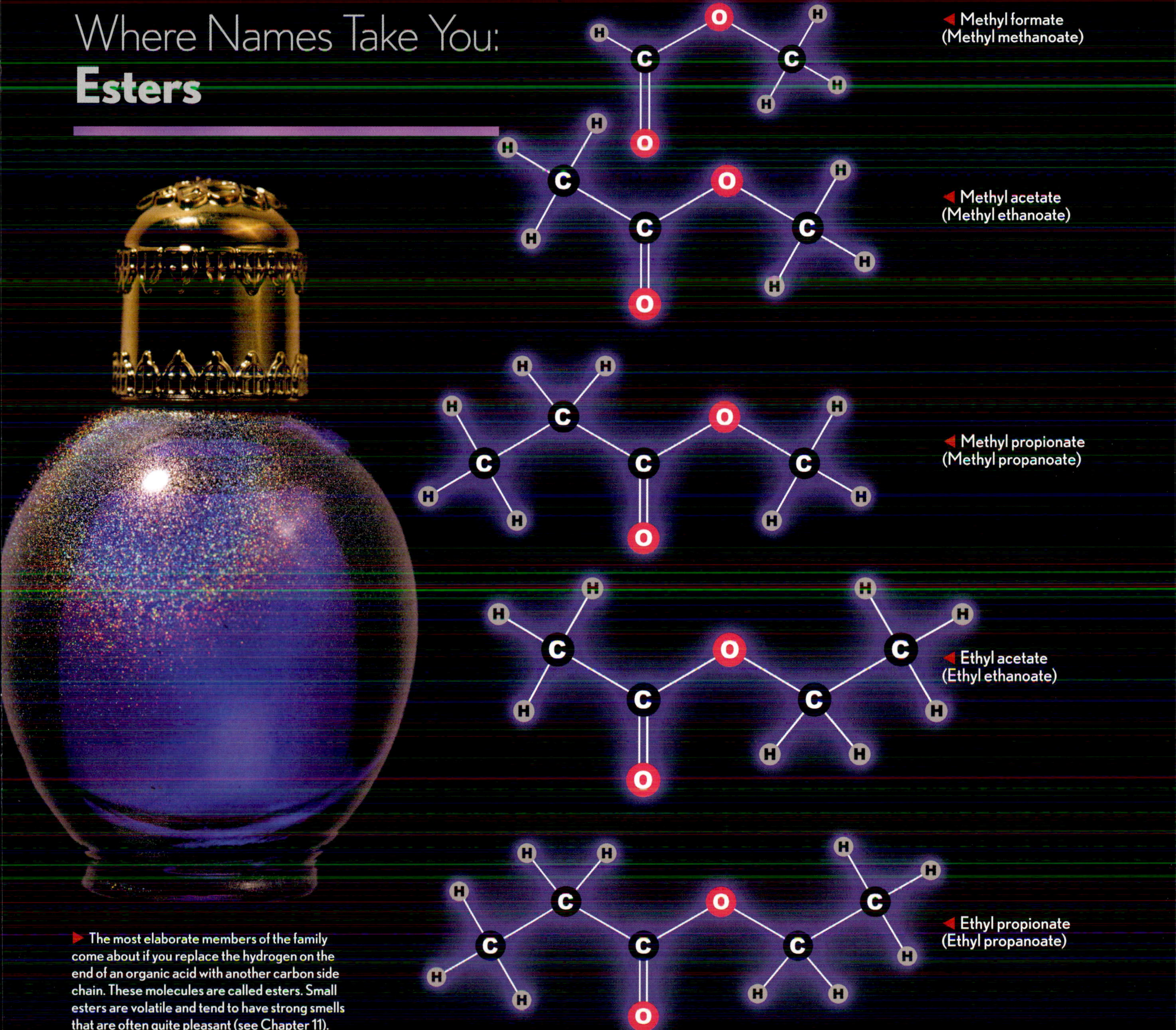

The most elaborate members of the family come about if you replace the hydrogen on the end of an organic acid with another carbon side chain. These molecules are called esters. Small esters are volatile and tend to have strong smells that are often quite pleasant (see Chapter 11).

Where Names Take You: **Esters**

▶ An ester with four carbons on the left and two on the right (called ethyl butyrate) is the smell of pineapple.

▶ An ester with four carbons on the left and five on the right (called pentyl butyrate) is the smell of apricots.

▲ Esters with long carbon side chains are the major constituents of natural wax. Beeswax, for example, is made largely of an ester with fifteen carbons to the right of the –COO– link and thirty to the left, a compound called triacontanyl palmitate. (See page 84 for more about waxes.)

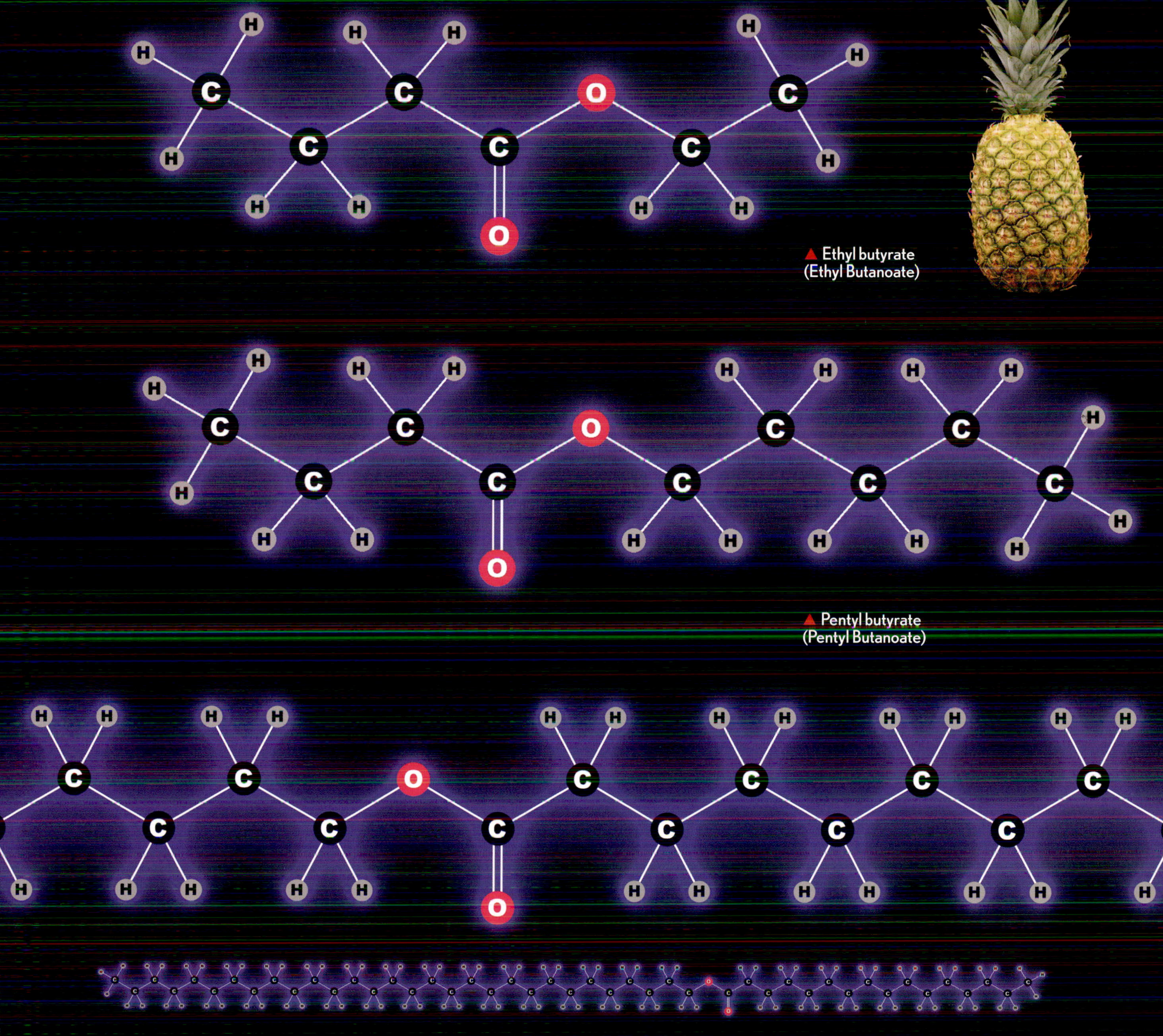

Ethyl butyrate
(Ethyl Butanoate)

Pentyl butyrate
(Pentyl Butanoate)

Triacontanyl palmitate
(Triacontyl Hexadecanoate)

Eiderdown, which grows on the belly of the Eider duck, is made of a complex organic molecule called keratin.

Chapter 3

Dead or Alive

THE WORLD OF CHEMICALS can be divided broadly into organic and inorganic compounds. The name *organic compound* intuitively feels soft, like something you might find growing in the garden. Indeed, many organic compounds are closely associated with life. On the other hand, the term *inorganic compound* sounds gritty, like a rock, and in fact rocks generally are inorganic. But this hard-versus-soft definition doesn't really work out—there are just too many exceptions.

So how, exactly, is the division between organic and inorganic defined?

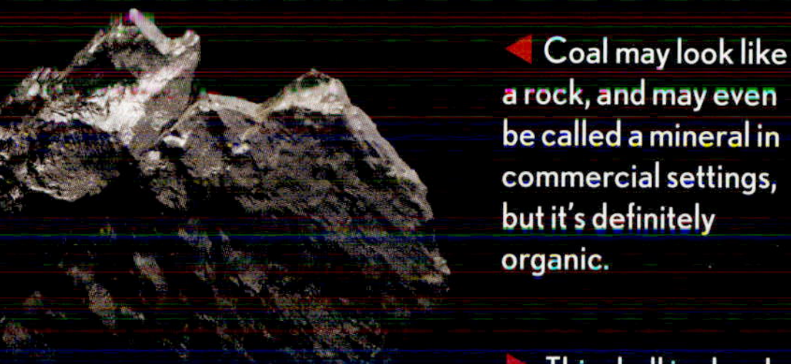

◀ Coal may look like a rock, and may even be called a mineral in commercial settings, but it's definitely organic.

▶ This skull is clearly from a living creature (a frilled dragon to be exact), but it's not an organic compound. It's made mostly of hydroxylapatite, a calcium phosphate mineral.

▶ Are these inorganic quartz crystals? No, it's actually crystals of menthol, an organic compound found in essential oils, cough remedies, and cigarettes.

▶ Even though some are called "mineral" oils, all oils are organic compounds.

▼ Asbestos is a lovely, soft fiber, not unlike wool in some ways—but definitely an inorganic compound.

What's an Organic Compound?

IF YOU SEARCH for definitions of "organic compound," many sources will tell you they are any compounds containing carbon. This is plainly and obviously wrong, and I can prove it with one word: limestone. There's no question that limestone is an inorganic compound. It's a chalky, gritty, hard thing—the bedrock underneath a garden, not something growing in it. But the chemical formula for limestone is $CaCO_3$, calcium carbonate. And limestone is far from the only obviously inorganic compound containing carbon.

Look a little longer, and you'll find organic compounds defined as containing carbon and hydrogen bonded to each other. And many of them do have this structure. But again, this definition is shot down by a single counterexample: Teflon. This wonderfully slippery stuff contains a backbone of carbon bonds that is absolutely classic organic chemistry, unambiguously and obviously an organic polymer. But it has no hydrogen whatsoever in it, and neither do any members of the large class of fully substituted fluoro- and chloro-fluoro-carbon compounds used in spray cans and as refrigerants (including the ones that are destroying the ozone layer, along with a host of less harmful ones).

Is there any clear definition of what's organic?

▼ Teflon is made from perfluoroethylene, which means ethylene (C_2H_4) in which all the hydrogens have been replaced with fluorine.

▶ The chemical name for Teflon is polyperfluoroethylene, meaning many repeats of ethylene with all its hydrogens replaced with fluorine. In effect, it is polyethylene, a very common plastic (see Chapter 7), but with fluorine in place of all the hydrogen. Because the carbon–fluorine and carbon–carbon bonds are both very, very strong, this material is almost impossible to attack chemically.

▲ My favorite organic compound that contains no hydrogen: Teflon. This huge cylinder of the stuff is so slippery you have to be careful picking it up!

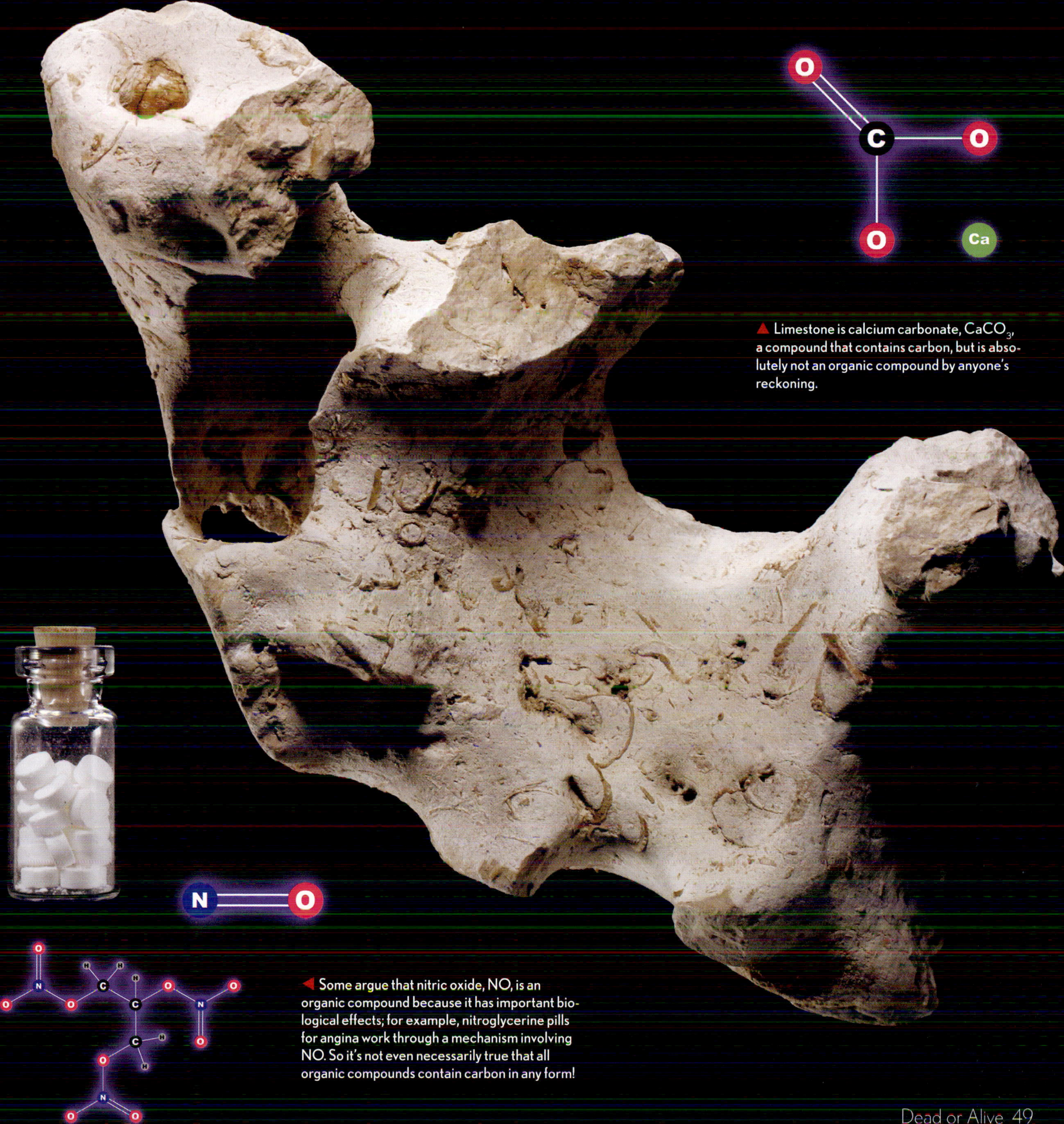

▲ Limestone is calcium carbonate, $CaCO_3$, a compound that contains carbon, but is absolutely not an organic compound by anyone's reckoning.

◀ Some argue that nitric oxide, NO, is an organic compound because it has important biological effects; for example, nitroglycerine pills for angina work through a mechanism involving NO. So it's not even necessarily true that all organic compounds contain carbon in any form!

The Compounds of Life

THE ORIGINAL DEFINITION of organic compounds was very clear: they were the compounds of life. In the early days of chemistry, many believed that living things contained a "vital force" without which certain chemical transformations were impossible. Organic compounds were those that originated in, and only in, living beings, created by the mystical life force within.

That definition, along with the whole idea of the vital force, came crashing down in 1828 with another simple counterexample: Friedrich Wöhler synthesized urea out of silver cyanate and ammonium chloride.

Urea was clearly known to be an organic compound; no one questioned that. Silver cyanate and ammonium chloride were definitely not organic compounds. It took a while for the magnitude of the synthesis of urea to sink in, but eventually educated people realized that this experiment, one of the most important ever done in any field, blew their whole worldview right out of the water.

If mere humans, through our own devices, could create the compounds of life, then perhaps life was not so mysterious after all. The synthesis of urea helped wash away remnants of alchemical mysticism and open people's minds to the possibility that all things could be understood eventually.

The synthesis of urea marked the beginning of the study of organic compounds as a real science, which is ironic because it simultaneously ruined the only good definition of "organic compound" there was.

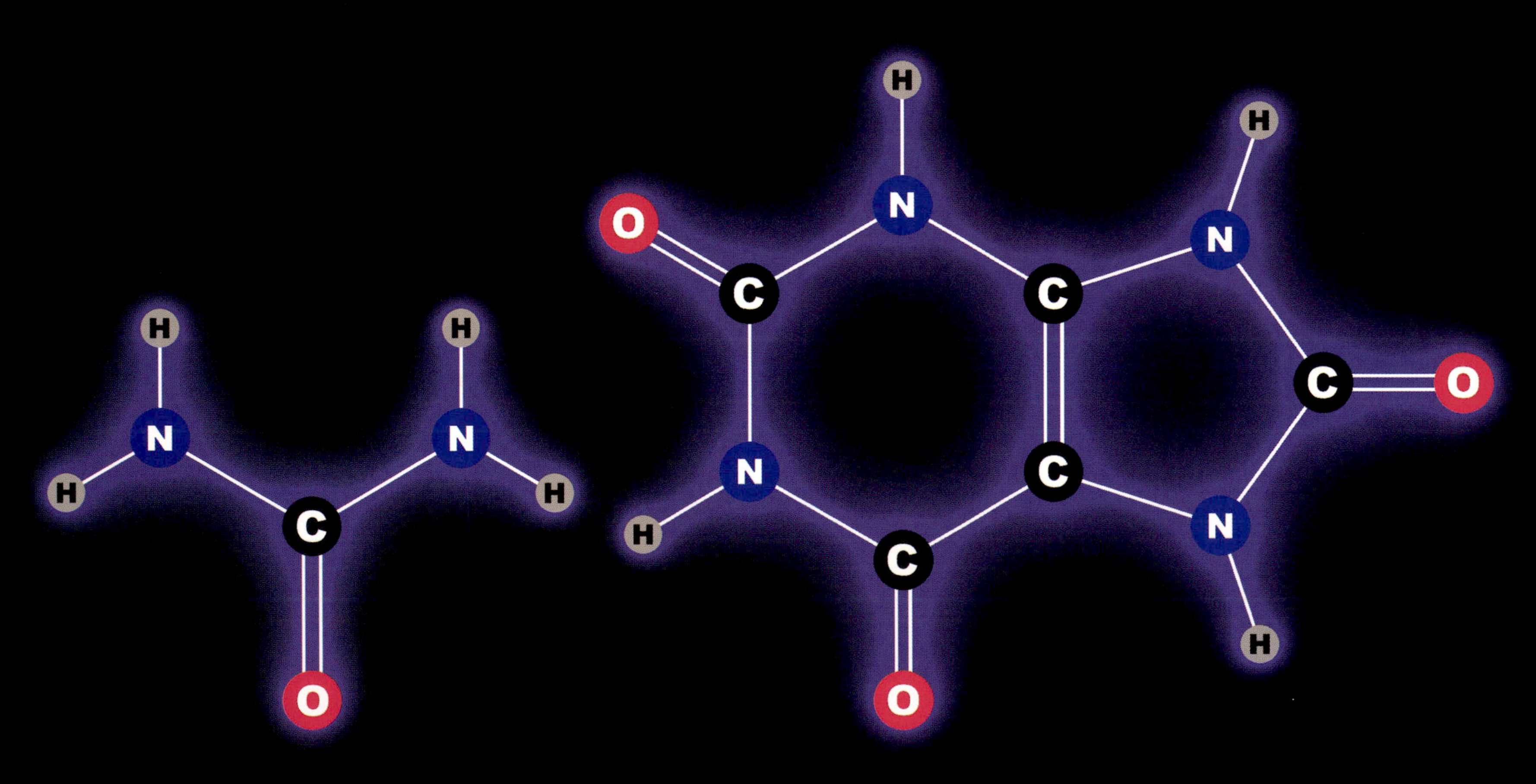

▲ The urea molecule is a fairly simple one.

▲ Uric acid is a relative of urea.

▶ Urea is definitely organic. Its name comes from urine; that's where it's found in large quantities. It plays an important role in many processes of life, and it does not normally occur naturally anywhere outside of living bodies (except when it's recently come out of one).

▼ The high value once ascribed by science and industry to snake poop was due to its extraordinarily high concentration of uric acid. This substance could not, at the time, be acquired in any other practical way.

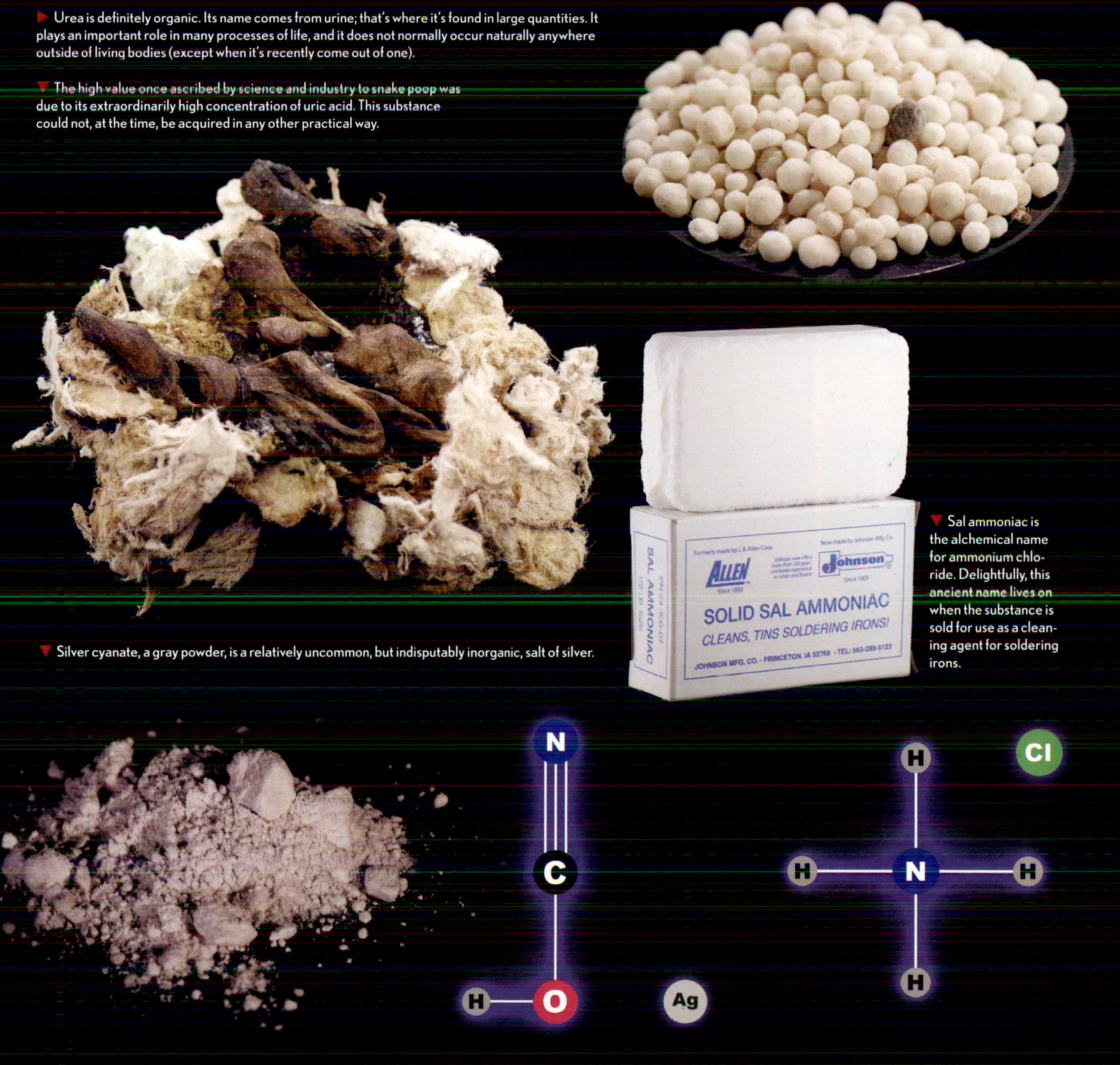

▼ Sal ammoniac is the alchemical name for ammonium chloride. Delightfully, this ancient name lives on when the substance is sold for use as a cleaning agent for soldering irons.

▼ Silver cyanate, a gray powder, is a relatively uncommon, but indisputably inorganic, salt of silver.

All Organic Chemical Free!

I CAN'T RESIST mentioning the least useful definition of the word *organic*, which is—big surprise— the one you see most often. Advertisements for food, nutritional supplements, cosmetics, and even hair dyes, push the "all natural," the "organic," and the "chemical free." This kind of thing makes chemical people tear their hair out; of course, *everything is made of chemicals, including the hair I just tore out* (see chapter 10). That organic apple you just bit into? Hundreds of chemicals.

There is no meaningful way to use the word *organic* to distinguish between good and bad, natural or unnatural, wholesome or big business. Chemicals are chemicals. There are only a couple of interesting questions to ask about any food, herb, or soft drink: What chemicals are meant to be in it, and are they good for you? What chemicals might be in it as contaminants, and are they bad for you? Where the chemicals came from makes exactly zero difference, except as guidance on what kind of contamination is most likely.

Let's move on, because you don't want to get me started on this. Suffice to say, do not use advertisements as a way to understand what organic might mean.

▼ Ephedrine

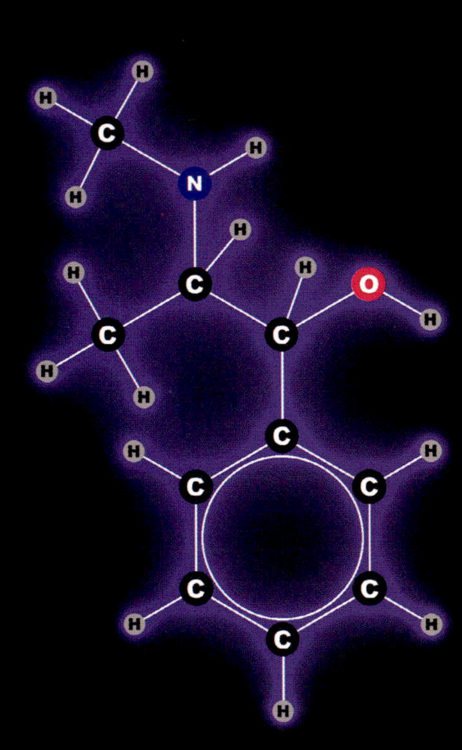

▼ Pseudoephedrine

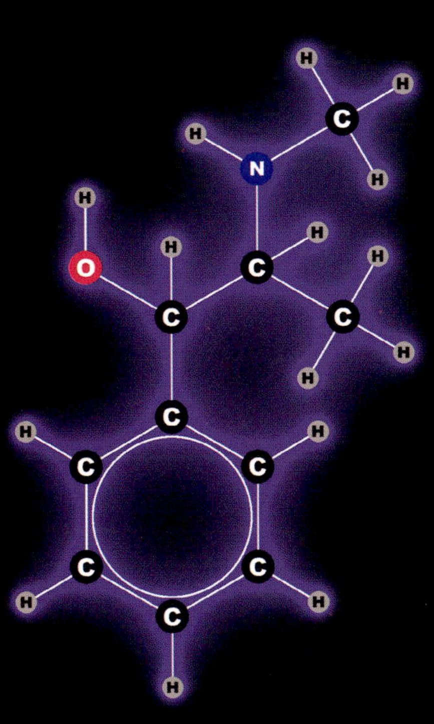

▼ Methamphetamine

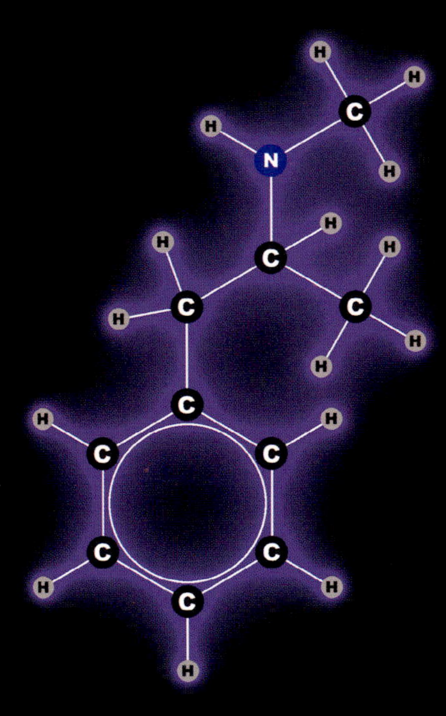

▲ The sale of herbal supplements containing ephedrine, known in Chinese medicine as *ma huang*, is banned in the United States. The chemical similarity between ephedrine, a natural herbal substance, and two synthetic drugs, pseudoephedrine (the active ingredient in Sudafed and other cold remedies) and methamphetamine, is striking to say the least. Natural ephedra is dangerous enough that it's been banned. The synthetic variant methamphetamine is really banned, and is clearly a more dangerous and unhealthy substance than its natural cousin. But the other synthetic variant, Sudafed, is a marvelous decongestant, safely sold over the counter for a generation (before restrictions were placed on it because people learned how to turn it into methamphetamine).

This package of indigo dye makes the bold claim that it contains "NO CHEMICALS." Oh, dear. Not only is indigo a chemical, it's one of the most important chemicals in the whole history of chemicals (see page 200). Indigo dye without chemicals is like the sandwich our foreign exchange student always orders: a BLT with no lettuce. Adding additional layers of irony, the powder arrives as a raw green leaf extract. In order to turn it the characteristic blue color of indigo, you have to heat it with water. That initiates a *chemical reaction* in which the *chemical* indican glycoside in the ground leaves is hydrolyzed into the *chemicals* indoxyl and glucose. Contact with air then *chemically* oxidizes indoxyl into the *chemical* indigo. Dye is *all about chemicals*. Argh. At least if they'd said it was ALL ORGANIC that would have been a true statement: indigo is an organic *chemical*. Darn, I said that word again.

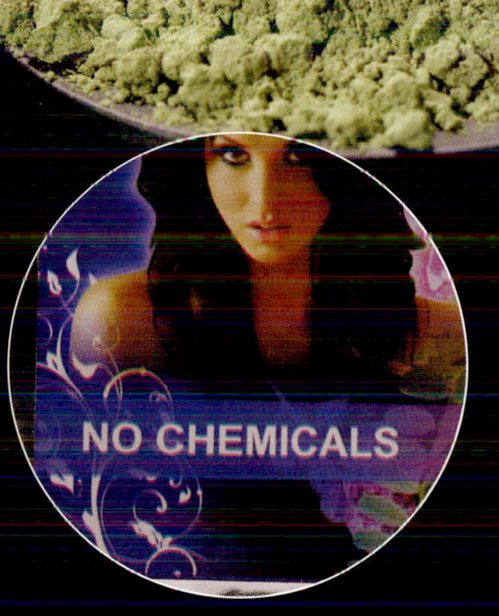

Not all companies shy away from admitting their products contain chemical compounds. In fact, the company who made this product claims that it contains not just any compound but the *ultimate* compound! It's a scratch-repair product for cars, and its superlative claim is based on a special blend of microabrasives.

Organic salt? Seriously?

So, Come On, What's the Answer?

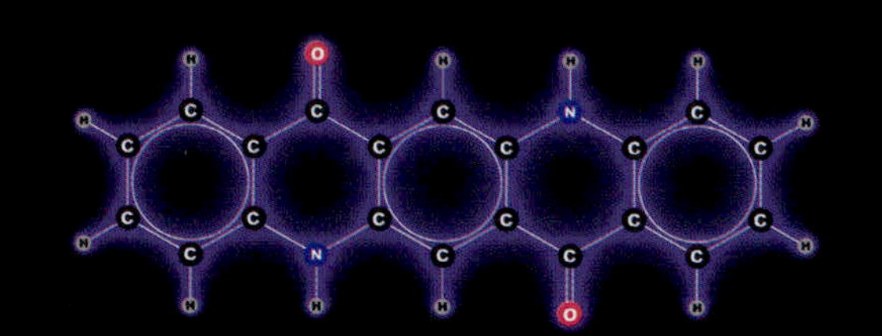

▲ Quinacridone

WHERE DOES ALL of this leave us today in terms of defining the term *organic compounds*?

The most widely accepted definition is that organic compounds are any compounds containing carbon, unless that carbon occurs in the form of carbonate (CO_3), carbon dioxide (CO_2), or carbon monoxide (CO). Or a cyanide group (CN). Or unless it's in a carbide, as in aluminum carbide (Al_4C_3). Or unless... The list of exceptions goes on a while longer and isn't very interesting.

The point this definition is trying to get at is that carbon is special. Carbon is the one and only element that can form highly complex chains, rings, trees, and sheets, bonding to itself in a way that not only permits but also encourages complex and varied three-dimensional structures. If you throw together random piles of elements including a sufficient amount of carbon and set up virtually any conditions that cause reactions to happen, you will end up with complex *organic* molecules—ones that play off of carbon's propensity to form into the chains and rings that are at the heart of what it means to be organic.

In the chapters that follow, we will meet some of these organic compounds, from the nastiest poison to the most adorably soft and fluffy polymerized ethylene phthalates.

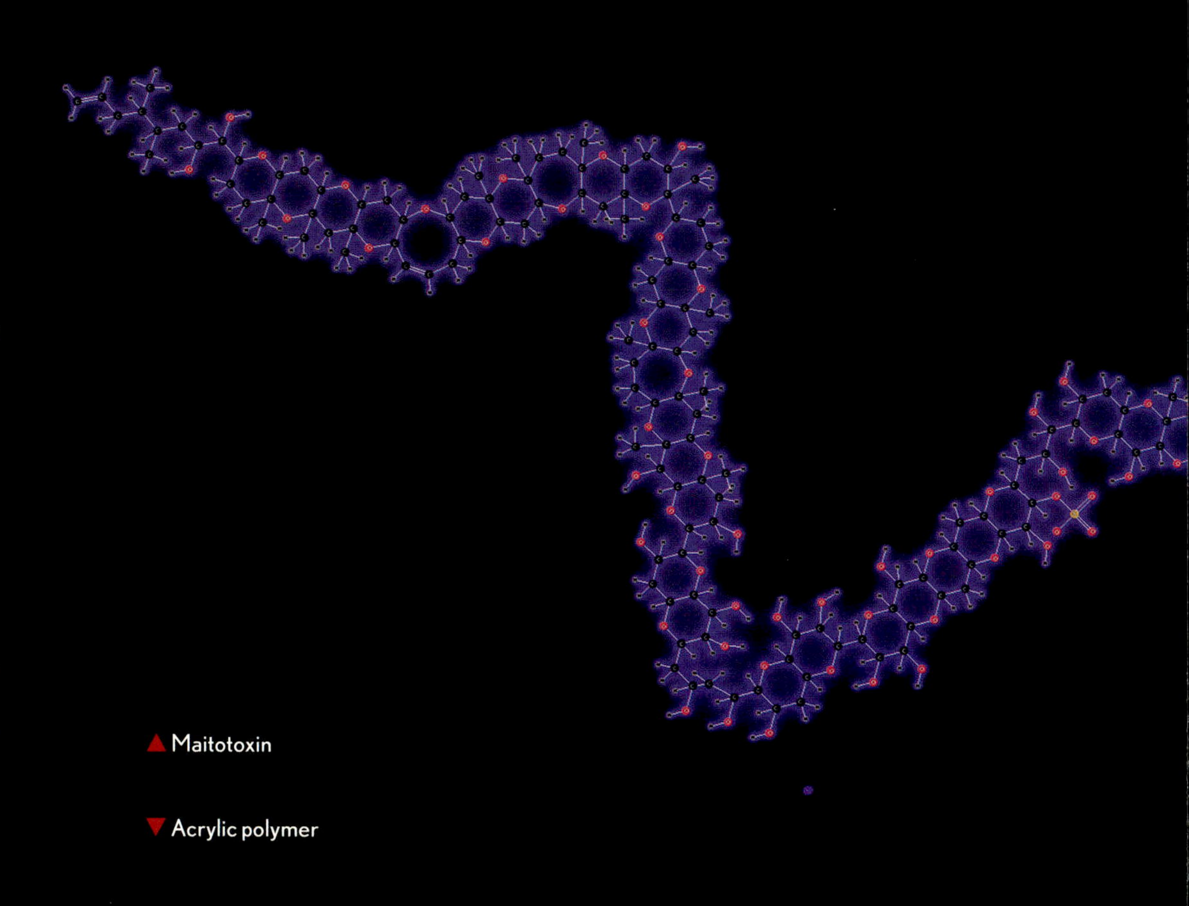

▲ Maitotoxin

▼ Acrylic polymer

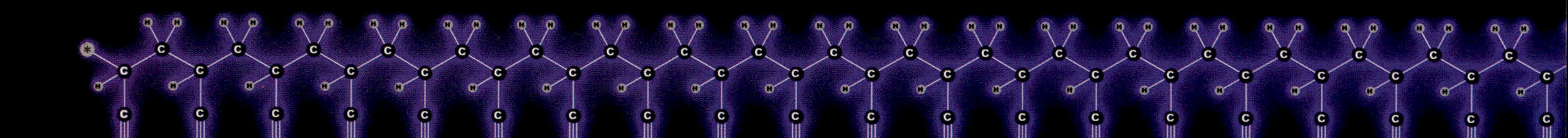

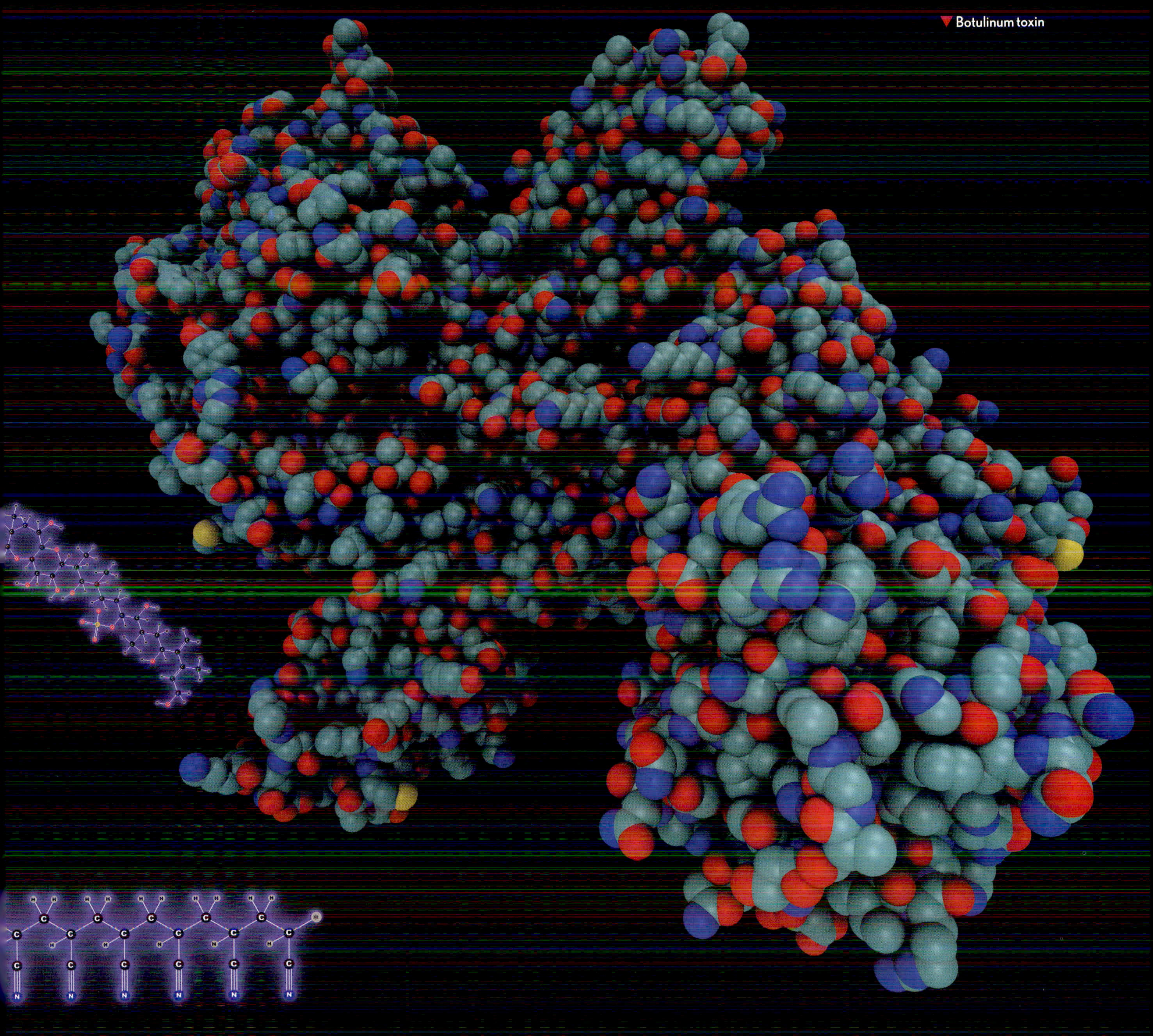
▼ Botulinum toxin

Chapter 4

Oil and Water

OIL AND WATER DON'T MIX, but why? And how is soap able to overcome this natural distrust? The answer to both questions lies in the way electric charge is distributed in the molecules of water, oil, and soap.

As we learned in Chapter 1, bonds between atoms are formed in two ways: either electrons move entirely from one atom to another (ionic bonds), or two atoms share some electrons between them (covalent bonds).

In the case of an ionic bond, the electric charge is unevenly distributed in the molecule. The molecule has "poles" of positive and negative charge (a bit like magnets have north and south poles) and is called a polar compound. Table salt, for example, is a polar, ionic compound.

Molecules with covalent bonds have a more uniform distribution of electric charge across all their atoms: they are "nonpolar." Oil is a common example of a nonpolar compound, and so are paint-thinner-type solvents such as hexane or kerosene, which don't mix with water any better than does oil.

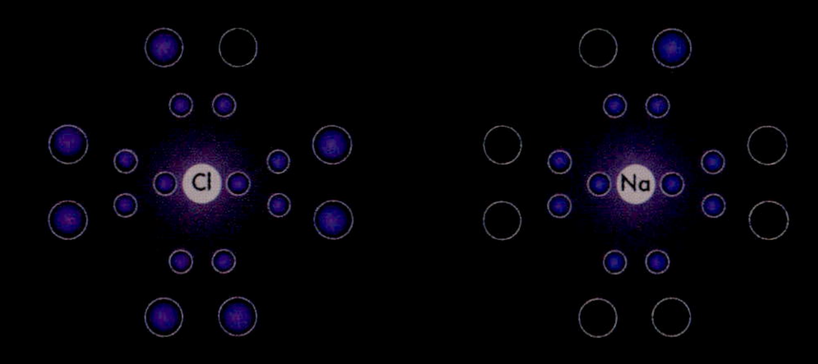

▲ As we saw in Chapter 1, when sodium and chlorine combine to form table salt, the result is a highly polar compound, with more negative charge concentrated on the chlorine atoms than on the sodium atoms. When atoms have an overall charge on them, they are called ions.

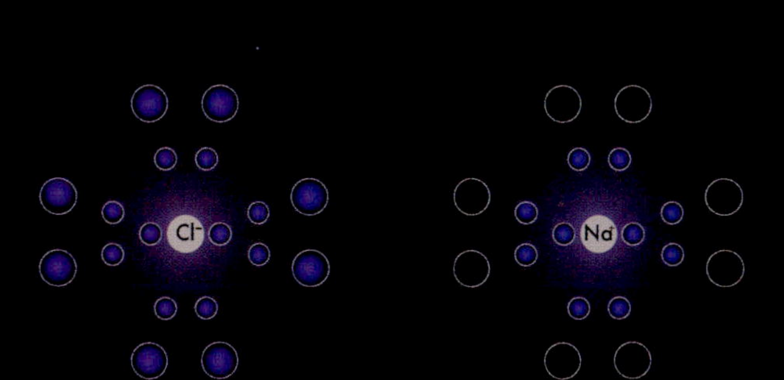

▲ Table salt is a combination of positive sodium ions and negative chlorine ions.

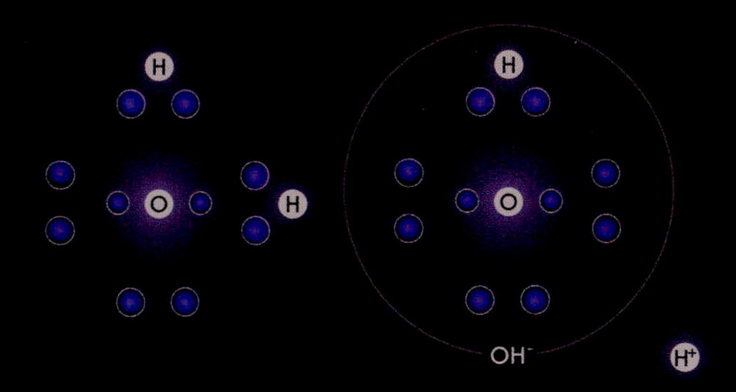

▲ Water, H_2O, isn't technically an ionic compound, but it is nevertheless polar, because the electrons holding it together are bunched up somewhat more on the oxygen atom than on the hydrogen atoms connected to it. And it fairly easily separates into two parts, a positively charged hydrogen ion (H^+), and a negatively charged hydroxide ion (OH^-). In pure water, about one in ten million water molecules is separated like this at any given time. (The hydrogen ion (H^+) is tiny because it has no electrons around it. An H^+ ion is in fact nothing more than a naked proton, and as such it is vanishingly small compared to any atom or ion that has electrons around it.)

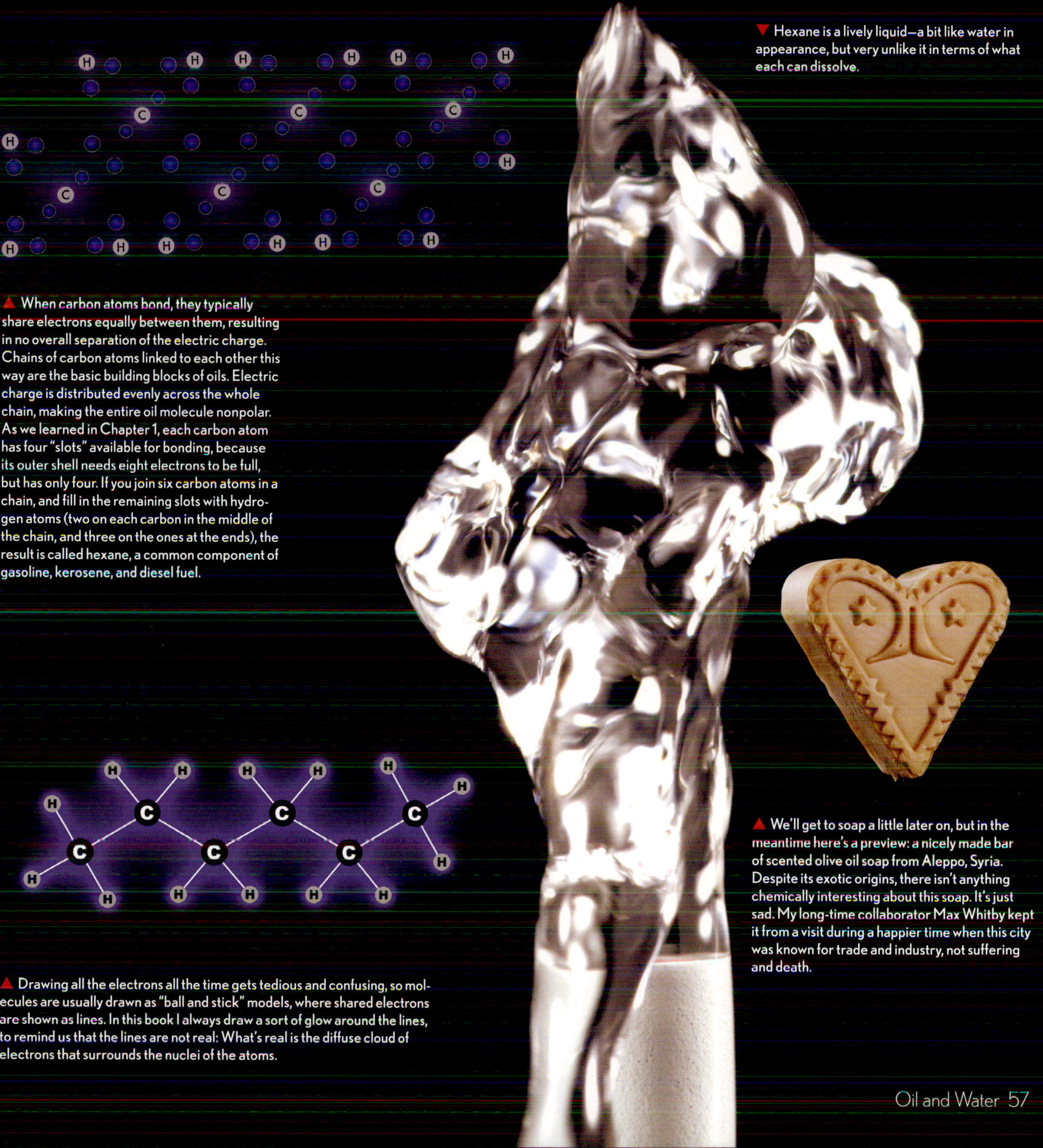

▼ Hexane is a lively liquid—a bit like water in appearance, but very unlike it in terms of what each can dissolve.

▲ When carbon atoms bond, they typically share electrons equally between them, resulting in no overall separation of the electric charge. Chains of carbon atoms linked to each other this way are the basic building blocks of oils. Electric charge is distributed evenly across the whole chain, making the entire oil molecule nonpolar. As we learned in Chapter 1, each carbon atom has four "slots" available for bonding, because its outer shell needs eight electrons to be full, but has only four. If you join six carbon atoms in a chain, and fill in the remaining slots with hydrogen atoms (two on each carbon in the middle of the chain, and three on the ones at the ends), the result is called hexane, a common component of gasoline, kerosene, and diesel fuel.

▲ We'll get to soap a little later on, but in the meantime here's a preview: a nicely made bar of scented olive oil soap from Aleppo, Syria. Despite its exotic origins, there isn't anything chemically interesting about this soap. It's just sad. My long-time collaborator Max Whitby kept it from a visit during a happier time when this city was known for trade and industry, not suffering and death.

▲ Drawing all the electrons all the time gets tedious and confusing, so molecules are usually drawn as "ball and stick" models, where shared electrons are shown as lines. In this book I always draw a sort of glow around the lines, to remind us that the lines are not real: What's real is the diffuse cloud of electrons that surrounds the nuclei of the atoms.

Polar Attraction

WE LEARNED THAT table salt (NaCl) is composed of positively charged Na^+ ions and negatively charged Cl^- ions. When a crystal of salt is placed in water (i.e., H_2O, which can also be written as HOH), the water molecules near it start separating into H^+ and OH^- ions. Some of the H^+ ions approach Cl^- ions in the salt and pair off with them, pulling them away from the crystal. Similarly, some of the OH^- ions approach Na^+ ions and pair off with them.

In this way, the salt crystal is systematically torn apart, ending up as separated Na^+ and Cl^- ions floating freely in the water, with loose, temporary associations forming between those ions and the ions resulting from the splitting of water. In other words, salt dissolves in water.

But if you try to dissolve salt crystals in a nonpolar solvent such as hexane, it doesn't work at all. Ions want to have other, oppositely charged ions to pair off with, and nonpolar molecules don't have any concentrated charges with which to attract them away from their mates.

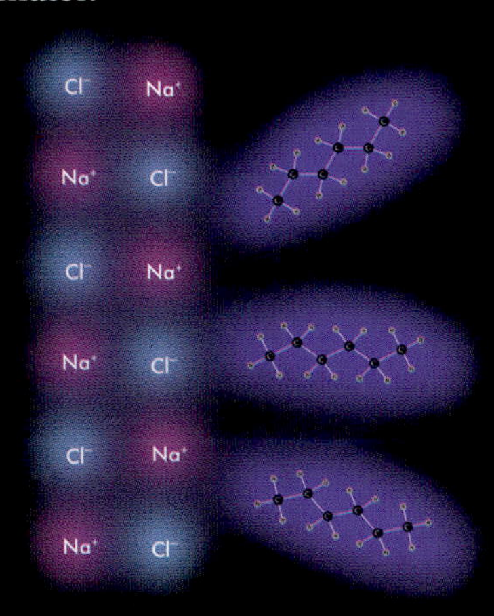

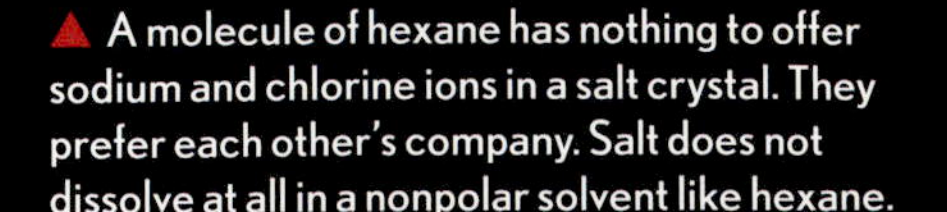
▲ A molecule of hexane has nothing to offer sodium and chlorine ions in a salt crystal. They prefer each other's company. Salt does not dissolve at all in a nonpolar solvent like hexane.

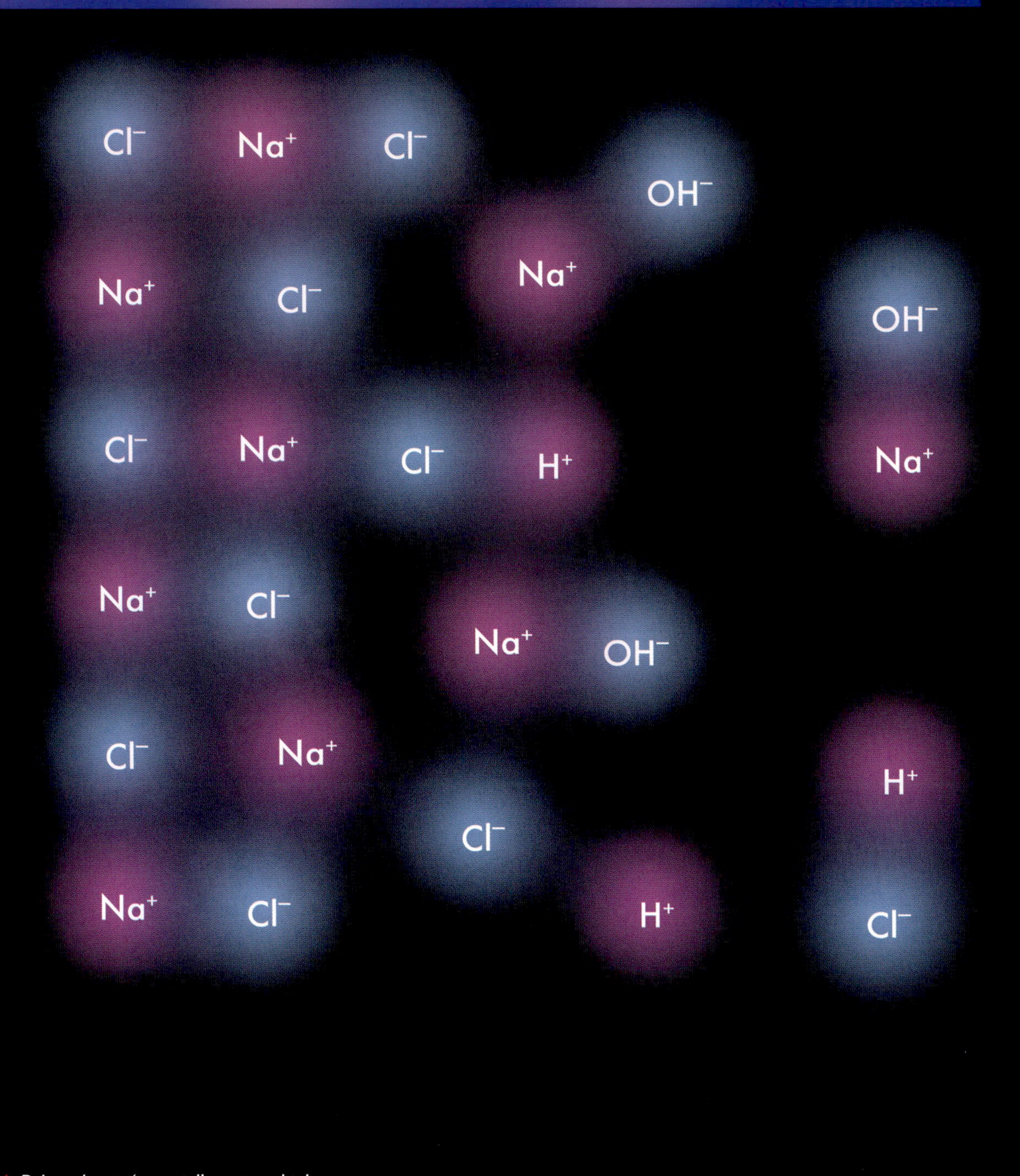

▲ Polar solvents (especially water, which partially dissociates into H^+ and OH^- ions) can insert themselves into polar compounds (like salt), which is why salt is highly soluble in water.

The Power of Being Nonpolar

DOES WATER'S POLARITY make it a stronger solvent than any nonpolar solvent? For dissolving ionic substances, water is in fact one of the harshest, most aggressive solvents known. But we think of it as life sustaining because it's not very good at dissolving our skin.

Being polar is an advantage only if you're trying to dissolve something that is itself polar. Dissolving is a mutual thing: each substance must be willing to mingle with the other. So asking whether polar water can dissolve nonpolar oil is the same as asking whether nonpolar oil can dissolve polar water. As we just learned, the answer is no: the polar molecules in water prefer to stick with each other.

If you have something nonpolar, such as oil or grease, the only kinds of molecules that will infiltrate it are other nonpolar ones, which is of course why hexane is a good solvent for oil.

And there you have it: oil and water, nonpolar and polar, each preferring to keep to itself, each finding that the other has nothing to offer it. This divide would be as fixed and eternal as the split between Protestant and Catholic, Mac and Windows, or dogs and cats, if not for the existence of soap.

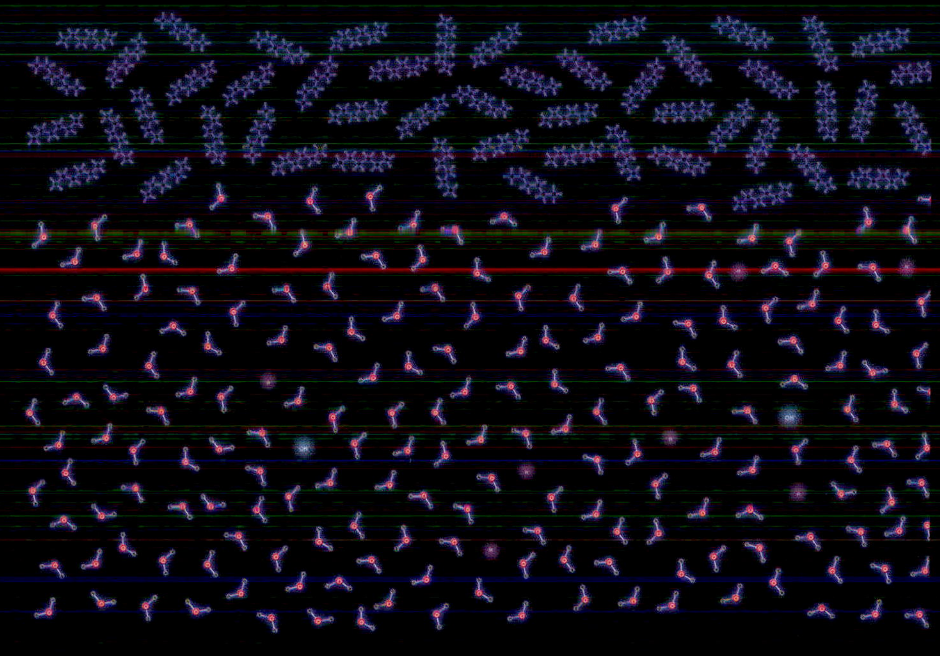

◀ Water is very happy playing with its own polar bonds and has no interest in getting between nonpolar oil molecules. Nor is water willing to allow those nonpolar molecules to get between its molecules.

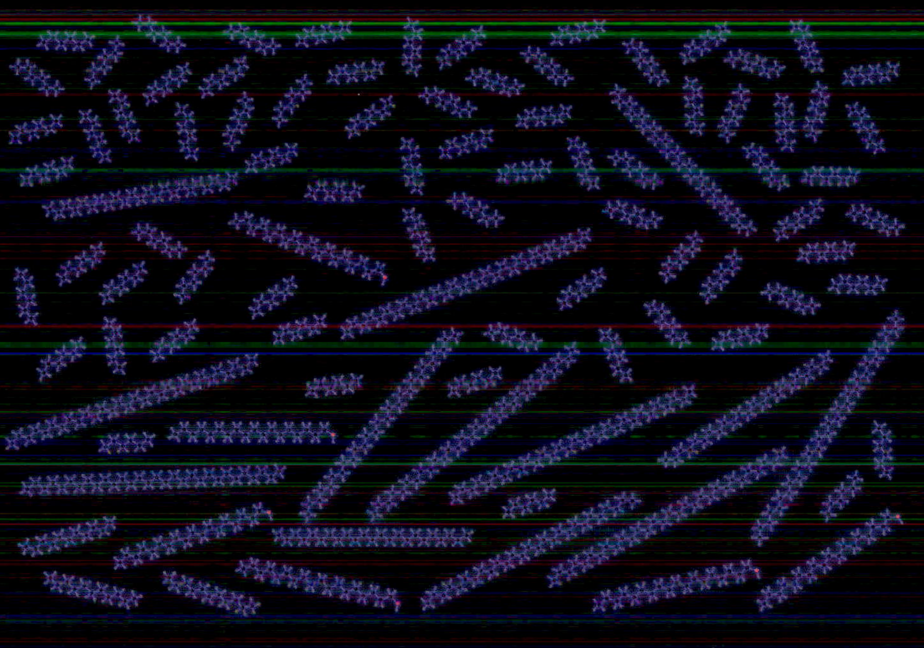

◀ Nonpolar hexane molecules are able to infiltrate similarly nonpolar, but longer-chain, oil molecules, which is another way of saying that kerosene dissolves oil.

▼ This soap is shaped like a teddy bear. It's here to remind you that this is, very soon, going to be a chapter about soap.

The Magic of Soap

SOAP DOES SOMETHING almost as remarkable as achieving world peace: getting oil to dissolve in water. It can do this because it's made of molecules that are both polar on one end and nonpolar on the other end, allowing one end to dissolve the oil and the other end to dissolve in the water.

How might you make such a molecule?

You could start with a nice long nonpolar carbon chain molecule—like, say, octadecane, which has eighteen carbon atoms in a row surrounded by thirty-eight hydrogen atoms. It's similar to hexane (six carbons), just longer and equally nonpolar. This molecule will readily insert itself into collections of oily molecules. It is, in fact, rather oily itself.

Then you could attach to it, at one end, a highly polar group of some kind. A good candidate would be a carboxylic acid group (carbon with two oxygen atoms bonded to it; see page 42). All acids are inherently polar because they dissociate to release a positively charged hydrogen ion.

Octadecane with an acid group at one end is called stearic acid, and it happens to be readily available in many animal fats. Stearic acid is one of the common fatty acids you hear so much about.

But stearic acid doesn't work as soap because, although it's an acid, it's a very, very weak acid that does not dissociate very much in water.

▼ Octadecane is a chain of eighteen carbon atoms in a straight line: octa means eight, deca means ten, and the suffix -ane signifies a carbon chain fully occupied (i.e., saturated) with hydrogen atoms. It's a squishy solid that melts just a bit above room temperature.

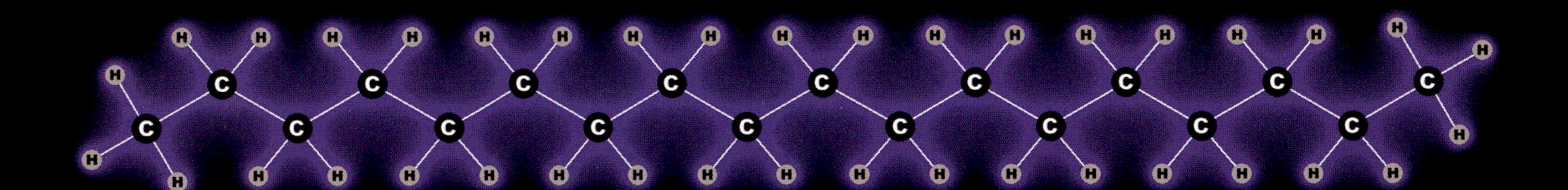

▼ Stearic acid, a fatty acid found in large amounts in many kinds of animal fat, is just like octadecane but with a -COOH group, called a carboxylic acid group, positioned at one end. Stearic acid is not very soluble in water; only about three milligrams will dissolve in a liter of water, making it useless as soap.

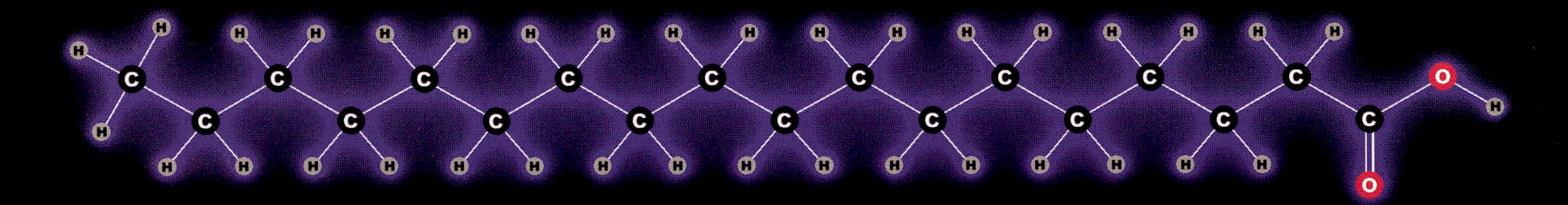

Making Soap Work

TO TURN STEARIC ACID into something that functions as soap, it needs to be made more soluble. This can be done by ripping off the hydrogen atom at its acid end and replacing it with something that is more easily separated from the rest of the molecule when it comes into contact with water.

Sodium hydroxide (caustic lye) can be used to replace the hydrogen atom with a sodium atom. This process is called "forming the salt of an acid"—in this case the sodium salt of stearic acid, which is called sodium stearate.

Sodium stearate is very soluble in water. It and similar salts of various other fatty acids are the major components, the active ingredients as it were, of all natural soaps.

So, how exactly does soap do its job?

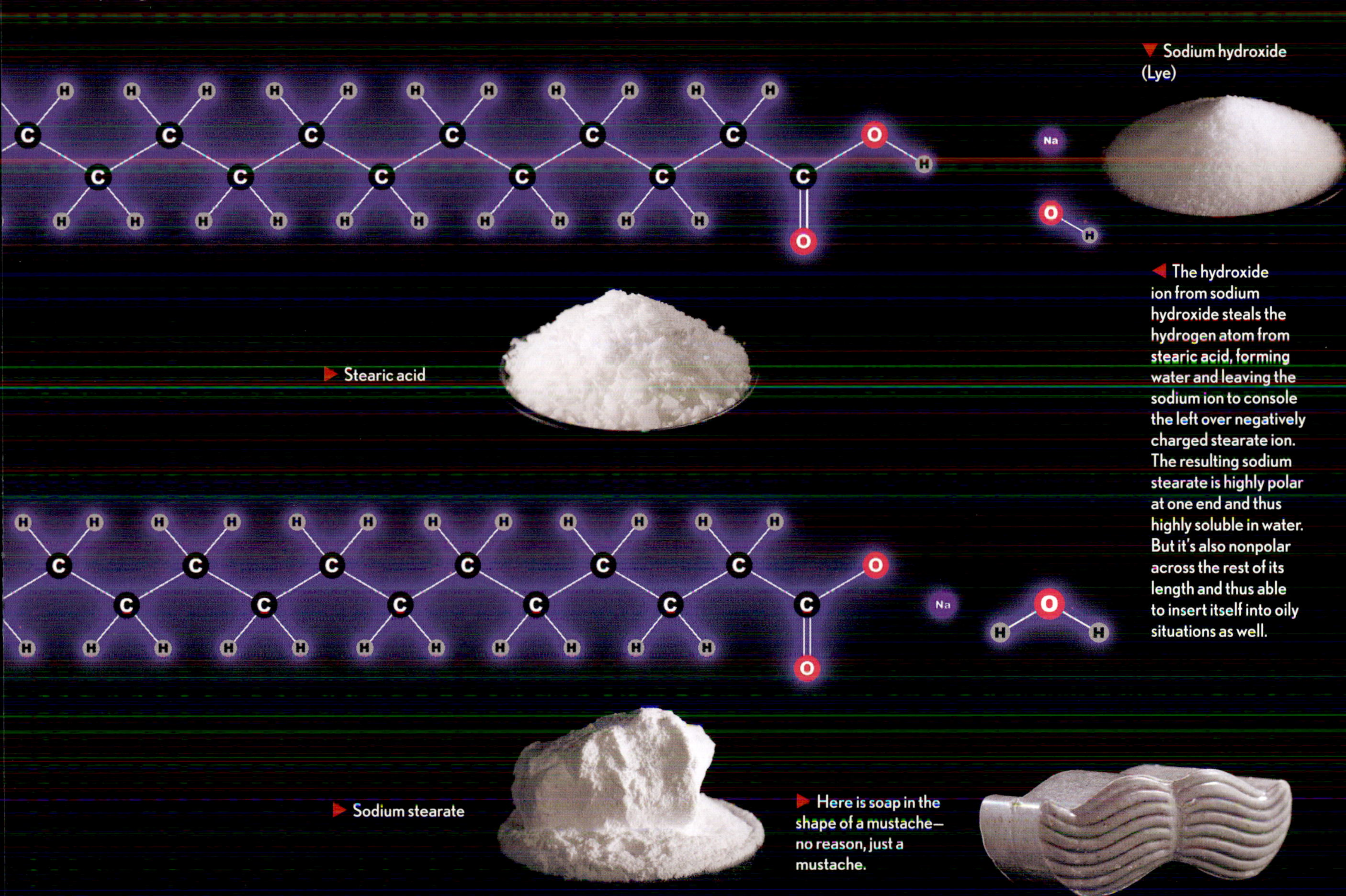

▼ Sodium hydroxide (Lye)

▶ Stearic acid

◀ The hydroxide ion from sodium hydroxide steals the hydrogen atom from stearic acid, forming water and leaving the sodium ion to console the left over negatively charged stearate ion. The resulting sodium stearate is highly polar at one end and thus highly soluble in water. But it's also nonpolar across the rest of its length and thus able to insert itself into oily situations as well.

▶ Sodium stearate

▶ Here is soap in the shape of a mustache—no reason, just a mustache.

The Mechanics of Soap

WHEN A SOAP MOLECULE, such as sodium stearate, is placed in the presence of both water and oil, it is literally pulled in two direct[illegible] The polar end of the molecule is attracted to the polar water molecules, while the nonpolar carbon chain is comfortable nestling in with the nonpolar oil molecules.

The nonpolar chains of the soap molecules slip into the surface of the oil and pull at it, breaking the oil up into microscopic blobs. Those blobs form spherical clusters (called *micelles*), with all the polar ends of the soap molecules pointing outward and all the nonpolar chains pointing inward.

▶ Liquid soaps and detergents are chemically the same as their solid counterparts. The only difference is that their ingredients are predissolved in water.

▶ Soap molecules cluster around oil to form micelles, which present themselves as water-loving polar objects. The nonpolar oil molecules are snugly hidden away in the core of the spheres.

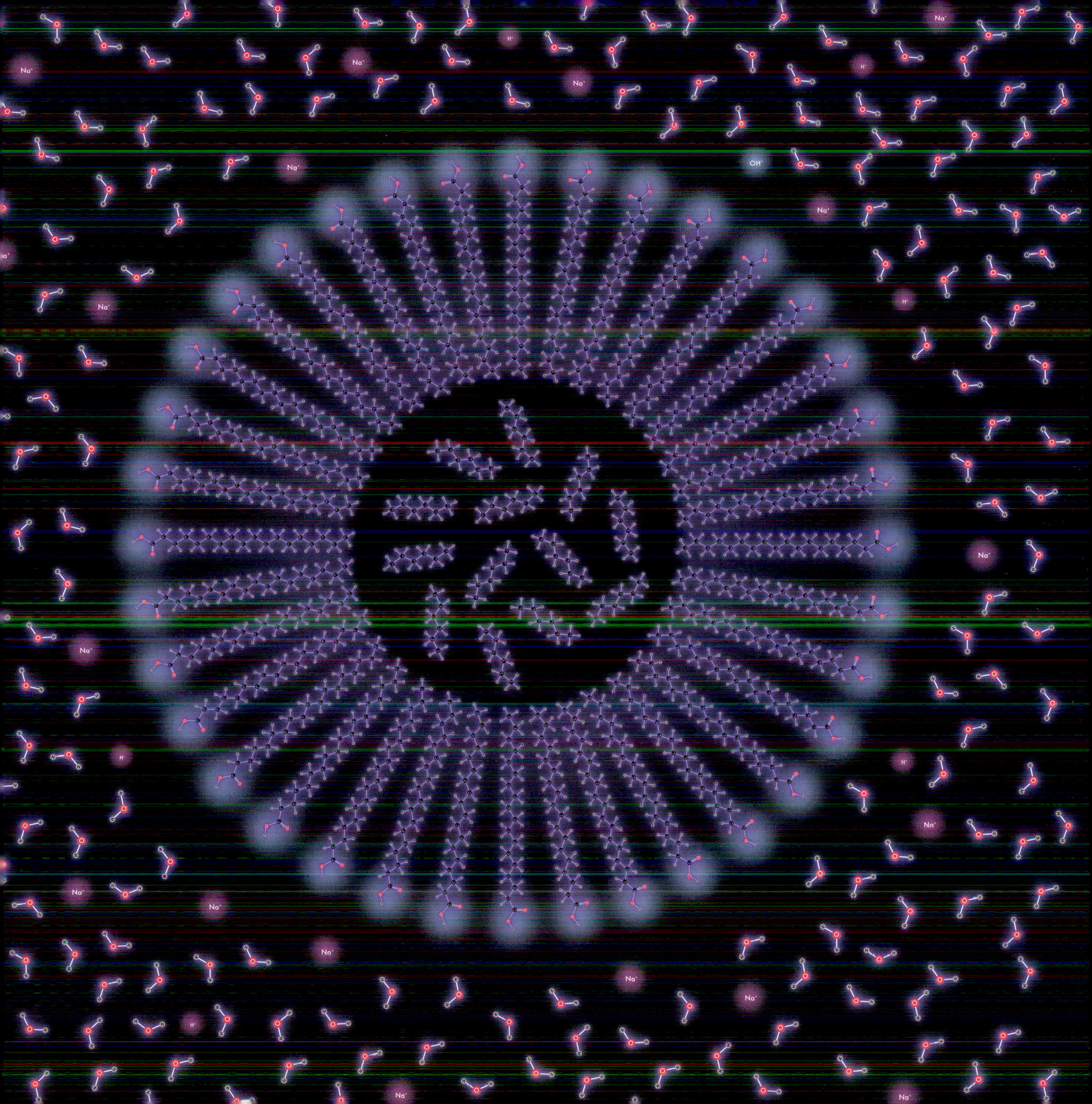

Making Natural Soap

SOAP-MAKING IS AN ancient industry, dating back at least as far as 2800 BC. It's also a remarkably simple one, practiced in kitchens and sheds all over the world. All you need is some animal or vegetable fat (made of fatty acids such as stearic acid) and some lye (sodium hydroxide). Historically, the lye was washed out of wood ashes, but today it's readily available in pure form as drain opener or as food-grade lye certified not to have any unsuitable impurities.

Fats are made of fatty acids, but there's a wrinkle I haven't mentioned. The fatty acids in animal fat and vegetable oil are not floating free. They are grouped together in the form of triglycerides, which consist of three fatty acids linked together by a glycerine backbone. (See page 79 for a more detailed explanation of how this works.)

When a triglyceride is processed with lye, the fatty acids are simultaneously torn off the glycerine backbone and turned into salts. The glycerine is left over. Most commercial soap manufacturers remove the glycerine, but makers of specialized "glycerine soaps" leave the glycerine in, and sometimes add even more glycerine, which results in a transparent soap that some people like the look and feel of.

▶ Glycerine soaps are more transparent because the fatty acids in them do not form crystals that scatter light. (It's a bit like water, which is transparent when it's in noncrystal, liquid form but becomes opaque when turned into many tiny irregular crystals, as in snow.)

▲ Beef tallow, made by boiling and straining the fat of cows, is almost pure triglyceride fat and is thus an ideal starting point for making soap.

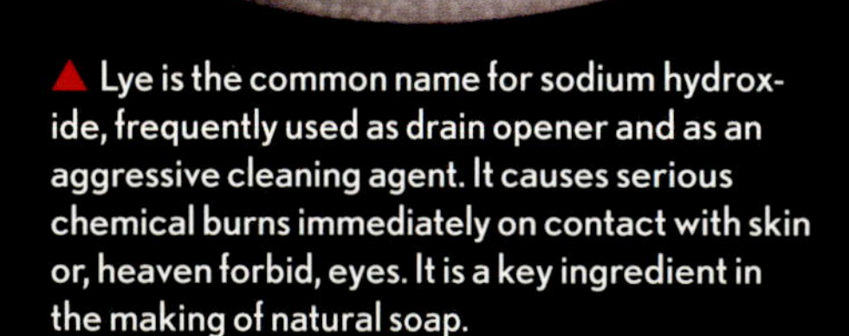

▲ Lye is the common name for sodium hydroxide, frequently used as drain opener and as an aggressive cleaning agent. It causes serious chemical burns immediately on contact with skin or, heaven forbid, eyes. It is a key ingredient in the making of natural soap.

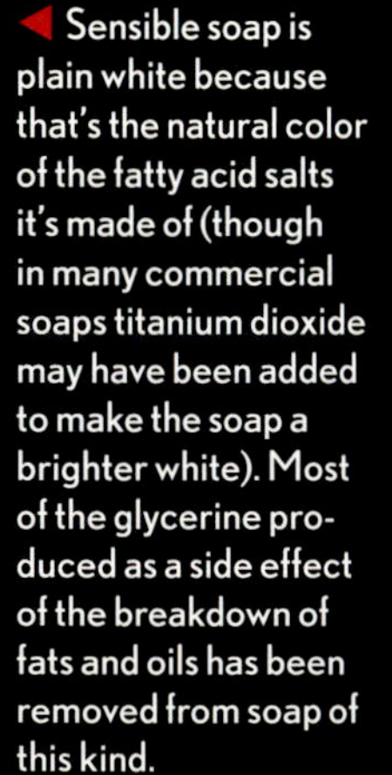

◀ Sensible soap is plain white because that's the natural color of the fatty acid salts it's made of (though in many commercial soaps titanium dioxide may have been added to make the soap a brighter white). Most of the glycerine produced as a side effect of the breakdown of fats and oils has been removed from soap of this kind.

▲ Since anything other than a plain white bar of soap is inherently silly, it's fair to say that one of the great advantages of glycerine soap is that it's transparent. That means you can embed any number of silly things inside of the soap and get someone to pay too much money for it. I paid $9 for this bar.

Artificial Soap

SOAP IS ANCIENT, but modern synthetic alternatives called detergents, which work in basically the same way, have largely replaced the traditional product.

A big problem with natural soaps is that in the presence of dissolved calcium, magnesium, or iron in the water, they tend to precipitate (form insoluble compounds). Water containing these sorts of ions is called "hard water" and is very common in many parts of the world. (You can tell how hard the water you're washing with is by how long soap feels slippery. If it seems to wash away rapidly and you need to use a lot of it, hard water is precipitating out the soap. If it feels slippery for a long time, you've got soft water.)

Detergents get around this problem by using a different polar group than natural soap does. Instead of salts of a carboxylic acid, they are typically salts of a sulfonic acid or a sulfate. For example, just as sodium stearate, with eighteen carbon atoms, is a common component of natural soap, sodium dodecylbenzenesulfonate, also with eighteen carbons, is a common component of synthetic laundry detergents.

▲ Dodecylbenzenesulfonic acid is a long name for a long molecule. The ring on the left contributes the "benzene" part of the name, and the sulfur atom with three oxygens bonded to it contributes the "sulfonic acid" part.

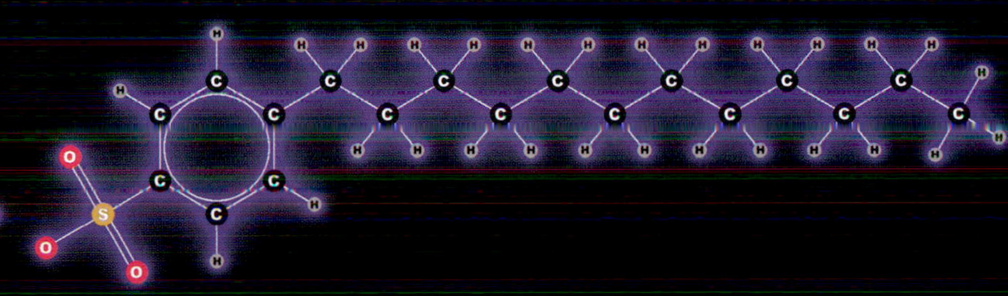

▲ Sodium dodecylbenzenesulfonate is the sodium salt of dodecylbenzenesulfonic acid. (OK, I admit it: I copied and pasted those names into my manuscript. There's no way to type them without making at least three typos.) As with the sodium salts that make up soap, this common ingredient in detergents is the salt of a weak organic acid.

◀ Linear (i.e., straight-chain) detergents are relatively easy for bacteria to break down, but branched-chain variations are not very biodegradable. Early branched-chain detergents led to widespread formation of foam on lakes and rivers in the 1950s and 1960s, one of the reasons we now have more biodegradable versions.

▼ Revolting pollution caused by nonbiodegradable, branched-chain synthetic detergents.

Artificial Soap

▶ For years, I kept seeing sodium lauryl sulfate and sodium laureth sulfate listed alternately on different shampoo labels. In my foggy little brain, I was never quite sure whether they were actually two different substances or I was just remembering the name wrong from the last time. Then I ran into a product with both.

▼ Sodium laureth sulfate is not sodium lauryl sulfate pronounced with a lisp. But they are chemically very similar: the laureth form has an ethyl ether group (see page 39) inserted between the polar sulfate group on the left and the nonpolar lauryl (twelve-carbon chain) group on the right.

▼ Sodium lauryl sulfate ▲

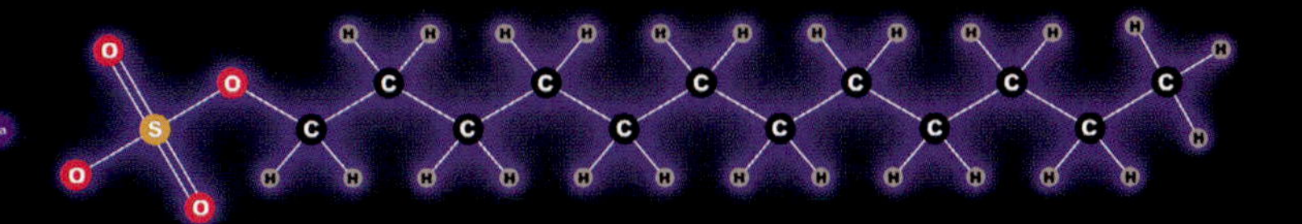

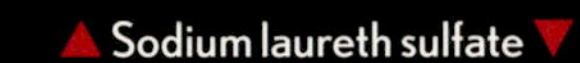

▲ Sodium laureth sulfate ▼

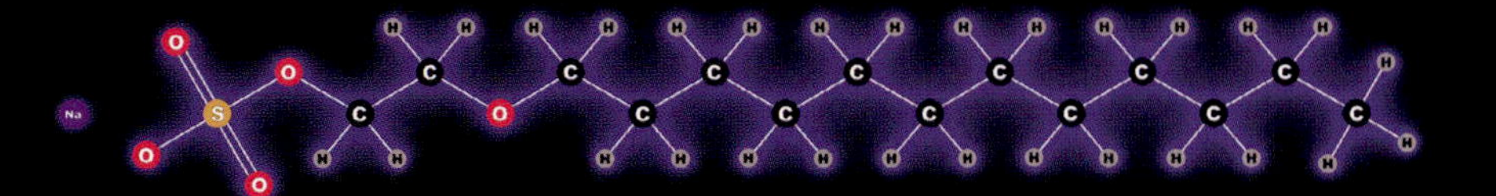

▼ Lauric acid

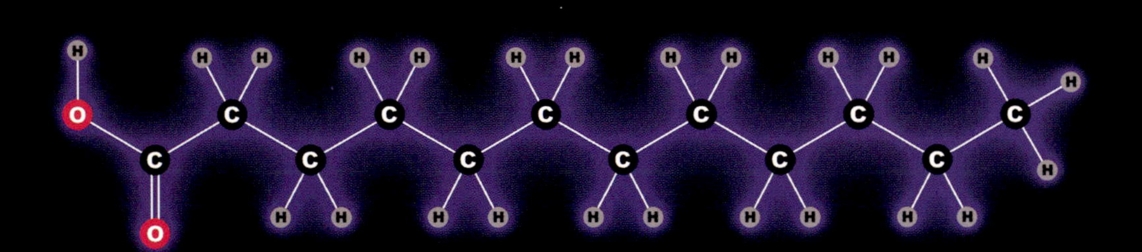

▲ Lauric acid is a common fatty acid found, for example, quite prominently in coconut oil. It is the starting point for making the common detergents and surfactants sodium lauryl sulfate and sodium laureth sulfate.

▶ "Sodium coco sulfate" is promoted by some detergent manufacturers as a safer, more natural alternative to sodium lauryl sulfate. It is made from pure coconut oil. The problem is that it's really just a relatively impure form of sodium lauryl sulfate under a different name. Pure coconut oil is anything but pure in a chemical sense. It's a complex mixture of different oils and fatty acids. But its principle component is lauric acid, so when it's processed into a sulfate salt, it becomes mostly sodium lauryl sulfate. This is great because sodium lauryl sulfate is a fine chemical. The only problem here is trying to sell sodium coco sulfate to people who don't like sodium lauryl sulfate. This kind of marketing jargon is really quite infuriatingly silly; if there is anything wrong with using sodium lauryl sulfate (which may or may not be the case), then using sodium coco sulfate would pose all the same problems because *it's the same chemical*. The only difference is that sodium coco sulfate also contains a bunch of other unknown or unspecified chemicals, which might or might not be bad for you and which you don't need to worry about if you're using simple, pure sodium lauryl sulfate.

Soap and the Origin of Life

WHEN SOAP BREAKS up oil, it does so by forming microscopic spheres of oil completely surrounded by a wall of soap molecules, all with their nonpolar tails pointing in and their polar heads pointing out (See page 62).

That's a pretty interesting structure: a spherical object with an interior filled with organic molecules that are protected by a wall of robust soap molecules. It sounds a lot like a biological cell. In fact, there are reasons to believe that exactly this kind of soap sphere played an important role in concentrating and protecting organic molecules during the long phase of chemical evolution that came before recognizable life emerged.

In any case, it's pretty intriguing that if you simply throw a random collection of organic molecules—some entirely nonpolar, some partially polar—into a pool of water, they will spontaneously self-assemble into structures that facilitate interaction between the organic molecules.

In other words, not only is soap fundamental in allowing modern humans to continue the process of natural selection, it may also have played a crucial role in getting the whole thing started in the first place.

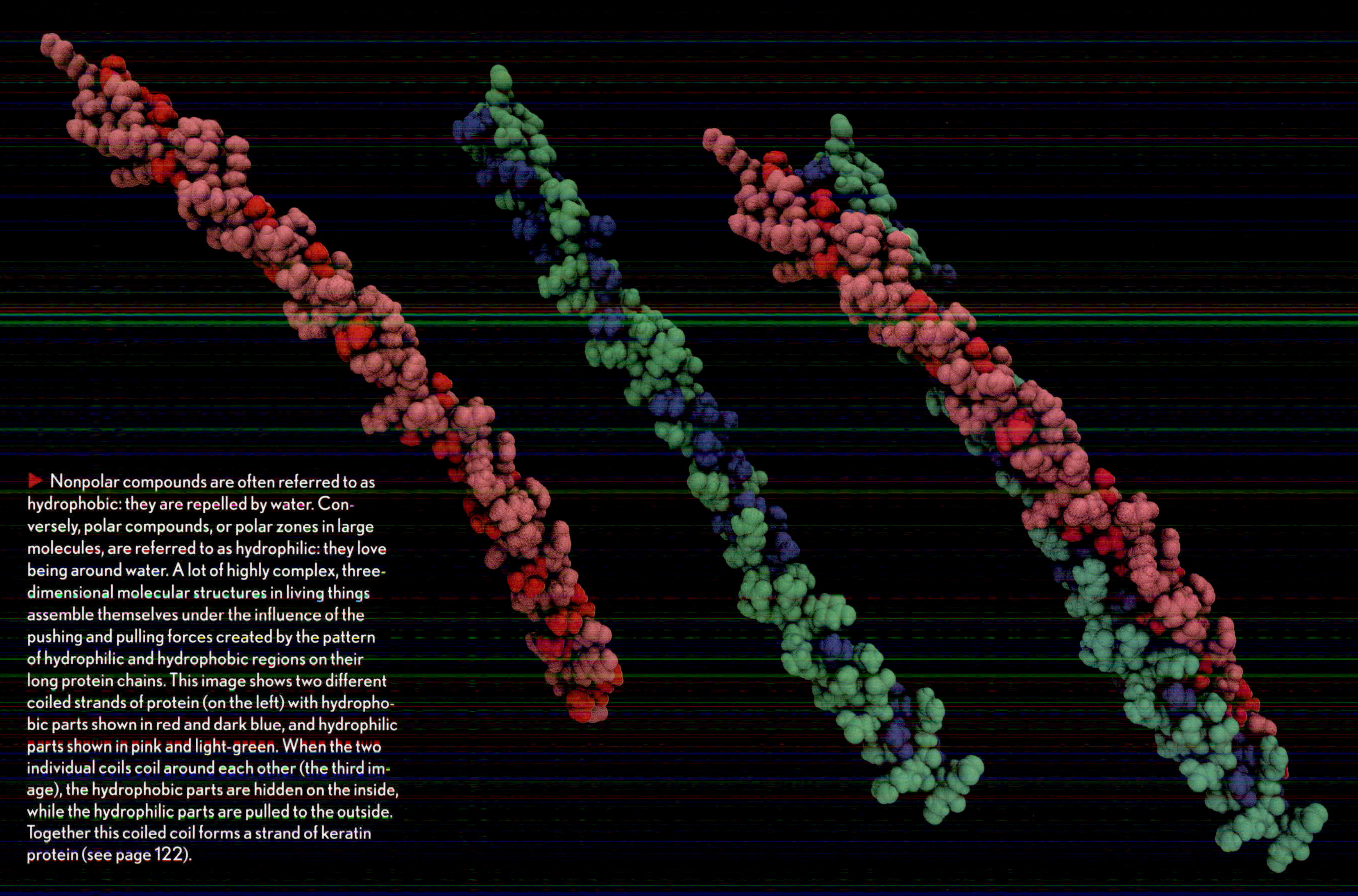

▶ Nonpolar compounds are often referred to as hydrophobic: they are repelled by water. Conversely, polar compounds, or polar zones in large molecules, are referred to as hydrophilic: they love being around water. A lot of highly complex, three-dimensional molecular structures in living things assemble themselves under the influence of the pushing and pulling forces created by the pattern of hydrophilic and hydrophobic regions on their long protein chains. This image shows two different coiled strands of protein (on the left) with hydrophobic parts shown in red and dark blue, and hydrophilic parts shown in pink and light-green. When the two individual coils coil around each other (the third image), the hydrophobic parts are hidden on the inside, while the hydrophilic parts are pulled to the outside. Together this coiled coil forms a strand of keratin protein (see page 122).

Too Many Soaps

SOAP, LIKE WINE, is basically all the same, so manufacturers go crazy making up variations to push back the pain of banality.

▲ Most soap is differentiated on the basis of what kind of oil or fat was used to make it—animal fat, olive oil, palm oil, and so forth. But this "African black soap" is special for the source of the lye (i.e., sodium hydroxide) used to process its oils (typically palm oil, palm kernel oil, or coconut oil). Lye was traditionally extracted from wood ash, but for this soap the ash of cocoa pods, coconut pods, or shea tree bark is used, and it's left in the soap rather than just used as a source of lye.

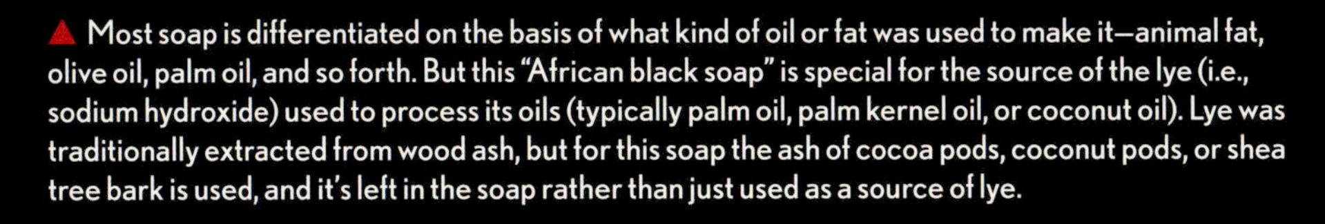

▲ Soap or candy? I'll never tell.

▲ Soap or wax? The wick gives this one away.

▶ This is, apparently, official government soap from Bangalore, India.

▼ Pine tar soap is made from the tar that results from heating pine wood under pressure. Rather than being made of straight-chain fatty acids, it contains mostly molecules with benzene (six-carbon) ring structures along their nonpolar chains. These aromatic compounds tend to absorb light, making the soap black.

◀ This olive oil soap is from Greece, home of many olives. Like pine tar, olive oil contains a wide range of complex, ring-structured molecules that can be turned into soap molecules.

▶ This sheep-shape soap (try saying that three times fast) is a combination of regular and glycerine soap. I got it at a yarn store, where it fit with the whole wool theme.

▶ Hand soap. Get it? For those reading in translation, "hand soap" in English means soap meant for washing hands. It's funny because this soap is shaped like hands.

▲ "Hotel soap" is a whole industry unto itself. These are ones I've collected over the years. I travel too much.

▼ Most soap is made from fat or oil, but any source of fatty acids is a potential foundation for soap. Beeswax is made mostly of fatty acids and fatty acid esters (not triglycerides like those in fat), so beekeepers sometimes get into soap-making to use their readily available raw ingredient. (It doesn't seem to be practical to make soap from only beeswax, so more conventional coconut, palm, or olive oils are included as well.)

Chapter 5

Mineral and Vegetable

◀ Methane is most commonly known as natural gas. It's what provides heat and cooking gas to a lot of people in many parts of the world, and it's the compound behind talk of gas exploration, "fracking," and other things to do with "gas" except in connection with motor vehicles. We'll get to the car kind of "gas" when we reach pentane (five carbons).

THERE ARE TWO VERY different kinds of oil and two very different kinds of wax. In each case, one kind originates from petroleum (crude oil pumped out of the ground) and the other from plants and animals.

The superficial similarities between mineral and vegetable oils, and between paraffin wax and beeswax, are only skin deep. Underneath, the chemistry is crucially different. For example, no living thing can digest mineral oil (with the exception of a few bacteria), but vegetable oils are high-caloric foods.

We'll start with the inedible side of things. Mineral oils are basically hydrocarbons; they are made from hydrogen, carbon, and nothing else. The systematic way to look at hydrocarbons is in order of the number of carbon atoms each one contains in the molecules that make it up. That number can range from one to many thousands.

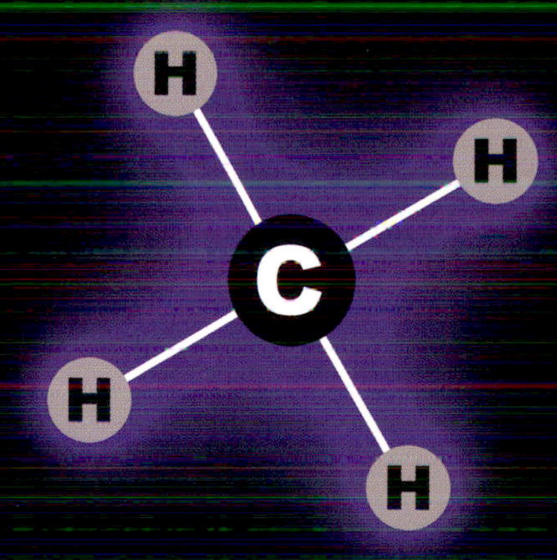

▲ The starting point for any discussion of hydrocarbons is always methane, the simplest hydrocarbon compound. It has one carbon with four hydrogens attached to it.

▶ The next-simplest hydrocarbon is ethane—two carbons and six hydrogens.

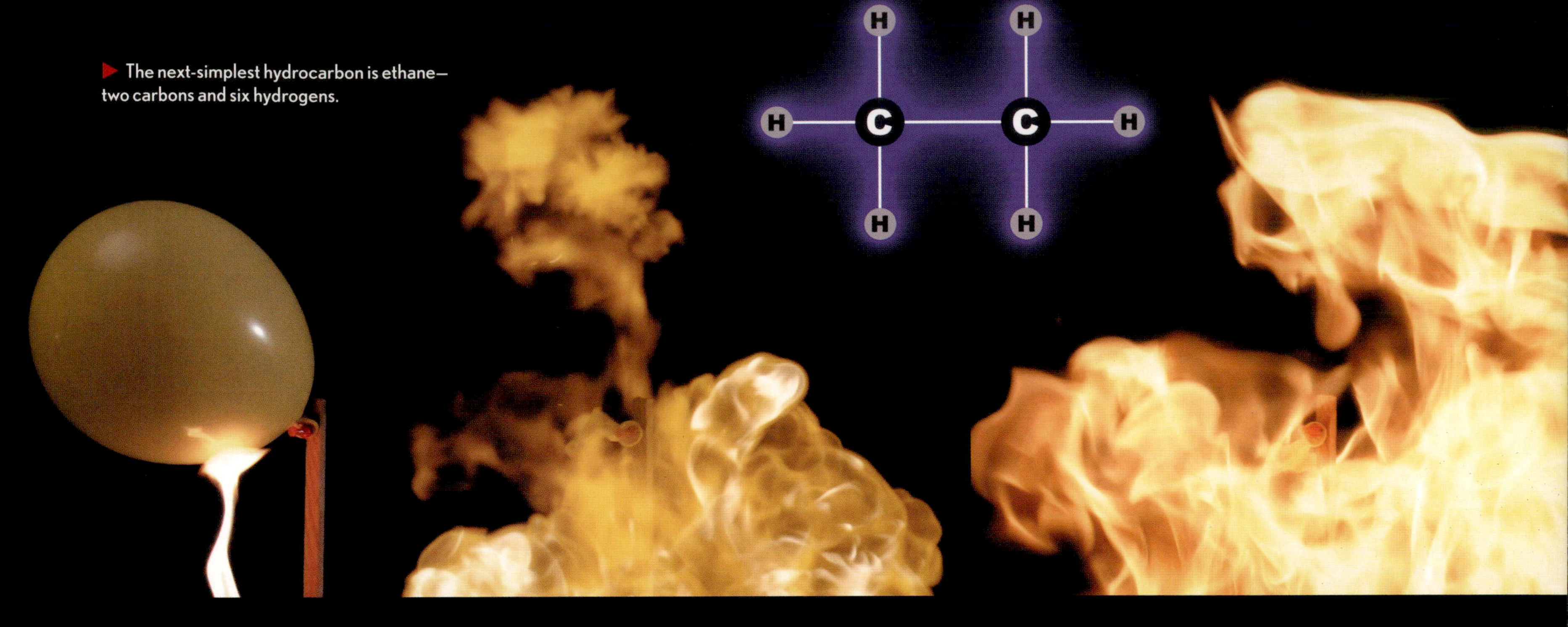

▲ Ethane is a gas similar to methane but slightly denser and with a somewhat higher boiling point. It makes nice fireballs when a balloon of it is lit.

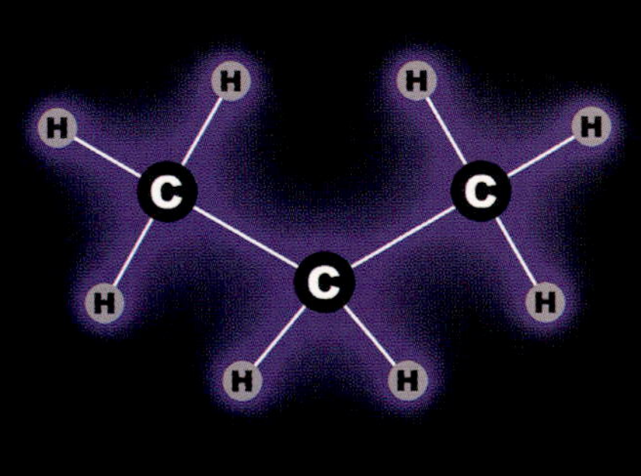

▲ Propane

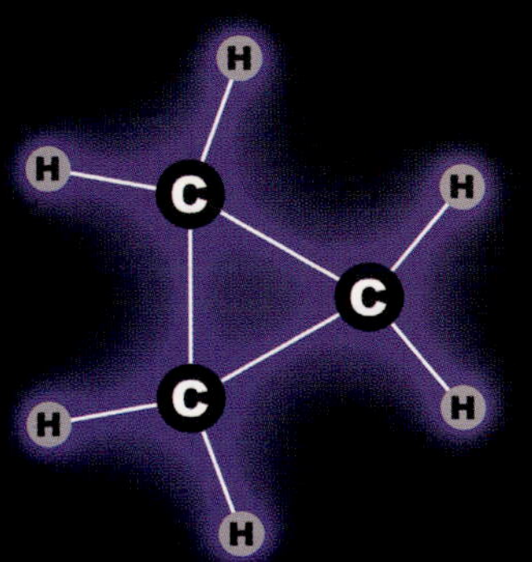

◀ Propane has three carbons and eight hydrogens. It's the simplest hydrocarbon that can be arranged in more than one way, either a straight chain or a ring called cyclopropane (which only has 6 hydrogens). Cyclopropane is a very strained molecule: Carbon–carbon bonds don't like to be at such sharp angles to each other, which makes cyclopropane explosively reactive, especially with oxygen. Its use as a sedating anesthetic was phased out due to the uncomfortable fact that patients needed to breathe oxygen at the same time they were breathing cyclopropane.

◀ Propane (the straight-chain kind) is handy because it turns into a liquid under a fairly modest amount of pressure. When a gas turns into a liquid, its volume goes down by a factor of many hundreds. Another way of saying this is that you can store vastly more liquid than gas at the same pressure in the same size container. This makes propane a practical fuel for portable gas torches like this one, used by insane people to kill weeds and by insaner people to heat-weld rubber roofs. The torch puts out about 500,000 BTU of heat, much more than the furnace in even a large house.

▼ Butane has four carbons and ten hydrogens. It can be arranged in quite a few different ways (see page 19). The more carbons there are in a hydrocarbon chain, the more ways there are of arranging them. For the sake of practicality, we are going to look mainly at the straight-chain hydrocarbons from here on out, though many of the substances we'll discuss actually contain mixtures of straight, branched, and cyclic molecules.

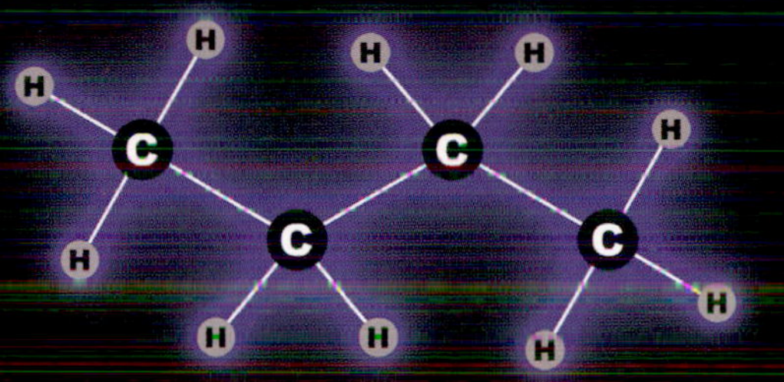

▲ Butane

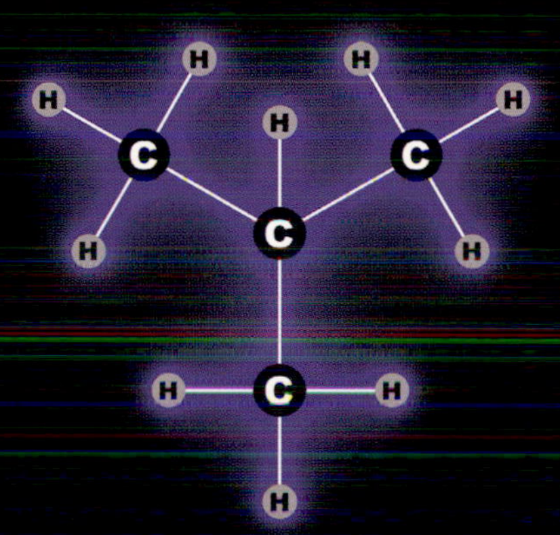

▲ Isobutane

▼ Cyclobutane

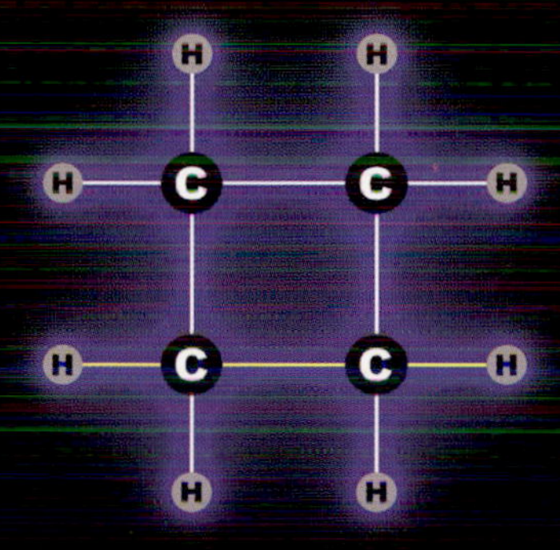

◀ Like propane, butane is a gas under normal conditions and can be turned into liquid under modest pressure. In the case of butane, the pressure required is so modest that a thin plastic container is all it takes. That's why we have cheap, disposable, plastic butane lighters that work about three times before breaking.

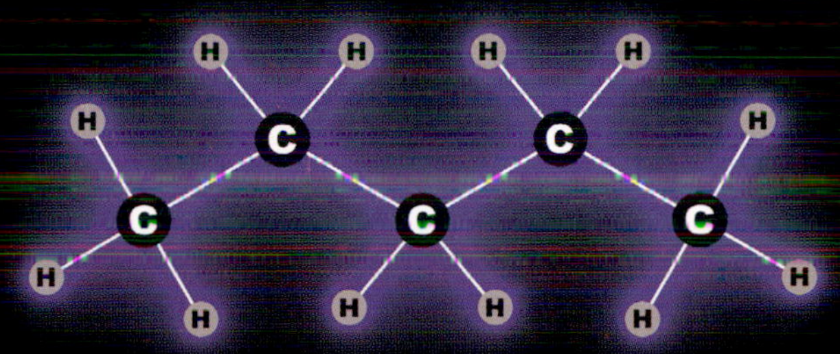

▲ Pentane has five carbons and twelve hydrogens.

▼ Pentane is the smallest hydrocarbon in the series that is liquid under ordinary conditions, but just barely (its boiling point is 97°F, 36°C). It's the lightest, most volatile substance that is normally found in the kind of "gas" used in cars, gasoline (called petrol in some countries). Pentane is partly responsible for the fact that gasoline vapors are explosive. It and other volatile components accumulate in explosive concentrations in the air above an open container. Gasoline is traditionally kept in red cans to warn people of this danger.

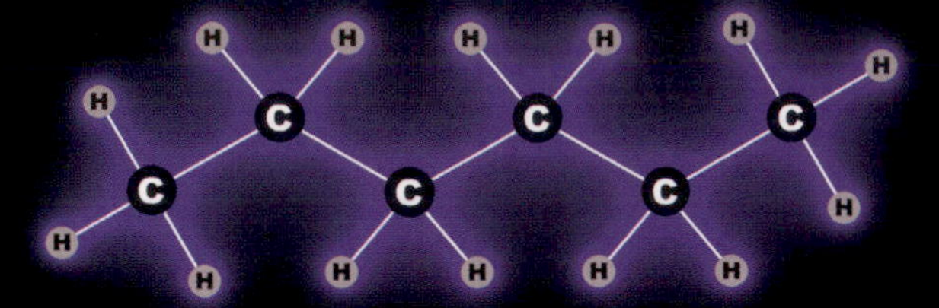

▲ Hexane has six carbon atoms and fourteen hydrogens.

◀ Heptane has seven carbons and sixteen hydrogens. This straight-chain version has a particular role to play as the "zero" standard on the octane scale for gasoline. Any hydrocarbon will potentially explode when it's compressed, which is a bad thing in a gasoline engine. The higher a fuel's octane rating, the more it can be compressed before exploding. Straight-chain heptane is prone to explode early, so it's defined as zero (i.e., bad) on the octane scale.

▼ Octane has eight carbons and eighteen hydrogens, whether it's straight or branched. The particular branched form shown here, called isooctane, is the "octane" referred to in the octane scale for gasoline. Pure isooctane has an octane rating of 100.

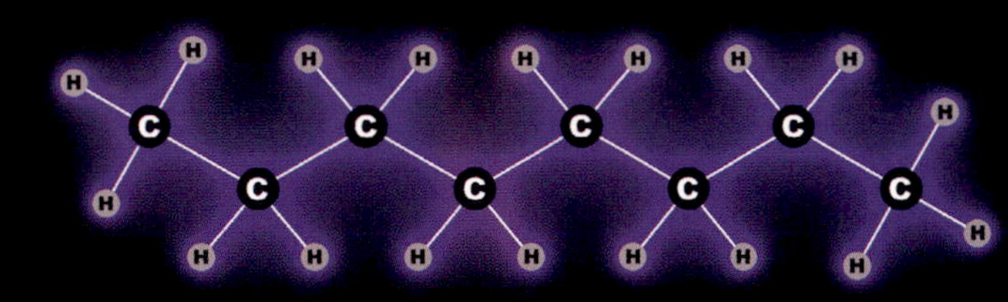

◀ Kerosene is a mix of hydrocarbons starting with hexane and including various straight and branched versions up to about sixteen carbons. It's an important property of kerosene that it does *not* include any lighter, shorter-chain, more volatile hydrocarbons than hexane, because that means no explosive vapor accumulates above it, making it much safer than gasoline. When crude oil was first pumped from the ground in the mid-nineteenth century, kerosene was the main product created from it. Inexpensive lamp oil allowed common people to stay up at night for the first time. Unfortunately, not all early refiners were careful to remove the hydrocarbons lighter than hexane, and death by kerosene lamp explosion was shockingly common. John D. Rockefeller named his company "Standard" Oil because he standardized kerosene and made it safer; he used a thermometer to measure the exact boiling point of his product instead of just calling any clear, distilled oil "kerosene." Today, kerosene is traditionally kept in blue containers to distinguish it from gasoline.

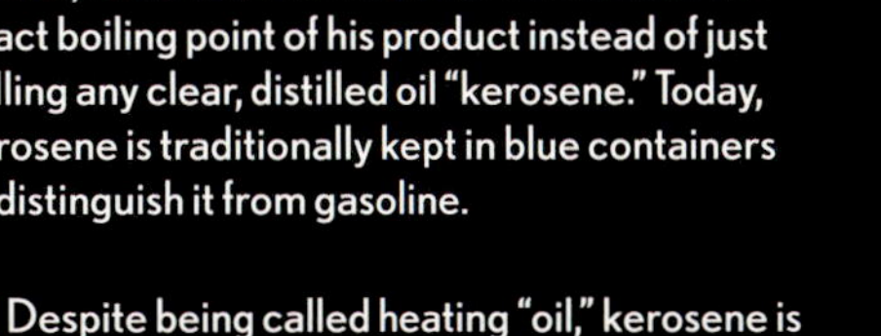

◀ Despite being called heating "oil," kerosene is a lively liquid, not viscous like heavier oils.

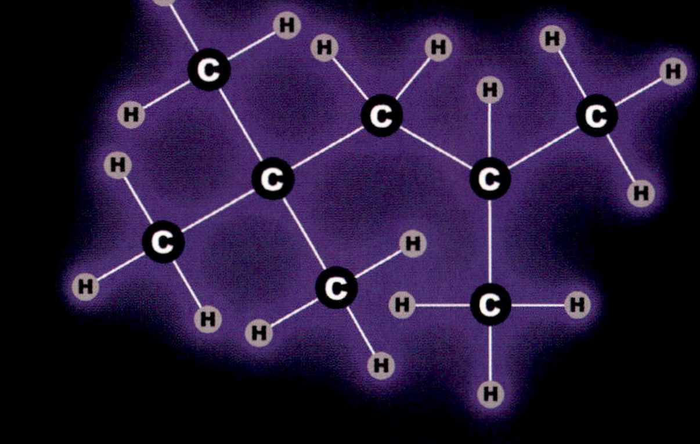

◀ Isooctane

▼ Decane

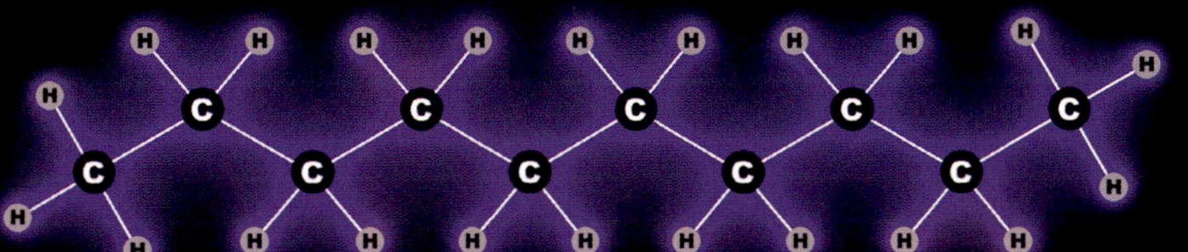

▶ As the number of carbons increases, hydrocarbons become increasingly "heavy," meaning both their boiling point and viscosity increase (i.e., more "oily" than "watery"). Decane has ten carbons and twenty-two hydrogens.

◀ Diesel fuel is a heavier mixture than kerosene, with most of it consisting of chains between ten and fifteen carbons long (straight chain, branched, cyclic, and some with carbon-carbon double bonds as well). Diesel is traditionally kept in yellow cans. (Confusing what fuel you're putting in which engine is a *bad thing*, hence all this color coding.)

▼ Undecane—a straight-chain hydrocarbon with eleven carbons—is, I swear, a moth pheromone. They use it to attract mates, in much the same way the human male uses it in sports cars (it's one of the higher-numbered hydrocarbons in gasoline).

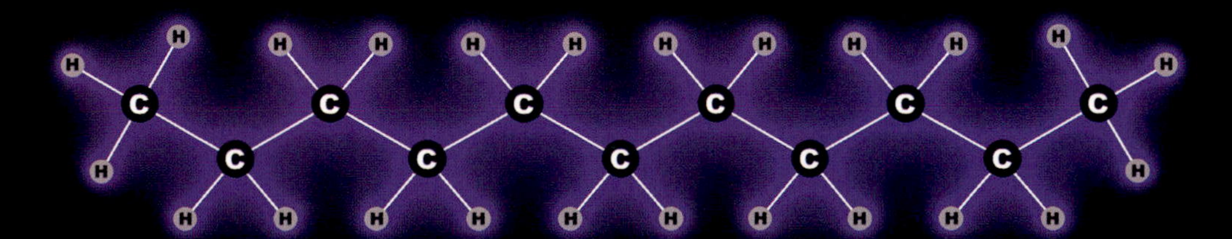

CROWN
THE BRAND PREFERRED BY PROFESSIONALS
LOW ODOR
MINERAL SPIRITS
DILUYENTE DE PINTURAS - BAJO DE OLOR
SCAN FOR INFO!
CLEANS UP PAINT
PREMIUM OIL-BASED THINNER
IDEAL FOR INDOOR USE
DANGER! COMBUSTIBLE LIQUID AND VAPOR. HARMFUL OR FATAL IF SWALLOWED. EYE, SKIN AND RESPIRATORY IRRITANT. VAPORS HARMFUL.
¡PELIGRO! LÍQUIDO Y VAPORES COMBUSTIBLES. NOCIVO O MORTAL SI SE INGIERE. IRRITANTE PARA LOS OJOS, LA PIEL Y EL SISTEMA RESPIRATORIO. VAPORES DAÑINOS.
LLOM4A - PS1112
See back panel for additional warnings.
Lea el panel del dorso para precauciones adicionales.
32 FL OZ (1 QT) 946 mL

▶ Mineral spirits are included in many kinds of solvents and paint strippers. This one, for example, contains mostly dichloromethane and methanol, plus some mineral spirits.

▶ Dichloromethane

▶ Methanol

▲There are a great many organic solvents used for different commercial and household purposes. Of these, the liquid known as "mineral spirits" is the closest to being a mixture of pure hydrocarbons. Mineral spirits and mineral oil, which we'll discus shortly, are closely related: both are gathered by distilling the same crude oil, with the "spirits" boiling off at a lower temperature than the "oil."

◀ The stuff called "mineral oil" and sold in drug stores instead of gas stations is a very clean grade of hydrocarbon containing almost entirely straight- and some branched-chain molecules with around fifteen to forty carbons per molecule (with a preponderance of shorter ones in that range, on average). You wouldn't want to eat mineral oil of any kind, but this food-grade version is promised to be free of harmful components, so it can be used on food preparation surfaces.

▶ Baby oil is just mineral oil with perfume added. It's not made from babies.

◀ "Light machine oil" is a fairly low-viscosity hydrocarbon oil—lighter than motor oil but more viscous than solvents and fuels. It differs from mineral oil in that it contains more of the sorts of additives, nonhydrocarbons, and unsaturated compounds (hydrocarbons containing double-bonds) that you find in motor oil, which aid in lubrication and make machine oil rather smelly.

▲ Trombone oil is used to lubricate trombone slides. It's basically a light machine oil, but it's also a wonderful metaphor. Trombone oil is very specialized. Musicians who need it will pay a high price for the best quality of trombone oil. And the total world market is a few gallons a year. (Actually, that's a made-up number, but this is a metaphor, so never mind.) The point is that no matter how excellent a trombone oil maker you are, no matter how fine your trombone oil is or how much you can charge for it, you're never going to make a really serious amount of money selling it because the total market is just too small. I find this to be an excellent analogy that can be applied to many situations, and now you can use it too.

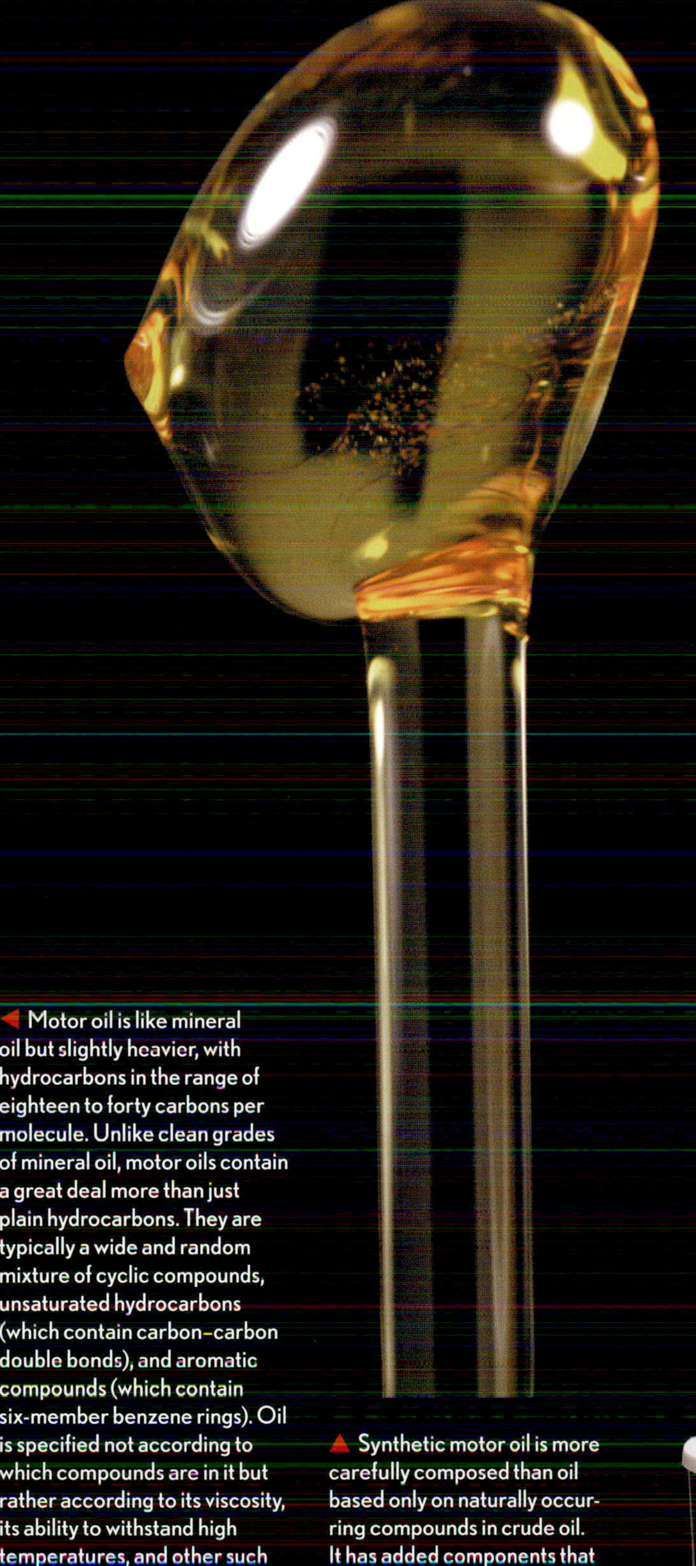

Motor oil is like mineral oil but slightly heavier, with hydrocarbons in the range of eighteen to forty carbons per molecule. Unlike clean grades of mineral oil, motor oils contain a great deal more than just plain hydrocarbons. They are typically a wide and random mixture of cyclic compounds, unsaturated hydrocarbons (which contain carbon–carbon double bonds), and aromatic compounds (which contain six-member benzene rings). Oil is specified not according to which compounds are in it but rather according to its viscosity, its ability to withstand high temperatures, and other such performance-related measurements. It's up to the individual manufacturers to decide what compounds to combine to meet these specifications.

Synthetic motor oil is more carefully composed than oil based only on naturally occurring compounds in crude oil. It has added components that give it a great deal of stickiness, allowing it to hold together and adhere to metal surfaces, thus fulfilling its function of protecting engines from wear.

All motor oils contain an assortment of special additives that improve quality by increasing the life of the oil, inhibiting rust in metals, scavenging contamination from the engine, and so on. For those of you who feel your oil doesn't have enough additives, you can buy additives in concentrated form in a range of oil supplements. These are often sold in exotic bottles with outlandish claims about the benefits they'll bring to your engine—not unlike, for example, olive oils or energy drinks.

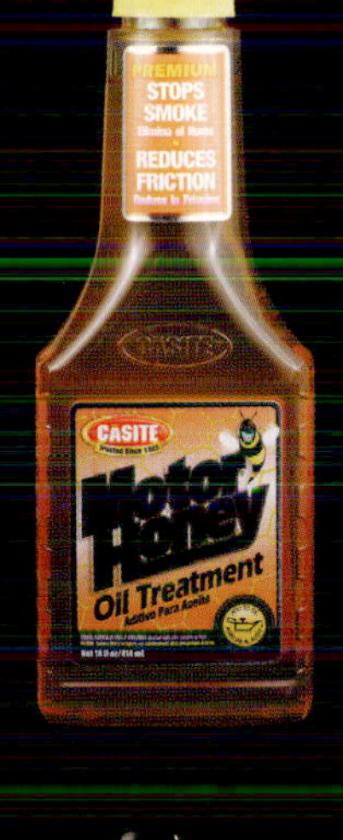

Drink it or pour it in your engine? Don't get these two confused! Both are designed to enhance the performance of an engine—one mechanical, the other biological. Both are all about marketing, which you can tell from the fact that they are sold at the checkout counter, sometimes alarmingly close to each other in a car parts store.

Oil can get pretty thick as the average length of its hydrocarbon chains grows. This extremely goopy material is oil meant for the gear boxes of locomotives. It comes in plastic bags, which are simply thrown into the crank cases of the giant machines. The gears chew up the plastic without a moment's hesitation.

As its average carbon chain length increases, the viscosity of a hydrocarbon eventually gets too high for it to be called oil anymore; it becomes grease. The point of grease is that it will stick in places where oil would just drip off.

Beyond grease is paraffin wax, with chains twenty to forty carbons long (tending toward the high end). In refined form, it is very little other than fully saturated hydrocarbon chains. (Confusingly, in some parts of the world the term *paraffin* is used for liquid mineral oils, while in the United States paraffin is always a solid. They are essentially the same kind of material, just with different average carbon chain lengths.)

Beyond paraffin wax lies polyethylene plastic, though it's a big jump. Polyethylene starts with chains a few thousand carbons long and reaches into the hundreds of thousands. Read more about the many uses of polyethylene in Chapter 7.

The mother of all mineral oils, solvents, greases, paraffins, and plastics is crude oil. This is the raw thing, pumped straight out of the ground in one of the historic oil fields of Pennsylvania. I used to always think of crude oil as extremely thick and sludgy, and some is. But this crude oil is almost the consistency of water. It's amazing how much chemistry is based on processing this stuff, and how soon we will be down to the last glass full of it.

Oils for Eating

OIL FROM PLANTS and animals may look and feel nearly the same as clean mineral oil, but there's a fundamental difference in chemical structure.

Like the mineral oils discussed in the last section, oils from animals and plants contain carbon chains, often in the range of fourteen to twenty carbons long. But the carbon chains in oils from living sources always have what is called an organic acid group at one end (see page 42 for more about what this means). These molecules are called fatty acids.

Their acid end makes it possible for fatty acids to link together in ways not possible for simple hydrocarbons, and they take advantage of this linking ability. In nearly all oils and fats from animals and plants, the fatty acids are linked together in groups of three on a glycerine backbone. These molecules are called triglycerides.

As with mineral oils, fatty acids vary in how long their carbon chains are, with longer chains yielding thicker, more viscous oils. But with fatty acids, we also worry a lot about the precise location and orientation of any carbon–carbon double bonds in the molecule. These are of tremendous interest because of their implications for health; it's these double bonds that people are talking about when they go on about how good omega-3 fatty acids are for you or how bad trans fats are.

▶ Glycerine

▲ Glycerine is a polyalcohol. As we learned in on page 38, an alcohol is any compound with a -OH group stuck on it somewhere. Glycerine has three of these, so it's a triple alcohol.

▼ Here's a typical fatty acid molecule, lauric acid. It looks superficially like the hydrocarbons in the last section, but notice the red oxygen atoms on the left: they make this a fatty acid. It is "fully saturated," which means that each one of its carbon atoms has the maximum two hydrogen atoms on it (except the last one, which ends the chain with one extra hydrogen). All the carbons are bonded to each other with single bonds. This molecule, and similar ones with slightly longer or slightly shorter carbon chains, are what make up saturated fat (when assembled into triglyceride units).

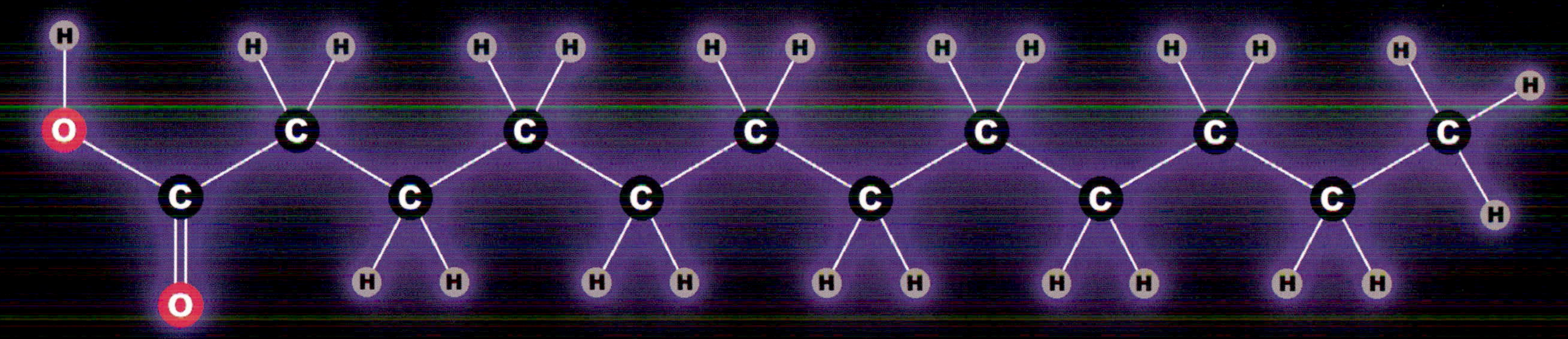

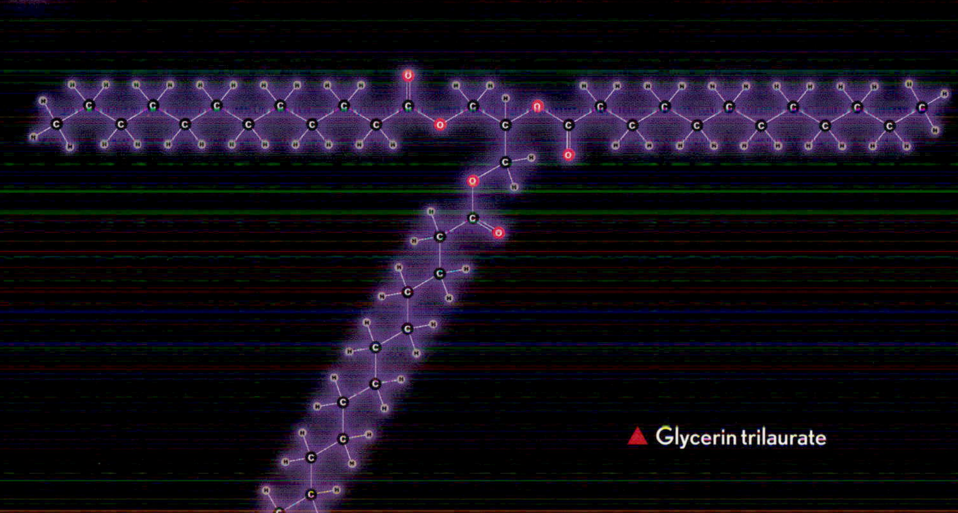

▲ Glycerin trilaurate

◀ When an organic acid, such as a fatty acid, is joined end to end with an alcohol, the result is called an ester (see page 43). Because glycerine has three alcohol groups, it can be linked to three fatty acids. If you do that, the result is called a triglyceride. This one is glycerin trilaurate, made from one glycerine molecule and three lauric acid molecules. All vegetable and animal oil and fat is composed primarily of this kind of triglyceride but with lots of variations in the individual fatty acids used to make up the triglyceride.

Oils for Eating

Here we have the same lauric acid molecule as before, but this time there is a double bond between two of the carbon atoms in the chain. As we learned on page 19, this means that an extra one of each of those two carbon's "slots" is taken up by the double bond. Each therefore has one fewer slot available for connecting to a hydrogen atom. So overall, this molecule has two fewer hydrogen atoms than lauric acid does. We say it's "unsaturated" because we could add more hydrogen to make it completely saturated with hydrogen. The double bond could occur between any pair of carbons, so a labeling system is used to make it easier to talk about specific examples. The carbon atoms are labeled with letters from the Greek alphabet starting with *alpha* α for the one closest to the acid end. Unfortunately, the fatty acids of human interest are special because of how far the double bond is from the *other* end of the chain, and the length of the chain varies. So we skip the middle and call the last carbon atom, no matter how long the chain is, the *omega* ω carbon, because ω is the last letter in the Greek alphabet. Then the double bond is labeled by how far away it is from this omega-carbon. Thus, this example is an omega-3 fatty acid. Did that ring a bell?

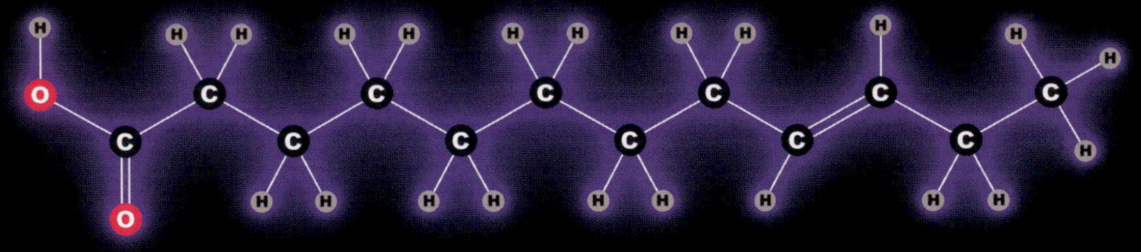

trans-Omega-3 lauric acid

There's another wrinkle! Carbon–carbon single bonds can rotate fairly easily around their axis, so molecules like this are rather floppy, and it doesn't really matter what angles you draw the bonds at. But double bonds are locked into a particular orientation. When the two sides of the chain leave the double bond on opposite sides (traveling in the same direction), it's called the *trans* configuration. When they leave on the same side (changing direction), it's called the *cis* configuration. So the example above is a *trans*-omega-3 fatty acid, while the example below is a *cis*-omega-3 fatty acid. That's what "*trans* fat" means. It's less healthy than *cis* fat. I think that's weird given how obscure this difference is, but the body is a subtle machine. It cares about this kind of thing.

- α alpha
- β beta
- γ gamma
- δ delta
- ω omega

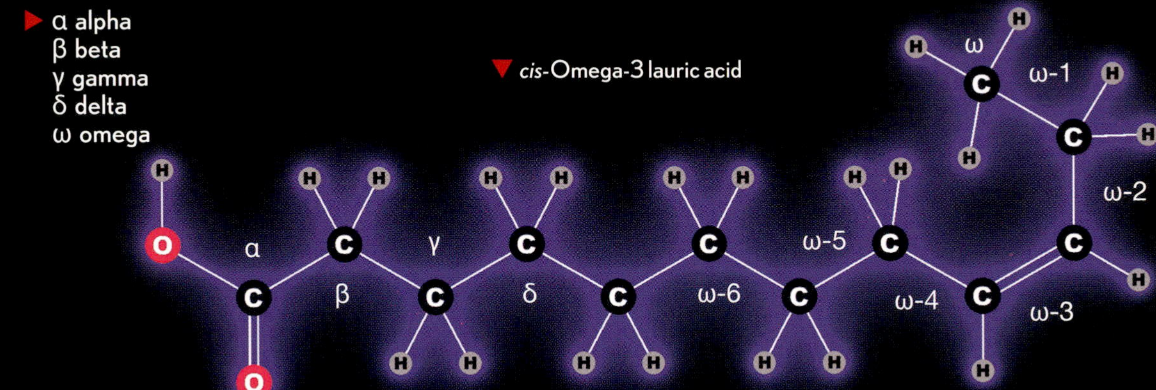

cis-Omega-3 lauric acid

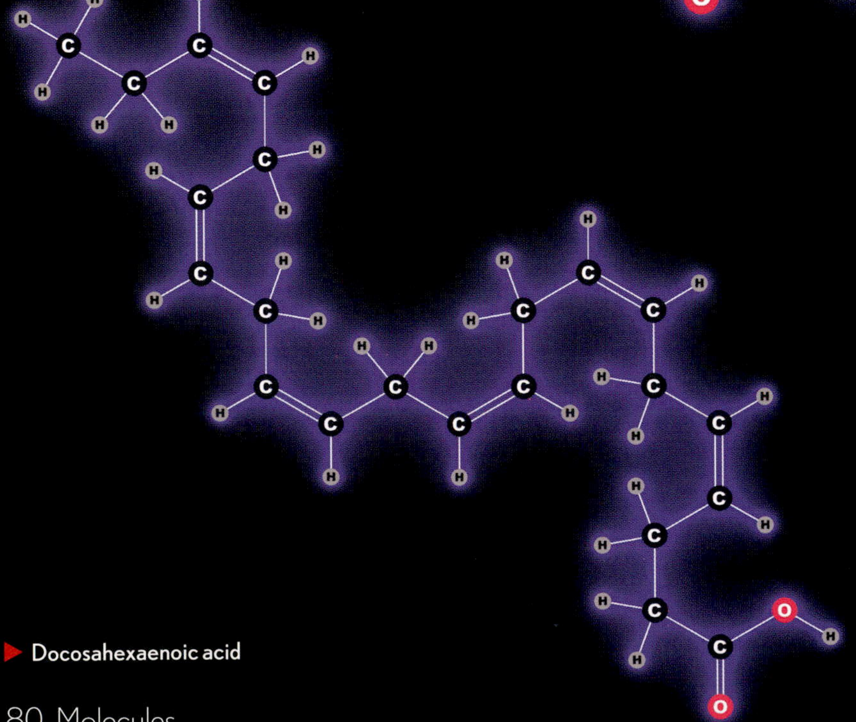

Docosahexaenoic acid

And one more complication! Our first example has just one double bond: it's monounsaturated. But you can have as many double bonds as you like. Anything more than one, and it's called "polyunsaturated fat." This seems to be better for you, or at least not as bad as, either monounsaturated or saturated fat. Because each double bond can be in either the *cis* or *trans* configuration, there are a lot of possibilities, and the body cares about each possible variation. It's only quite specific patterns of *cis* and *trans* that occur in plants and animals. This particular polyunsaturated fatty acid, with its unique combination of *cis* and *trans* double bonds, is a major structural component of the brain, the retina, and other important body systems. It's found commonly in seafood, but if you don't get enough from fish, your body can make it from other fatty acids.

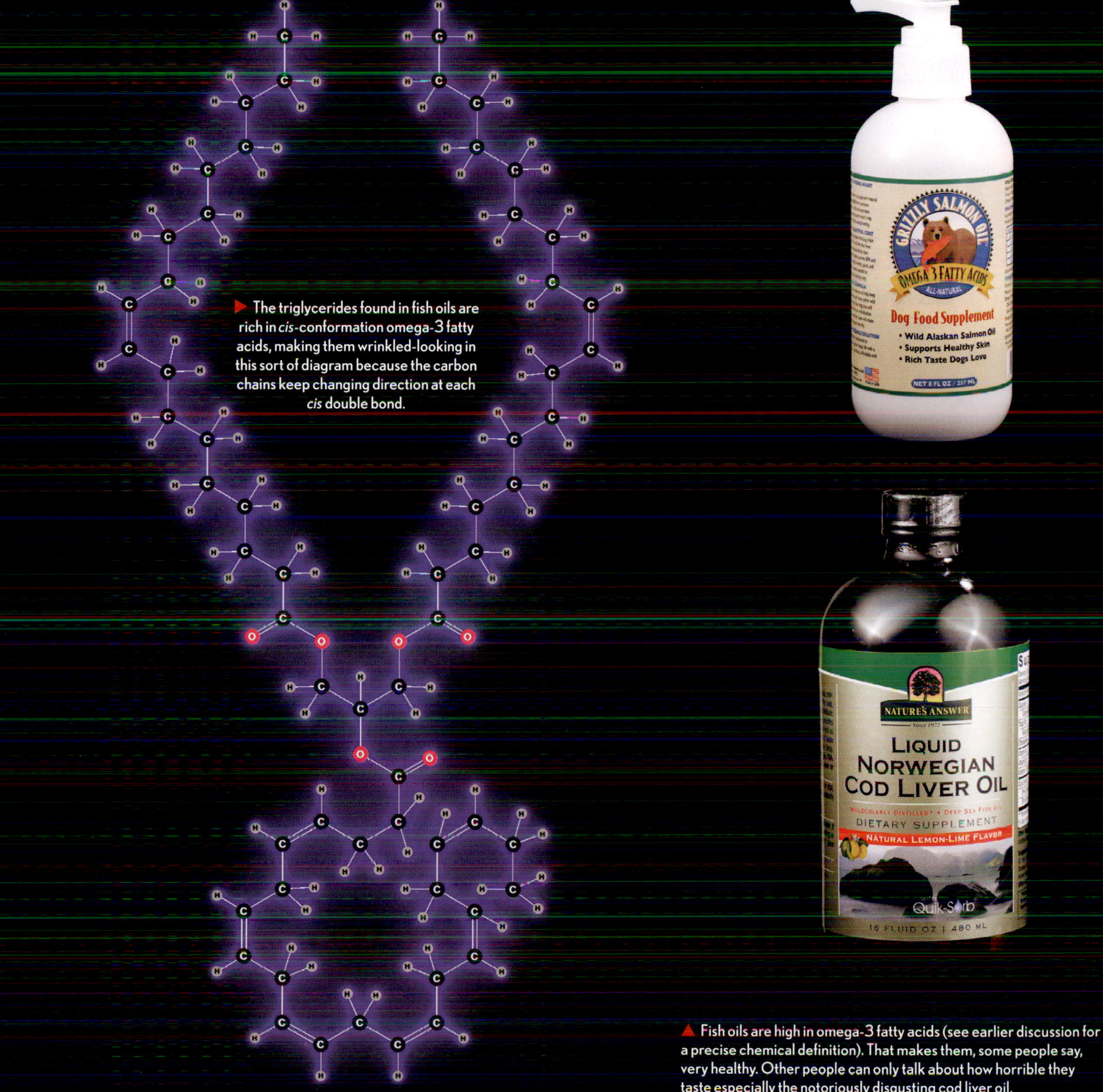

▶ The triglycerides found in fish oils are rich in *cis*-conformation omega-3 fatty acids, making them wrinkled-looking in this sort of diagram because the carbon chains keep changing direction at each *cis* double bond.

▲ Fish oils are high in omega-3 fatty acids (see earlier discussion for a precise chemical definition). That makes them, some people say, very healthy. Other people can only talk about how horrible they taste especially the notoriously disgusting cod liver oil.

Oils for Eating

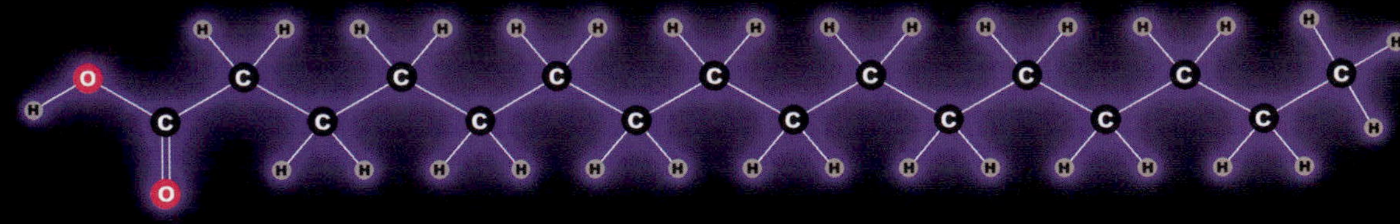

▲ Palmitic acid

▲ This fully saturated fatty acid is called palmitic acid, betraying its origin in palm trees.

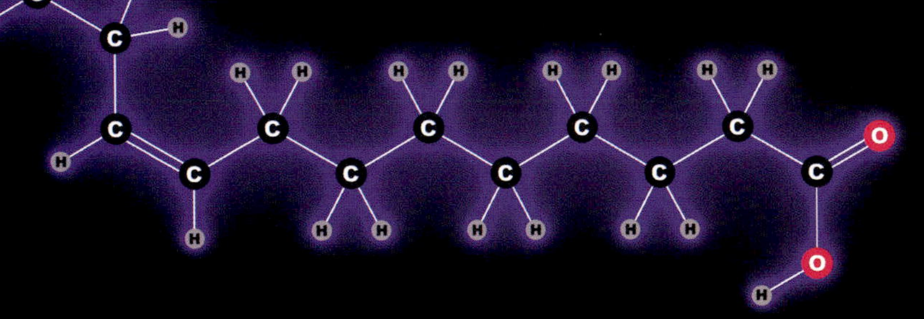

◀ Linoleic acid

◀ Omega-6 fatty acids have a double bond at the sixth carbon from the end (see earlier discussion for more detail). This example is linoleic acid, a polyunsaturated fatty acid that has double bonds at the six and nine positions from the end. It's found in many vegetable oils and is considered an essential dietary fatty acid. Like vitamins (see page 184), you cannot live without consuming at least some of this acid, but unlike vitamins, it's basically impossible not to get enough in any plausible diet.

▼ Palm oil

▼ Three units of linoleic acid assembled on a glycerine backbone give a very common triglyceride found in the majority of vegetable oils, with the highest concentration occurring in safflower oil.

▼ Typical vegetable triglyceride

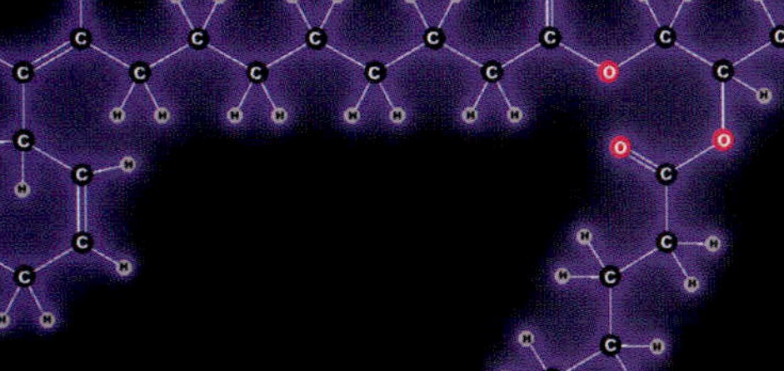

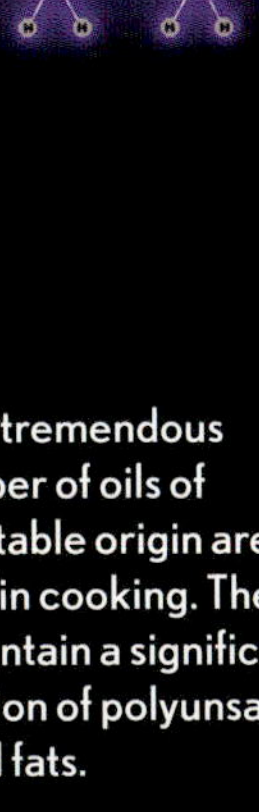

▶ A tremendous number of oils of vegetable origin are used in cooking. They all contain a significant fraction of polyunsaturated fats.

Some vegetable fats and most animal fats have a terrible reputation because they contain a high fraction of saturated fatty acids (see the earlier explanation of what "saturated" means). The tropics harbor a lot of this unhealthy stuff in the form of coconut, palm, and palm kernel oils. The more saturated a fat is, the higher its melting point, so these highly saturated fats tend to be solids or pastes at room temperature. In terms of their saturated fat content, they are basically equivalent to pure animal fat.

Baby oil isn't made from babies, Girl Scout cookies aren't made from Girl Scouts, but neatsfoot oil actually is made from feet. Specifically, the feet and shin bones of cattle ("neat" being an Old English name for ox, cow, or other cattle). As an oil of animal origin, it is made of triglycerides.

Wax

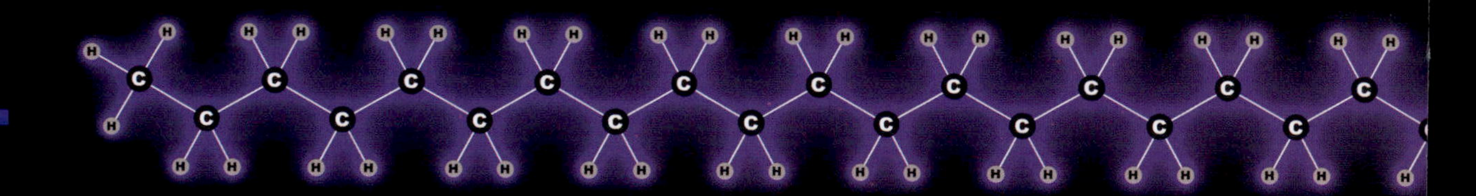

WE ENCOUNTERED paraffin wax, a pure hydrocarbon derived from petroleum, earlier in this chapter. But proper wax is closely related to both soap (see Chapter 4) and to the fats and vegetable oils discussed in this chapter. While vegetable oils are esters with three fatty acids joined to one glycerine polyalcohol, waxes are esters with one fatty acid joined to one long-chain alcohol.

▶ The color of beeswax depends on whether the wax comes from comb the bees used strictly for honey (light colored) or comb that contained brood and pollen (darker). Additional variation reflects the number of times the cells were filled and emptied by the bees before the wax was removed and purified; older wax is darker than single-season wax. The colors come from small amounts of impurities: refined beeswax is almost entirely waxy esters, and always pale in color.

▲ Beeswax is made (by bees) largely of an ester with fifteen carbons to the left of the -COO- link and thirty to the right, a compound called triacontanyl palmitate.

▶ Carnauba wax comes from the leaves of the carnauba palm tree and contains a more complex mixture of compounds than beeswax, including not only simple esters but also di-esters and some long-chain alcohols.

▶ A lot of wax is sold for particular purposes. Because waxes from different sources contain many different combinations of carbon chain lengths, they can be combined and blended with each other and with solvents to create an almost infinite variety of waxy products.

◀ Carnauba wax is widely admired for being especially hard and shiny, but in many products it's softened with a solvent that turns it into a paste so it can be applied easily. When the solvent evaporates, what's left is a hard wax surface that can be polished to a high shine. People apply wax to bowling lanes, cars, and other things they want to have smooth and shiny. (Carnauba wax is also known as Brazil wax, but not all Brazilian wax is carnauba. Some is a blend of beeswax and paraffin wax.)

Special purpose wax.

Chapter 6

Rock and Ore

COMPOUNDS ARE MADE up of elements, so it's logical to think that in order to get compounds, you would put together the necessary elements. But in practice, the opposite is true.

What you find in nature, for the most part, are elements already put together into various compounds. If you want elements, you have to take those compounds apart. For example, if you want the element iron, you're not going to find much of it in the wild. The only naturally occurring iron metal is from meteors, and there aren't many of those around (at least that haven't completely rusted away already).

So you have to look for iron *ore* instead: material from which you can extract iron. The word *ore* is an economic description, a name based on what something is useful for. So "iron ore" means anything, regardless of its specific composition, that is useful as a source of iron metal.

The ore from a given mine will always be made up of particular minerals. Unlike *ore*, the word *mineral* means a specific chemical compound, or at least a defined, reasonably consistent mixture of specific compounds. When a mineral is beautiful, we call it a crystal or even a gemstone. When it's ugly, we call it a rock.

Iron ore typically contains a mixture of the minerals hematite (Fe_2O_3), magnetite (Fe_3O_4), pyrite (FeS_2), or any of a number of other iron-containing compounds.

◀ Magnetite is a hard, shiny material that looks quite a bit like metal when polished, even though it's an oxide (Fe_3O_4). Its name comes from the fact that it can be magnetized, which explains why the skull in this photograph is able to hang from a magnet and *may* explain why it is described as being highly charged with psychic energy.

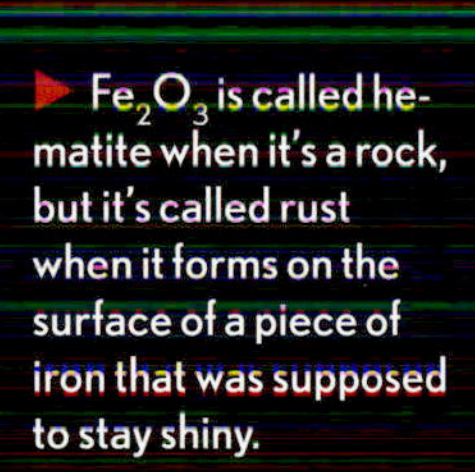

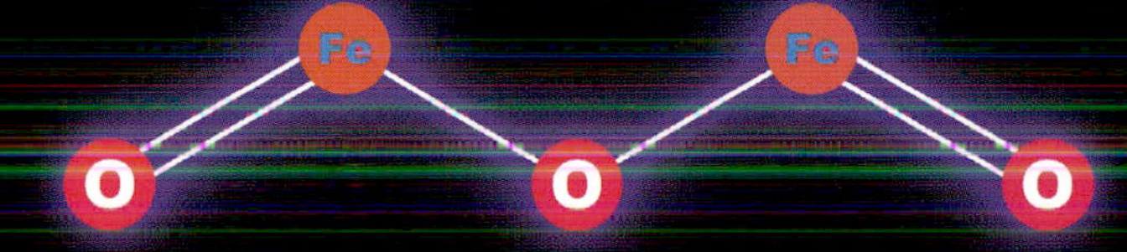

▶ Fe_2O_3 is called hematite when it's a rock, but it's called rust when it forms on the surface of a piece of iron that was supposed to stay shiny.

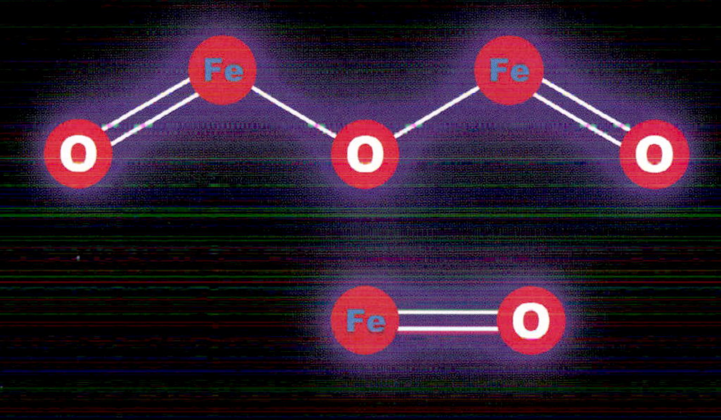

▶ Fe_3O_4 is a mixed oxide of iron. It's not a mixture of two different chemicals in arbitrary proportions. Instead, Fe_2O_3 and FeO units are bound together in exact one-to-one proportions, resulting in a perfect three-to-four ratio of iron to oxygen atoms overall. This kind of mixture is fairly common in minerals.

▼ Hematite, Fe_2O_3, is one of the two main iron ores processed in huge quantities in steel mills all over the world. It's also what iron rusts into, and in this sample you can see the reddish color that is so characteristic of rusted iron. The smelting process that converts iron ore into iron metal is the chemical opposite of rusting. In other words, while we think of rust as coming from the rusting of iron, in reality iron comes from the un-rusting of rust.

▲ Imagine grinding up beautiful stuff like this to make truck axles. But ore is ore, and no one's looking at most of it. This tiny chip of hematite was lucky enough to catch the eye of a collector and thus was saved from axle oblivion.

▲ These balls are sold on eBay as cheap slingshot ammunition, but that's not what they were made for. These are iron ore feedstock ready to be loaded into blast furnaces for reduction into iron metal. They are made by the megaton and shipped in giant barges and freight trains, which explains why they're so cheap if you want only a thousand or two for your slingshot! They start as the mineral taconite, which is ground up and separated to isolate its magnetite content, and is then heated and formed into these convenient balls. During the heating process, the magnetite (Fe_3O_4) is oxidized further into hematite (Fe_2O_3).

◀ Chunks of magnetite are historically known as lodestone. They sometimes become magnetized naturally, leading to the discovery that small pieces floated on a cork would always point north. This led to the first compass.

◀ This is a reproduction of a nineteenth-century lodestone compass. Modern compasses use much more powerful magnets, but even a very weak lodestone can work as a compass if it's carefully balanced.

▲ The mineral martite is a peculiar form of hematite. It's called a pseudomorph because, although chemically it is hematite (Fe_2O_3), it has the crystal structure of magnetite (Fe_3O_4). Pseudomorphs happen in one of two ways: either a chemical crystallizes in its natural form and is then converted into a different chemical by some reaction, or one chemical leaches out and replaces another chemical, occupying the same space and shape. This is an example of the former, and occurs when magnetite is further oxidized into hematite without changing its overall shape.

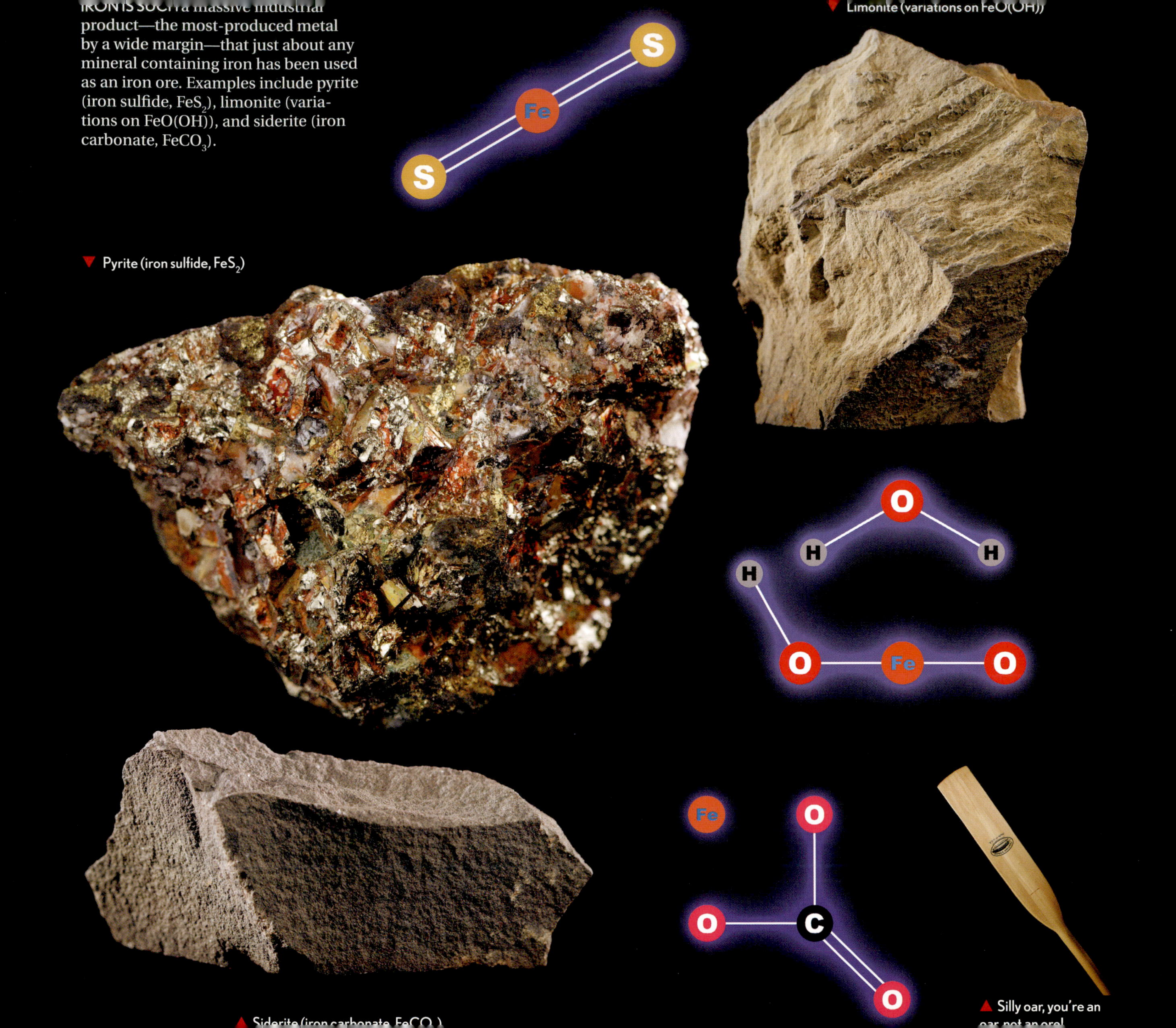

IRON IS SUCH a massive industrial product—the most-produced metal by a wide margin—that just about any mineral containing iron has been used as an iron ore. Examples include pyrite (iron sulfide, FeS_2), limonite (variations on FeO(OH)), and siderite (iron carbonate, $FeCO_3$).

▼ Limonite (variations on FeO(OH))

▼ Pyrite (iron sulfide, FeS_2)

▲ Siderite (iron carbonate, $FeCO_3$)

▲ Silly oar, you're an

Processing Ores

HOW ORE IS TURNED into elements depends a lot on the ore. Sometimes the hardest part of getting an element isn't finding the ore, it's figuring out how to refine it.

Iron is relatively easy: it can be smelted into iron metal simply by heating it in the presence of coke (a baked, hardened form of charcoal largely composed of the element carbon). People figured out how to do this almost three thousand years ago. (I say relatively easy because it's easier than most metal extractions from ore, but it's not easy in an absolute sense. The temperatures required are very high, and great skill is needed to maintain the right conditions. It took 150 generations after people started living in large cities before we first figured out how to smelt iron.)

But smelting iron is a whole lot easier than converting aluminum ore into aluminum metal. It turns out that this can be done in a practical way only if you have a lot of electricity, so aluminum remained exotic until electricity became available in bulk from generators rather than in dribbles from chemical batteries. Today a lot of aluminum is refined in Iceland simply because they have lots of cheap geothermal electricity. The ore arrives on huge barges, and the aluminum leaves in container ships. It visits Iceland only for the electricity.

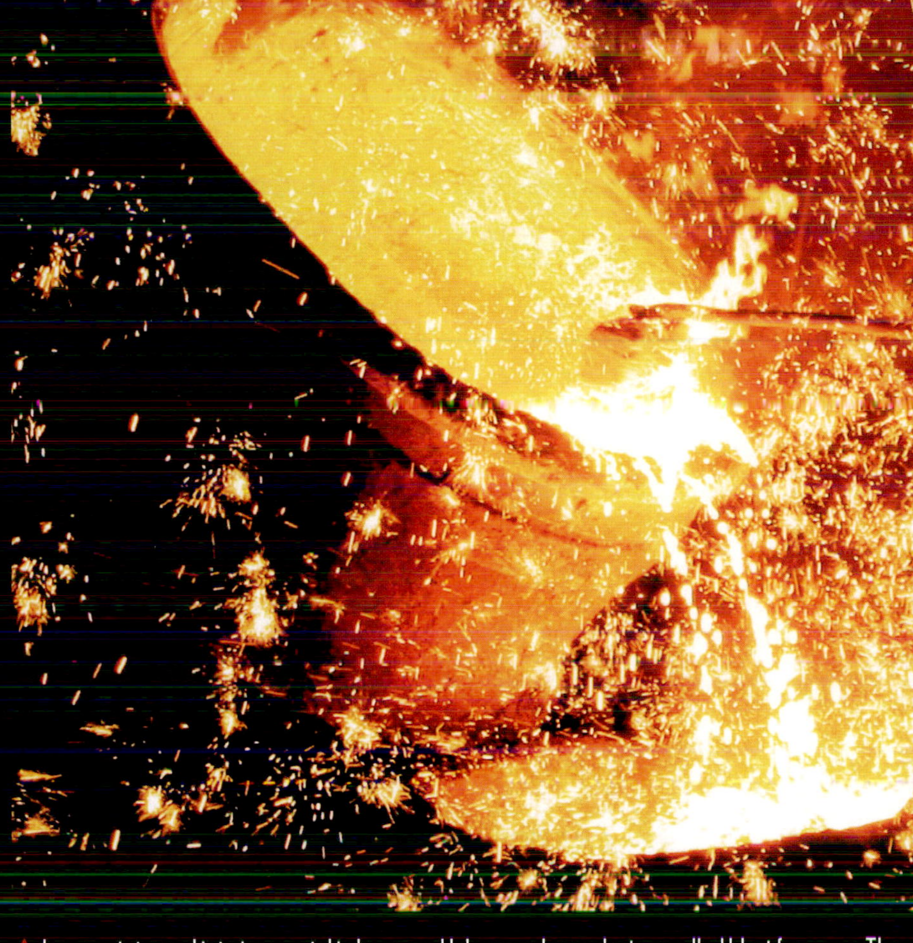

▲ Iron ore is turned into iron metal in huge, and I do mean huge, devices called blast furnaces. The iron ore is packed into the furnace together with coke (made mostly of carbon), and the whole thing is lit on fire and then fed from below with pressurized air (the *blast* in blast furnace). The carbon in the coke steals oxygen from iron oxides in the ore, creating carbon dioxide and at the same time freeing iron metal from the ore. The iron runs out the bottom of the furnace as a white-hot liquid.

◀ Aluminum can be extracted from aluminum ore by chemical means, but it's very difficult to accomplish and requires using elements that are even harder to isolate than aluminum itself. But with electricity in vast amounts, you can do it. Aluminum oxide is extracted from the ore bauxite, mixed with cryolite (another mineral containing aluminum), and then melted in large cells. Each cell has a pair of electrodes through which are run several hundred thousand amperes of current (at about three to five volts). Aluminum metal collects on the negative electrode and flows to the bottom of the cell where it is siphoned off periodically. Here we see molten ore being added to a cell. Notice the *unbelievably* thick electrical cables on the right side of the picture.

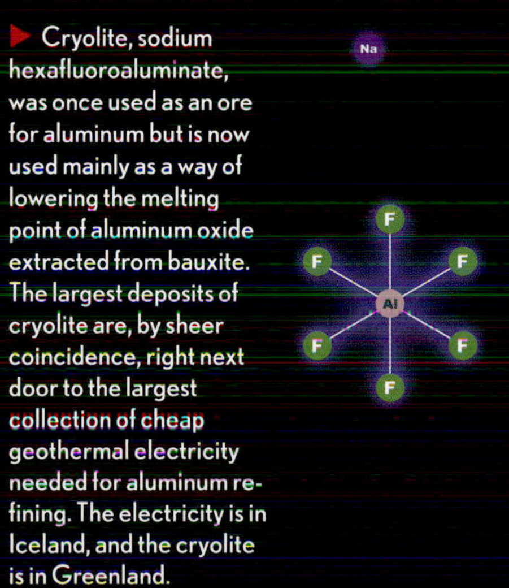

▶ Cryolite, sodium hexafluoroaluminate, was once used as an ore for aluminum but is now used mainly as a way of lowering the melting point of aluminum oxide extracted from bauxite. The largest deposits of cryolite are, by sheer coincidence, right next door to the largest collection of cheap geothermal electricity needed for aluminum refining. The electricity is in Iceland, and the cryolite is in Greenland.

Processing Ores

▶ Bauxite is the principal ore of aluminum. It's a mixture of several specific minerals that tend to occur together.

▶ The minerals that make up bauxite include $Al(OH)_3$, called gibbsite, and two different crystal structures of AlO(OH), called boehmite and diaspore. While bauxite is always a lumpy mass, these pure minerals can occur in crystalline form. (Diaspore even occurs in the form of gemstones that can be cut and polished.) That's no accident, because crystals can form only out of relatively pure substances. By the way, I include molecular diagrams for these kinds of inorganic compounds because they are a pretty way of showing what elements go into each substance. But often the diagram doesn't tell you much more than you could learn from the simple chemical formula, like for example, $Al(OH)_3$ for gibbsite. This is in stark contrast to the same kind of diagram applied to organic molecules, as you see in most of the rest of this book. The chemical formulas of organic molecules are often nearly useless, simply listing some number of carbons, hydrogens, and maybe oxygens, without telling you anything about how they are put together. This difference is testimony to carbon's truly unique status as the *only* element capable of consistently producing structures of such logical complexity that a picture is the only way to describe them.

▶ Gibbsite

▶ Boehmite

▼ Diaspore

▲ Diaspore octagon

More Ore

FOR EVERY METAL, there are one or more ores from which it is derived. Metals mined in large quantities have the ore named after them: iron ore, copper ore, aluminum ore, and so on. Other metals are hangers-on. For example, the main ore for gallium, a fairly obscure metal, is bauxite, the same ore we just talked about as being aluminum ore. Gallium is just a minor impurity in the bauxite and is extracted purely as a side effect of aluminum refining.

Similarly, the "minor platinum group metals," osmium, iridium, rhenium, rhodium, and ruthenium, are for the most part extracted only as a side effect of platinum mining, occurring as impurities in what is considered, from a commercial point of view, to be platinum ore. This makes their prices extraordinarily volatile. When demand for rhodium goes up, the supply doesn't, as it's not economical to mine more platinum just because people are paying a lot for a tiny rhodium impurity in the ore. When the demand for platinum goes up, the price of rhodium crashes because, as platinum mining is ramped up, there's more rhodium available, whether anyone wants it or not.

▲ Gold ore is generally pretty boring stuff. Sure, you find the occasional pure gold nugget, but most gold comes from thoroughly non-gold-like rock similar to these core samples taken from a potential gold mining site.

▶ Chalcopyrite, known as peacock ore, is the most important copper ore, and it's also very pretty because of the dichroic effect of the oxide layers that form on its surface. But no matter how pretty it is, the mining companies grind it up anyway. There's too much money to be made selling the copper squeezed from it.

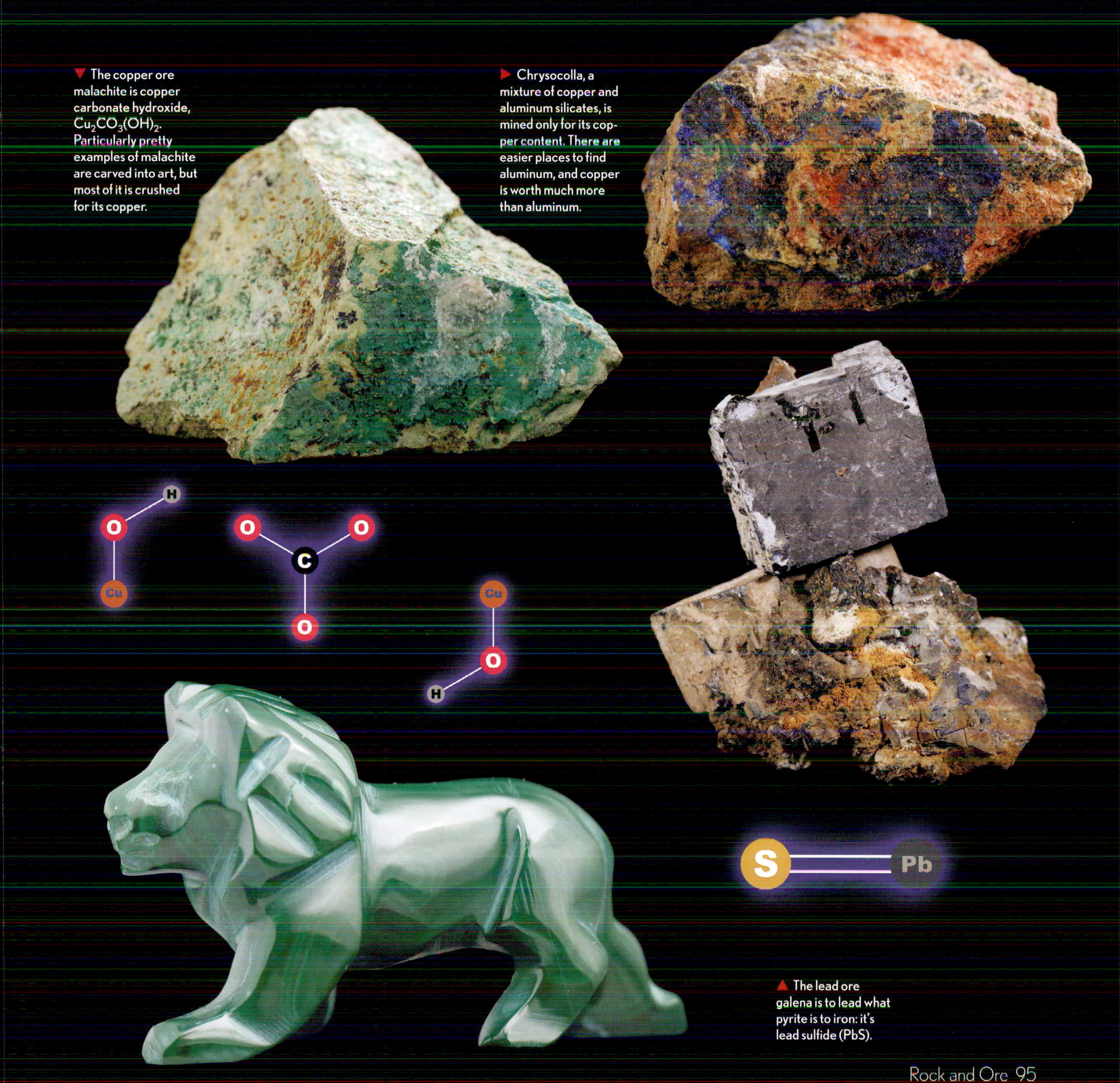

▼ The copper ore malachite is copper carbonate hydroxide, $Cu_2CO_3(OH)_2$. Particularly pretty examples of malachite are carved into art, but most of it is crushed for its copper.

▶ Chrysocolla, a mixture of copper and aluminum silicates, is mined only for its copper content. There are easier places to find aluminum, and copper is worth much more than aluminum.

▲ The lead ore galena is to lead what pyrite is to iron: it's lead sulfide (PbS).

More Ore

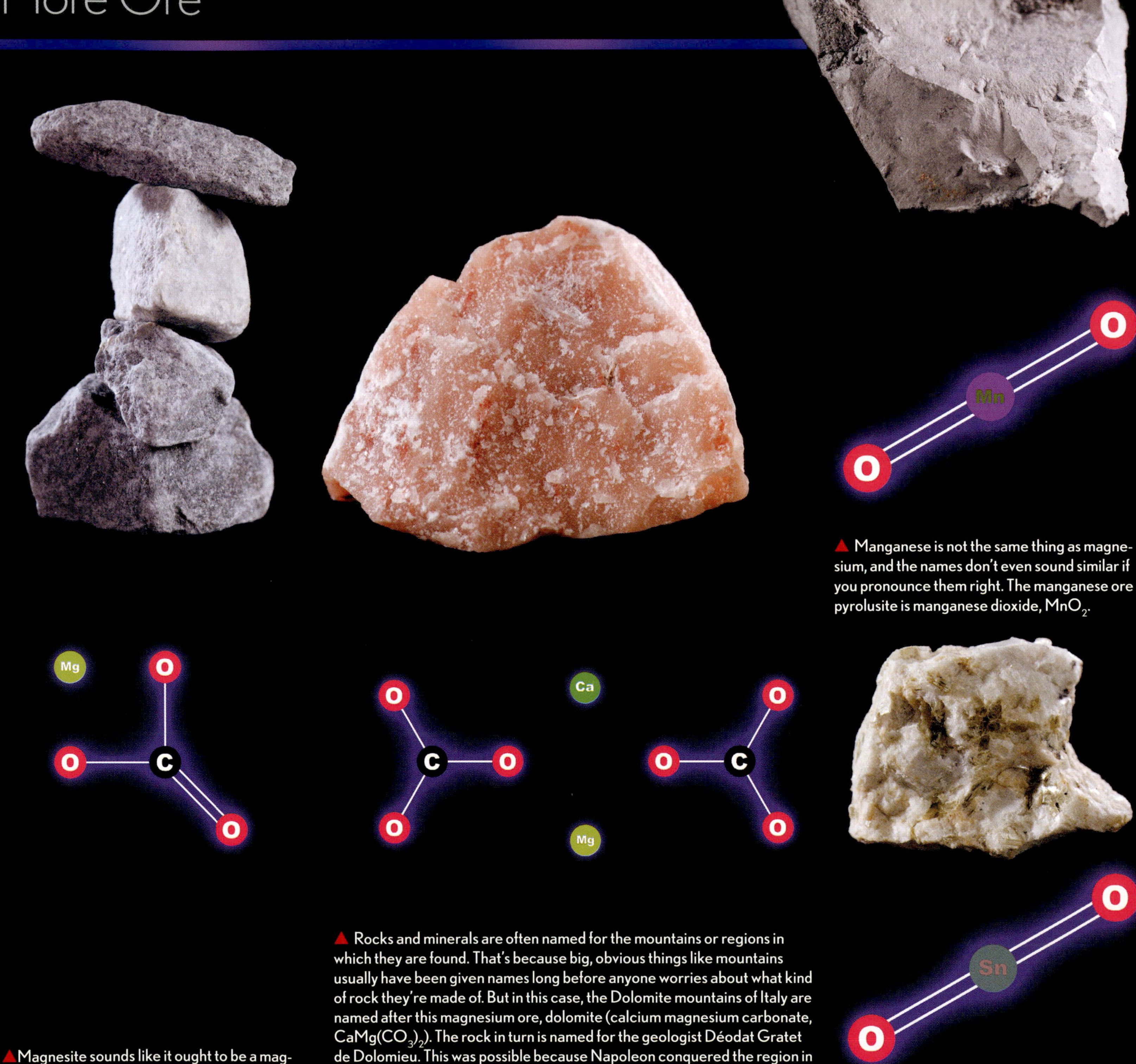

▲ Manganese is not the same thing as magnesium, and the names don't even sound similar if you pronounce them right. The manganese ore pyrolusite is manganese dioxide, MnO_2.

▲Magnesite sounds like it ought to be a magnetic mineral and thus probably an iron ore, but in this case the "mag" comes from magnesium. Magnesite is magnesium carbonate ($MgCO_3$).

▲ Rocks and minerals are often named for the mountains or regions in which they are found. That's because big, obvious things like mountains usually have been given names long before anyone worries about what kind of rock they're made of. But in this case, the Dolomite mountains of Italy are named after this magnesium ore, dolomite (calcium magnesium carbonate, $CaMg(CO_3)_2$). The rock in turn is named for the geologist Déodat Gratet de Dolomieu. This was possible because Napoleon conquered the region in 1800, allowing an Italian mountain range to be named after a French geologist (who, incidentally, had been held captive by the Italians for nearly two years). Politics.

▲ The tin ore cassiterite is tin oxide, SnO_2.

The zinc ore sphalerite is zinc sulfide, ZnS, which is usually contaminated with some FeS.

The beryllium ore beryl is beryllium aluminium cyclosilicate, $Be_3Al_2(SiO_3)_6$. When it's in the form of a pure transparent crystal, it's a moderately valuable gemstone. Ugly chunks of it get chewed up in rock crushers and turned into missile parts. Lesson: if you're a rock, try to be beautiful.

When beryllium ore is especially attractive and green from impurities, it's called an emerald.

Ores Are Not Just For Elements

THE TERM *ore* is used specifically to refer to rocks that get turned into metals (which are elements in pure form). Many other things are dug, pumped, or harvested from the earth and turned directly into other useful chemical compounds instead of into elements.

Most compounds you're familiar with are made by transforming one compound into another in sequences that may be dozens of reactions long. Sometimes an element or two gets introduced along the way, but that's the exception rather than the rule. (And the most common example by far is when the element oxygen gets introduced to oxidize or burn something.)

Coconut fibers, for example, are processed to extract one component (cellulose) and transform it into a pure, isolated chemical compound known as rayon. Drugs come from plants and snails, soap from pigs and trees, pigments from plants and minerals, perfumes from whales and wildflowers, and a little of everything from crude oil.

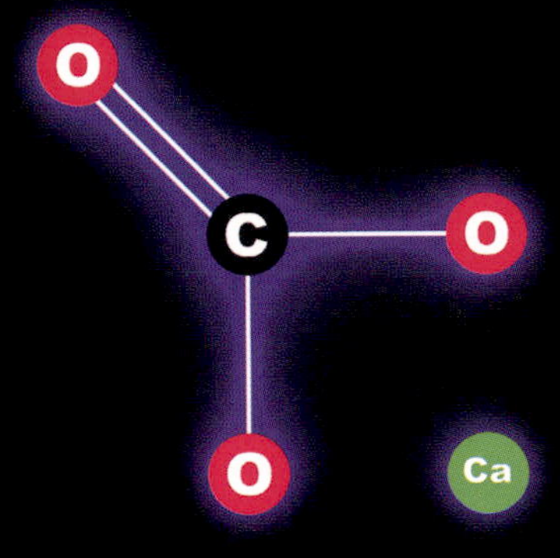

▲ Limestone is calcium carbonate, $CaCO_3$. It could, in principle, be used as an ore for calcium metal, but instead it is used far more often in the form in which it is dug out of the ground and, for example, crushed into gravel for roads. It's also the source of agricultural lime (finely ground calcium carbonate) and Portland cement (calcium oxide mixed with silicon-, iron-, and magnesium oxides).

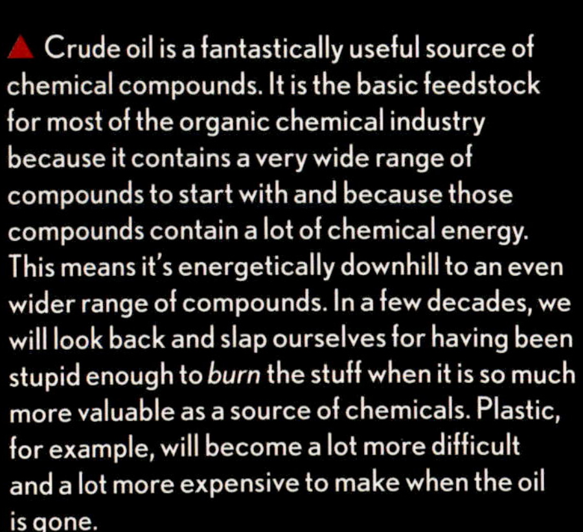

▲ Crude oil is a fantastically useful source of chemical compounds. It is the basic feedstock for most of the organic chemical industry because it contains a very wide range of compounds to start with and because those compounds contain a lot of chemical energy. This means it's energetically downhill to an even wider range of compounds. In a few decades, we will look back and slap ourselves for having been stupid enough to *burn* the stuff when it is so much more valuable as a source of chemicals. Plastic, for example, will become a lot more difficult and a lot more expensive to make when the oil is gone.

◀ Concrete and cement are not the same thing. Cement, more specifically Portland cement, is a very fine powder consisting of a specific mixture of calcium oxide (commonly known as quicklime) and oxides of silicon, iron, and magnesium. It turns rock hard a few hours after being mixed with water. Concrete consists of Portland cement combined with sand and small rocks (called aggregate). The cement is the glue that holds the aggregate together to make concrete.

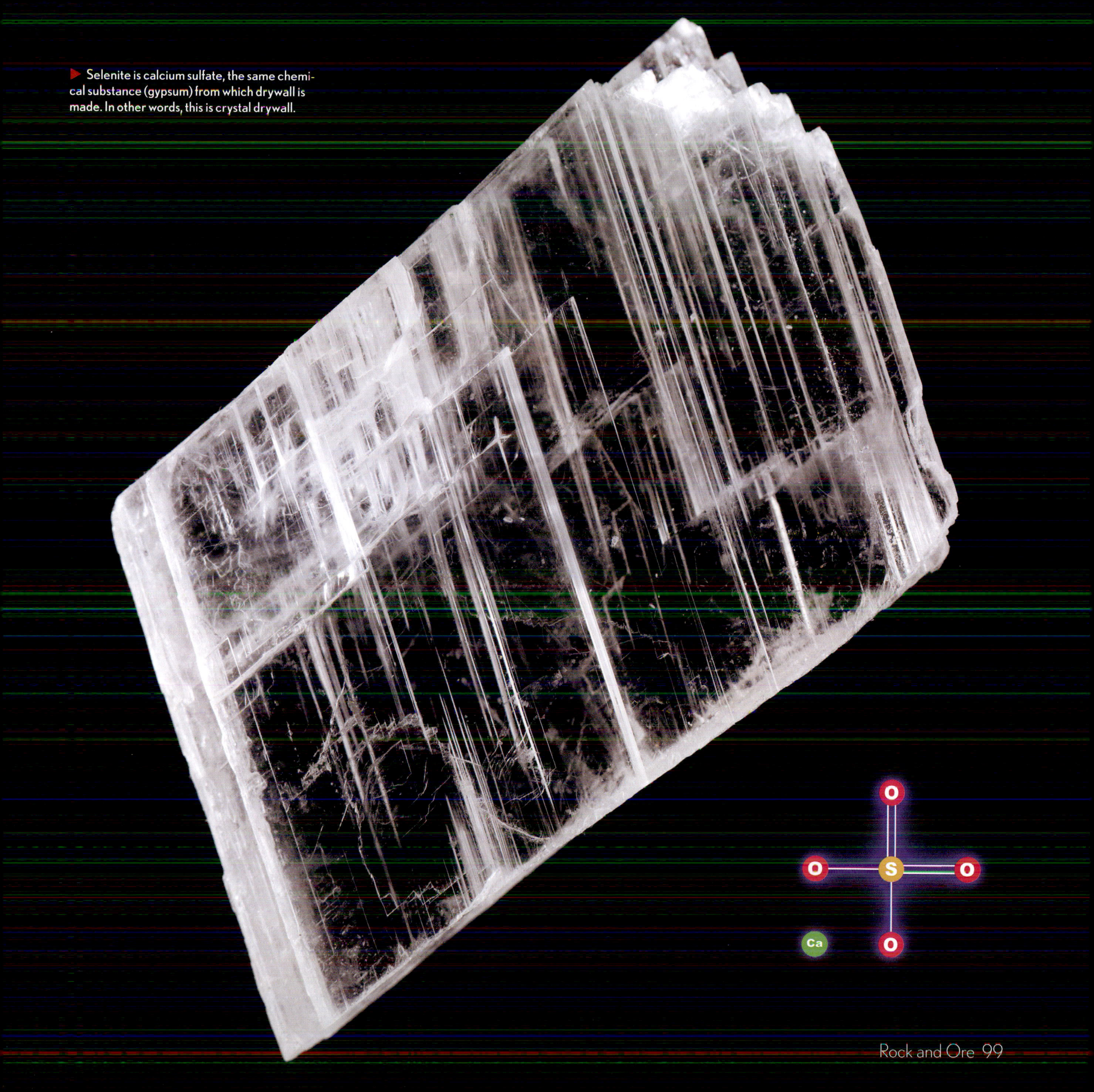

▶ Selenite is calcium sulfate, the same chemical substance (gypsum) from which drywall is made. In other words, this is crystal drywall.

Ores Are Not Just For Elements

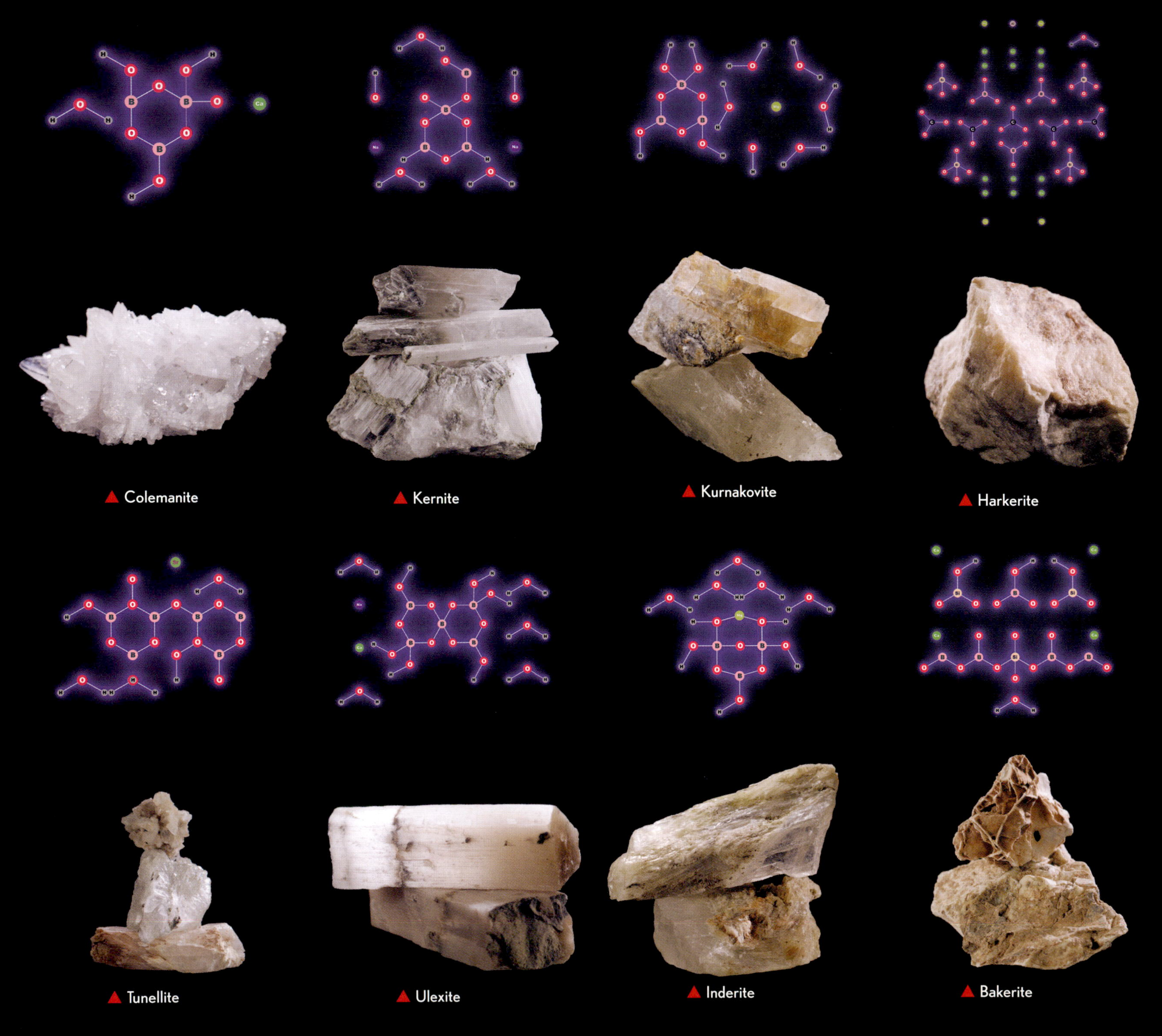

Colemanite

Kernite

Kurnakovite

Harkerite

Tunellite

Ulexite

Inderite

Bakerite

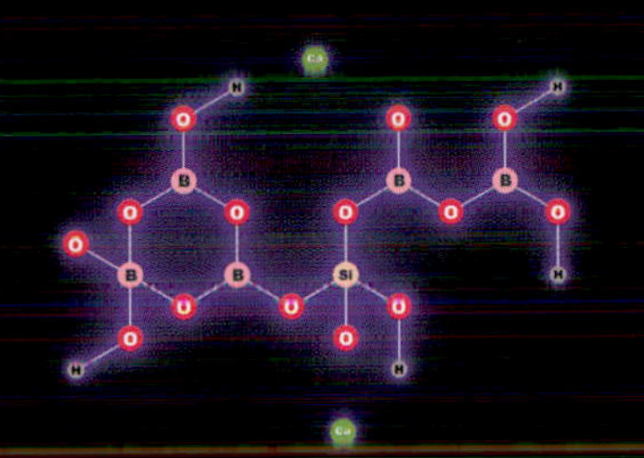

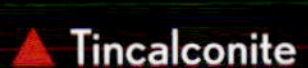

▲ Tincalconite

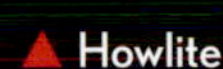

▲ Howlite

◀ All these minerals could be used as ores for boron but are instead mostly used as a source for borax and other compounds containing boron, not for elemental boron itself. Pure boron has a few applications but is difficult and expensive to prepare. If you need boron atoms in a compound, it's always cheaper and easier to get to the final compound by transforming one compound into another without ever isolating the boron.

▲ Borax (sodium borate) is usually made from other minerals, but it also occurs in natural form, as in this crystal.

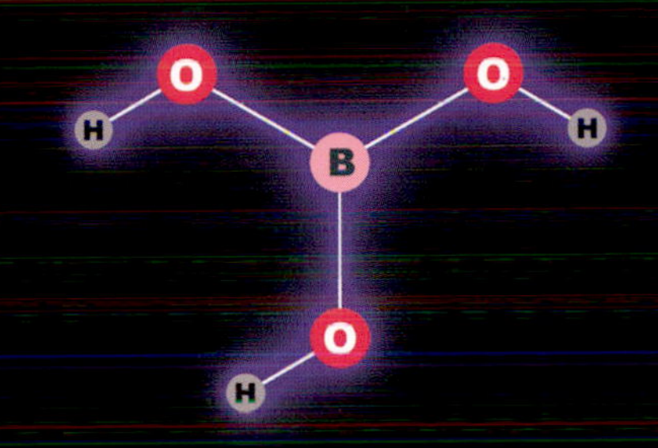

▲ Boric acid is H_3BO_3. If you've read Chapter 2, on the topic of names, you may wonder why this is called an acid and not an alcohol: if the boron atom were a carbon atom instead, this would indeed be a trialcohol (it also couldn't exist, because carbon won't bond to three oxygen atoms). But boron isn't carbon, and the distribution of electrons in this molecule means that the hydrogen atoms are bound loosely enough to dissociate in water, which is the fundamental property of an acid.

▲ Probertite

▲ Hydroboracite

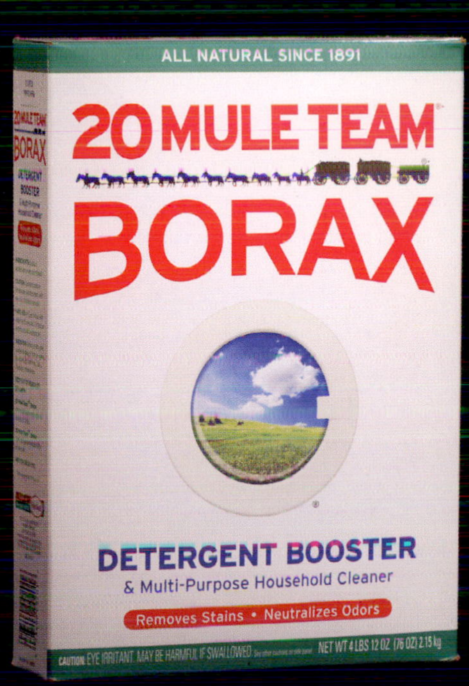

◀ The most common place you'll see a boron compound is in the borax used in laundry products, both pure and mixed with detergents. This classic 20 Mule Team Borax is an iconic form of boron compound derived from the minerals on this page.

▶ The borax used for washing is the sodium salt of boric acid. Boric acid itself (H_3BO_3) is used as an insecticide.

Chapter 7

Rope and Fiber

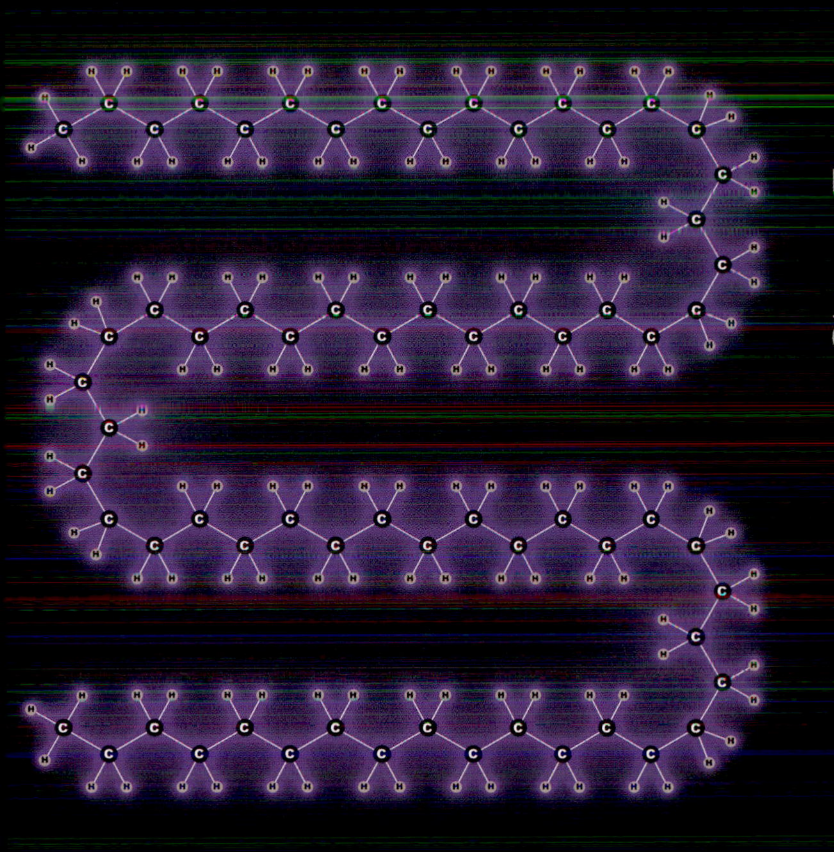

IF YOU HAVE an intuitive feeling about something to do with the world of atoms and molecules, it's probably wrong. Electrons don't exist in any one place; light is both a wave and a particle at the same time; it's all just very weird. So I find it a bit of a surprise that, despite the idea's being completely obvious, most fibers really are made of long, thin molecules. And the fibers are strongest when those molecules are all lined up in the right direction. Under the right conditions, you can even directly feel with your hands when the molecules in a fiber come into alignment.

These long molecules are called polymers because they consist of *many* (from the Greek *poly*) repeated *units* (from the Greek *meros*). Just about the simplest polymer is polyethylene, which is made from many repeating units of ethylene.

Polyethylene molecules are simply very, very long chains of carbon atoms, each with two hydrogen atoms attached. It's the same structure we saw in the mineral oils in Chapter 5. When you start linking carbons together, first you have gases, then solvents, then light oils, then heavier oils, then grease, and then paraffin wax. The end point of that process, once you reach many thousands of carbon atoms in a row, is polyethylene.

◀ This two-inch nylon rope consists of alternating units of hexamethylene diamine and adipic acid.

▲ Polyethylene is a floppy molecule that is able to twist and turn quite freely. There is a small barrier to rotation around the carbon–carbon bond, but not much.

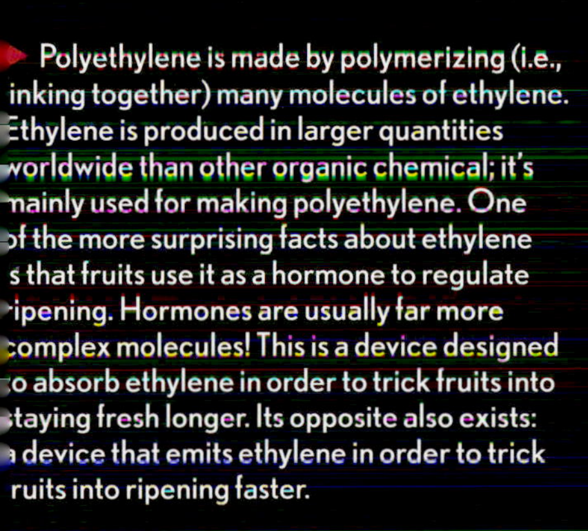

▶ Polyethylene is made by polymerizing (i.e., linking together) many molecules of ethylene. Ethylene is produced in larger quantities worldwide than other organic chemical; it's mainly used for making polyethylene. One of the more surprising facts about ethylene is that fruits use it as a hormone to regulate ripening. Hormones are usually far more complex molecules! This is a device designed to absorb ethylene in order to trick fruits into staying fresh longer. Its opposite also exists: a device that emits ethylene in order to trick fruits into ripening faster.

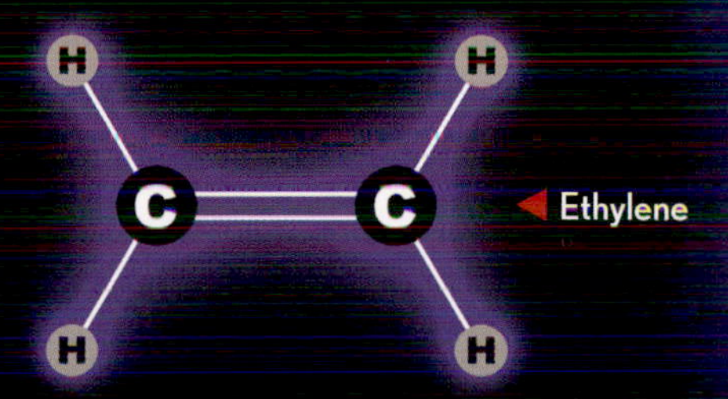

◀ Ethylene

The Simplest Polymer

POLYETHYLENE IS USED to make a great many things, but its properties are most easily seen in the humble plastic grocery bag. The carbon chains that make up these flimsy bags are a few thousand atoms long, and they are arranged quite randomly—some curled up, some snaking around others. The material deforms easily because the individual chains are not bonded to one another through anything but the very weakest of forces (called Van der Waals attraction, see page 12). There is a lot of freedom for the molecules to bend, straighten, and slide against each other.

You can easily tear the bag or stretch it in different directions. But if you continue to stretch it out longer and longer, at some point it will suddenly stop stretching and become dramatically stronger, cutting your finger where before it was yielding easily. That's the point at which all the molecules have lined up in the direction you were stretching and can't go any further. The force you are feeling is the strength of the carbon–carbon bond.

Fancier versions of polyethylene have much longer, prestretched carbon chains. But in all forms of polyethylene, these separate large molecules are not connected to each other. So why don't the molecules just slide apart when you pull them? For the same reason short fibers stay together in a long rope.

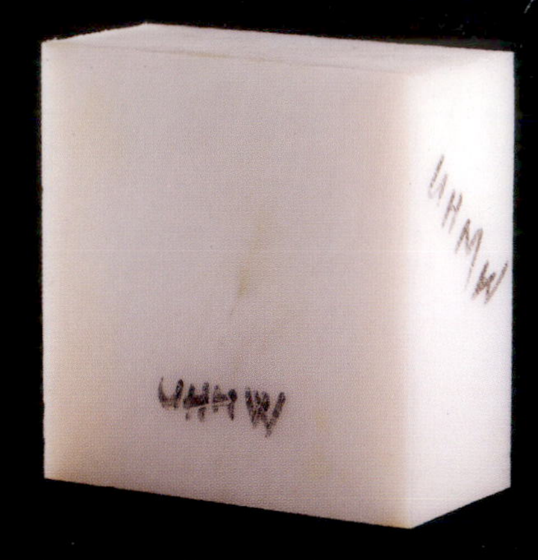

▲ This satisfyingly solid and slippery block is made of ultra-high-molecular-weight (UHMW) polyethylene. The molecules in it are on the order of several hundred thousand carbon atoms long, compared to only a thousand or two for ordinary polyethylene. A 500,000-carbon-long polyethylene molecule is about $^{1}/_{500}$ inch ($^{1}/_{20}$ millimeter) long, which is *very* long for a molecule!

▼ Anyone who has tried to tear apart a plastic grocery bag knows there's a right way and a wrong way to do it. If you ever let a patch of it stretch into a filament, all is lost: it suddenly becomes painfully strong.

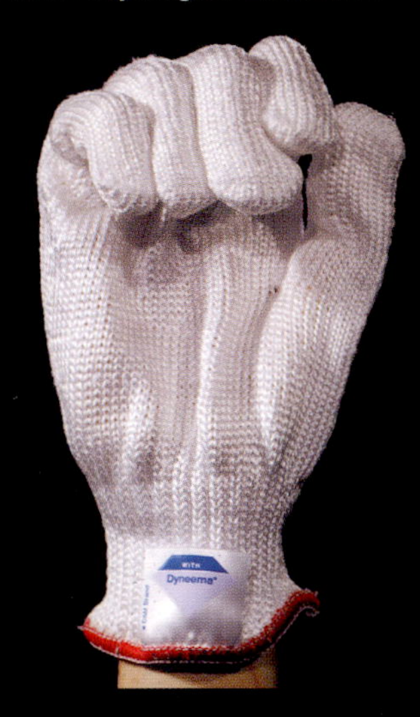

▲ Dyneema is a brand name of very strong UHMW polyethylene fiber used to make ropes and cut-resistant gloves like these. In these fibers, the molecules can be as much as 95 percent aligned in the direction of the fiber.

▼ The molecules in polyethylene can slide around more and more easily as the temperature increases. That means polyethylene has a melting point low enough to allow it to be remelted and cast, pressed, rolled, injection molded, or otherwise formed into new shapes after it has been synthesized. Beads like these have no function other than to be melted down and made into something else.

Polyethylene is a humble and ubiquitous material. I paid for someone to ship me these polyethylene packing blocks, but only because I wanted to be sure they were actually polyethylene and not one of the alternative materials such things can be made of. Chances are I've thrown away a hundred similar blocks of polyethylene, which are used to protect relatively heavy equipment during shipment.

Twisted into a Thread of Great Strength

COTTON FIBERS ARE ONLY about an inch (two and a half centimeters) long. In a three-mile spool of cotton thread, the individual fibers are *still* only an inch long. They are not glued to each other in any way. The thread is strong only because many fibers overlap each other, with the twist of the thread locking individual fibers one to another through their rough surfaces.

In exactly the same way that short cotton fibers lock together to create a long string of cotton thread, many long molecules can lock to each other by their random overlapping, and by the way they twist and turn into each other. Even though the individual forces between atoms in neighboring molecules may not be very strong, when molecular chains thousands of atoms long are lined up close to each other, it's very hard for those molecules to slide against each other.

This phenomenon is also what holds human civilizations together over great depths of time. Our individual lives may last only a few short decades, but we are spun into communities that draw strength from the overlapping of generations, each of us locked by the twists and turns of our lives to those who came before and those who will come after. So far, the inches of our lives have spun a thread of nearly a thousand feet, since first we sat together at a camp fire.

▲ These are cotton fibers after they have been processed in a cotton gin, which separates the fibers from the seeds. Before 1800, separating seeds from fiber required around a day's human labor per pound. The cotton gin reduced that labor by a factor of fifteen. If you ever catch yourself thinking that humans are clever and progress inevitable, consider that all the technology to build a cotton gin existed for at least a thousand years before anyone actually made one. During all that time, people sat around picking the seeds out by hand, one by one, day after day, year after

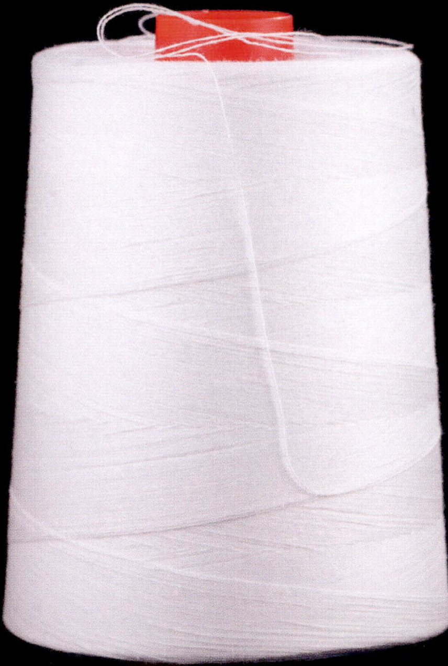

▲ This spool (called a cone) contains 6,000 yards of three-ply cotton thread. That's about 3.4 miles (5.4 kilometers)—or 10 miles (16 kilometers) if you count all three individual filaments. The individual cotton fibers making up this thread are each about 1 inch (2.5 centimeters) long, and they are held together only by the twist of the

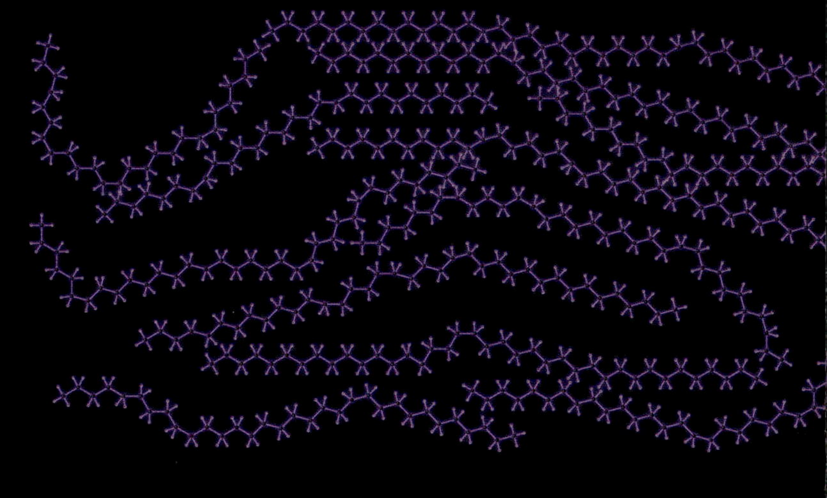

▲ The bonds between carbon atoms in these polyethylene chains are very strong, but the chains are held to each other only by the way they are twisted together and by the weak Van der Waals forces that act between neighboring molecules.

▲ If you untwist cotton thread, it can be pulled apart without breaking any of the individual fibers. In multi-ply thread, this untwisting is made more difficult by the fact that the strands are twisted together in the opposite

▼ This is cotton exactly how it comes off the plant. The fibers grow around the seeds in a cotton boll, protecting them and helping them disperse in the wind and on the backs of animals. That's "boll" with an "o," not "ball" with an "a." The cotton balls you buy in the drug store are different, even though they are just about the same size. They are highly processed cotton fibers shaped back into a ball.

A Shoe-Shaped Molecule

IN POLYETHYLENE, each long chain molecule is completely separate, not chemically bonded to any of the others. But with other kinds of polymers, made of similarly long molecules, those individual molecules can be connected to each other by chemical bonds through a process called cross-linking. Cross-linking makes the material stronger and resistant to melting at higher temperatures. It also prevents "creep," the slow deformation that happens when a material like polyethylene is subject to constant stress and the individual molecules very slowly slide past each other.

In a way, cross-linking turns the material into a single, giant molecule whose parts can't slide around anymore regardless of the temperature. So once a material has been cross-linked, it can't be melted, which means the cross-linking has to be done after the material has been molded into its final shape. (Or the final shape has to be created by machining a block of cross-linked material.)

Vulcanized rubber is an early example of a cross-linked polymer. The term *vulcanized* comes from the sulfur, heat, and pressure used in the process, which interconnects rubber molecules with chains of several sulfur atoms. (Anything involving sulfur and high temperatures is liable to be named after Vulcan, the god of volcanoes, because volcanoes are where you find heat, sulfur, and its characteristically acrid, choking smell.)

Today, there are many families of artificially created cross-linked polymers.

▲ Natural latex rubber as it comes from the plant (purified, of course) is widely used in medical and scientific applications. This latex tubing is more stretchy than pretty much any synthetic alternative.

▼ Vulcanized rubber

▲ Cross-linked, vulcanized rubber can be made so hard (by adding more sulfur cross-links) that it feels like solid plastic, not at all like what you ordinarily think of as rubber. Ebonite, which is what this electrical insulator and clarinet bell are made of, can be as much as 30 percent sulfur!

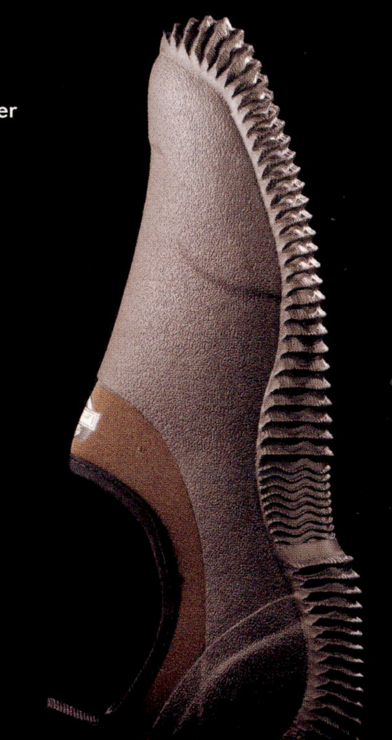

▶ The soles of these shoes are made of vulcanized rubber. They will not melt or dissolve because they are, in a sense, a single huge molecule created in the shape of a shoe. When heated, this molecule will char and burn rather than melt.

▲ Liquid latex, used for special effects makeup, is unvulcanized natural rubber. Its lack of cross-links between molecules allows latex to be dissolved in various solvents. Latex is good for simulating scars and peeling skin because when a pool of it dries, a hardened "skin" forms over the still-liquid pool, allowing for all kinds of gross manipulations.

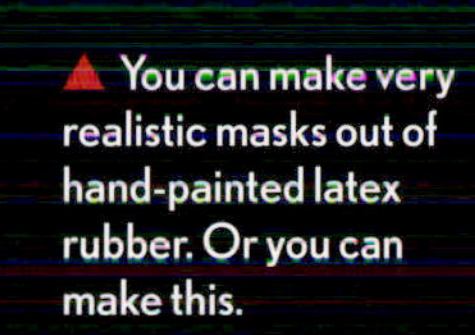

▲ You can make very realistic masks out of hand-painted latex rubber. Or you can make this.

▲ Latex is not just for serious medical applications. These artificial flowers are made of dyed natural latex rubber.

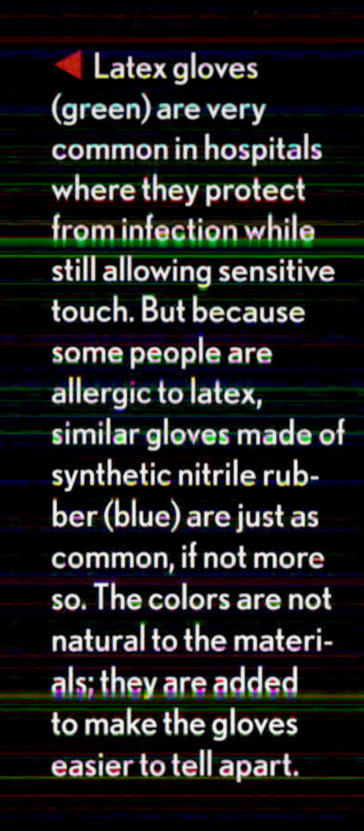

◀ Latex gloves (green) are very common in hospitals where they protect from infection while still allowing sensitive touch. But because some people are allergic to latex, similar gloves made of synthetic nitrile rubber (blue) are just as common, if not more so. The colors are not natural to the materials; they are added to make the gloves easier to tell apart.

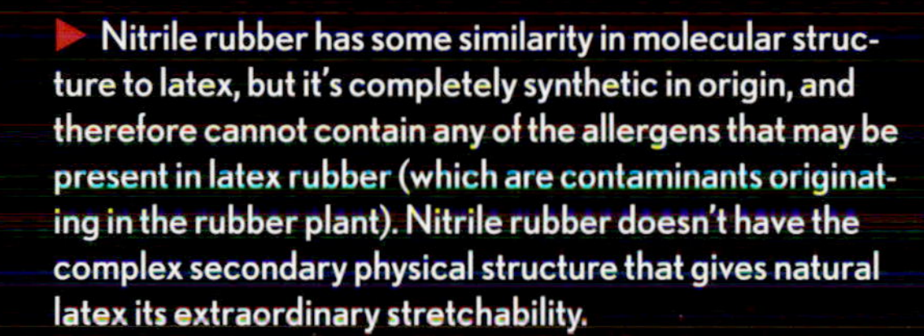

▶ Nitrile rubber has some similarity in molecular structure to latex, but it's completely synthetic in origin, and therefore cannot contain any of the allergens that may be present in latex rubber (which are contaminants originating in the rubber plant). Nitrile rubber doesn't have the complex secondary physical structure that gives natural latex its extraordinary stretchability.

▶ Nitrile monomer

▲ Nitrile polymer

Gutta-percha

▼ Gutta-percha is a close chemical relative of natural latex rubber, but its slightly different chemical structure causes it to be a hard, plastic-like solid even without cross-linking. This gutta-percha photo holder looks and feels like hard plastic.

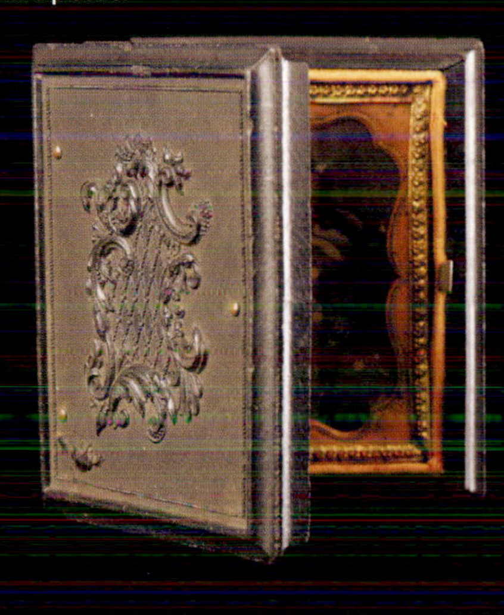

▶ Gutta-percha is an antique material made from *Palaquium* trees. Even the name sounds antique. But I've got some in one of my teeth, as it is used even today to pack the void after dentists scrape out the dead nerve and blood vessels in a damaged tooth root. Gutta-percha (dyed red on the ends of these tiny rods) is worked down into the root, preventing infections from lodging there beyond the reach of the immune system. Other fillers have been tried but none have proved better.

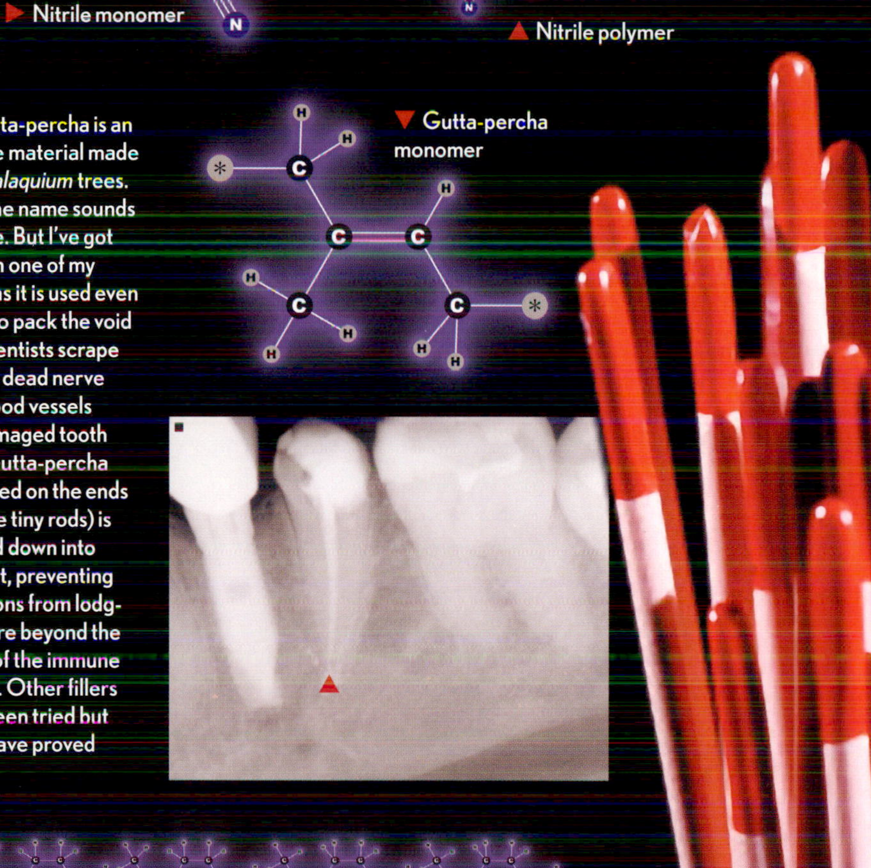

▼ Gutta-percha monomer

▲ Gutta-percha polymer

Sexy Synthetics

FIBER ENGINEERING is a high-tech business. Breaking strength is one measure of how fancy-pants a fiber is, but not the only one. For example, under just the right conditions, carbon fibers are stronger than any other fiber known (and carbon nanotube fibers may be orders of magnitude stronger still). But carbon fibers are quite brittle, so other fibers beat them in many applications.

Kevlar, a brand name of para-aramid fiber (whose full chemical name is poly-paraphenylene terephthalamide) is strong, like carbon fiber, but it's also very tough, which means it can absorb a lot of energy before it breaks, making it useful for bullet-proof vests and spear-fishing line. And it's resistant to abrasion, which makes it durable for ropes and protective gloves.

Other fibers are desirable because they float on water, resist rot, or feel extra soft on the skin. By modifying their chemical structure (i.e., the molecules they are made of) and/or their physical forms (e.g., how fine the fibers are, whether they are straight or kinked, etc.), synthetics can be engineered to meet a wide range of needs.

In many cases, the goal of these modifications is to mimic, more cheaply or more humanely, the feel of natural fibers, because natural fibers have some pretty remarkable properties.

Nylon

▼ The polymer chains of nylon consist of alternating units of hexamethylene diamine and adipic acid, making nylon what is called a copolymer.

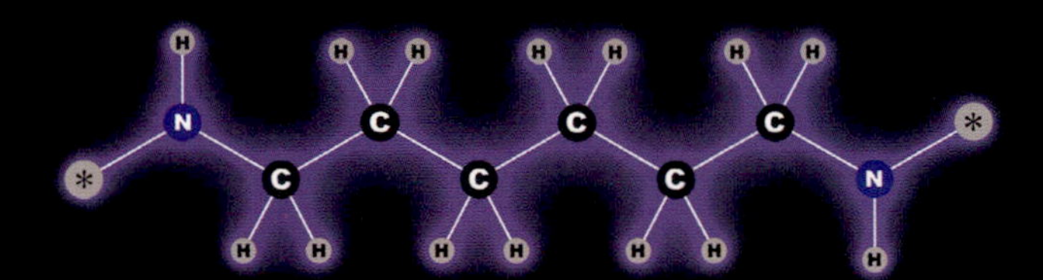

▲ Nylon monomer Hexamethylenediamine

▼ Nylon monomer Adipic Acid

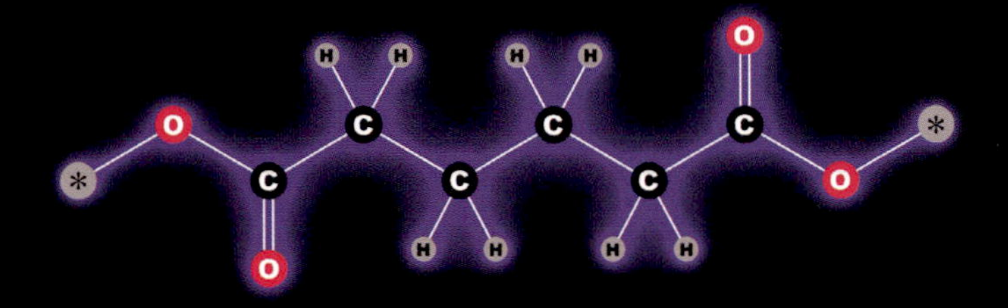

▼ Nylon 66 polymer

Acrylic

▶ This faux (fancy French word for fake) fur blanket is hands-down the softest, most amazingly cuddly warm blanket you can imagine. You can sell your cats because you won't be needing them anymore. I'm not sure which is more remarkable: that an acrylic engineered fiber can feel this close to real fur or, given that acrylic polymers have been known for quite some time, that it took until 2013 for fake fur manufacturers to achieve this level of purr-fection.

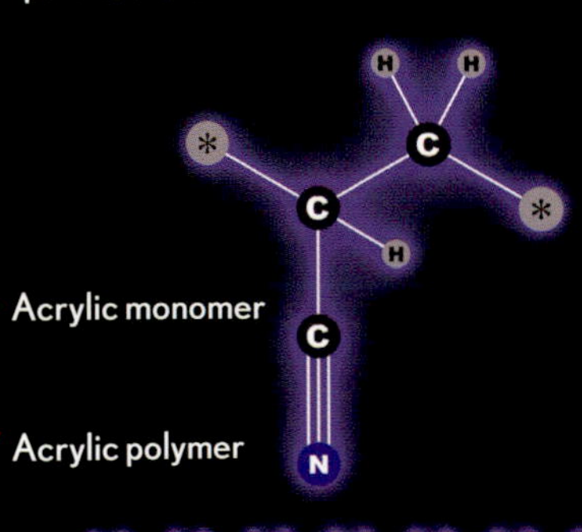

▶ Acrylic monomer

▼ Acrylic polymer

▶ Stockings were greatly improved by the invention of nylon. In fact, nylon hosiery was an early success story for synthetic fibers in general—one of the first times the public could immediately experience for themselves the superiority of an artificial material over any natural alternative.

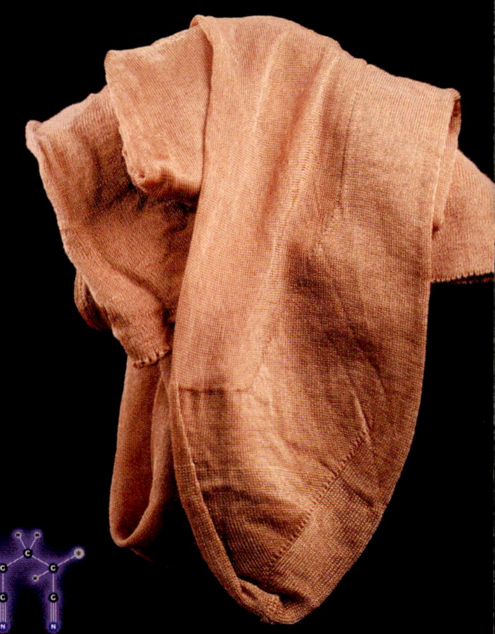

▲ Nylon proved to be revolutionary to the stocking and pantyhose industry. Today, it's taken for granted, but when it made its debut, it was a big deal.

▲ The reason nylon works well for stockings is because it's very strong, making even the super-thin threads used in sheer hosiery resistant to breaking. Make the thread thicker, and you end up with things like this monofilament fishing line with a breaking strength of 250 pounds (110 kilograms).

Kevlar

▼ The repeating unit in Kevlar is a quite a complex structure, which allows neighboring polymer molecules to coordinate and stick to each other particularly well.

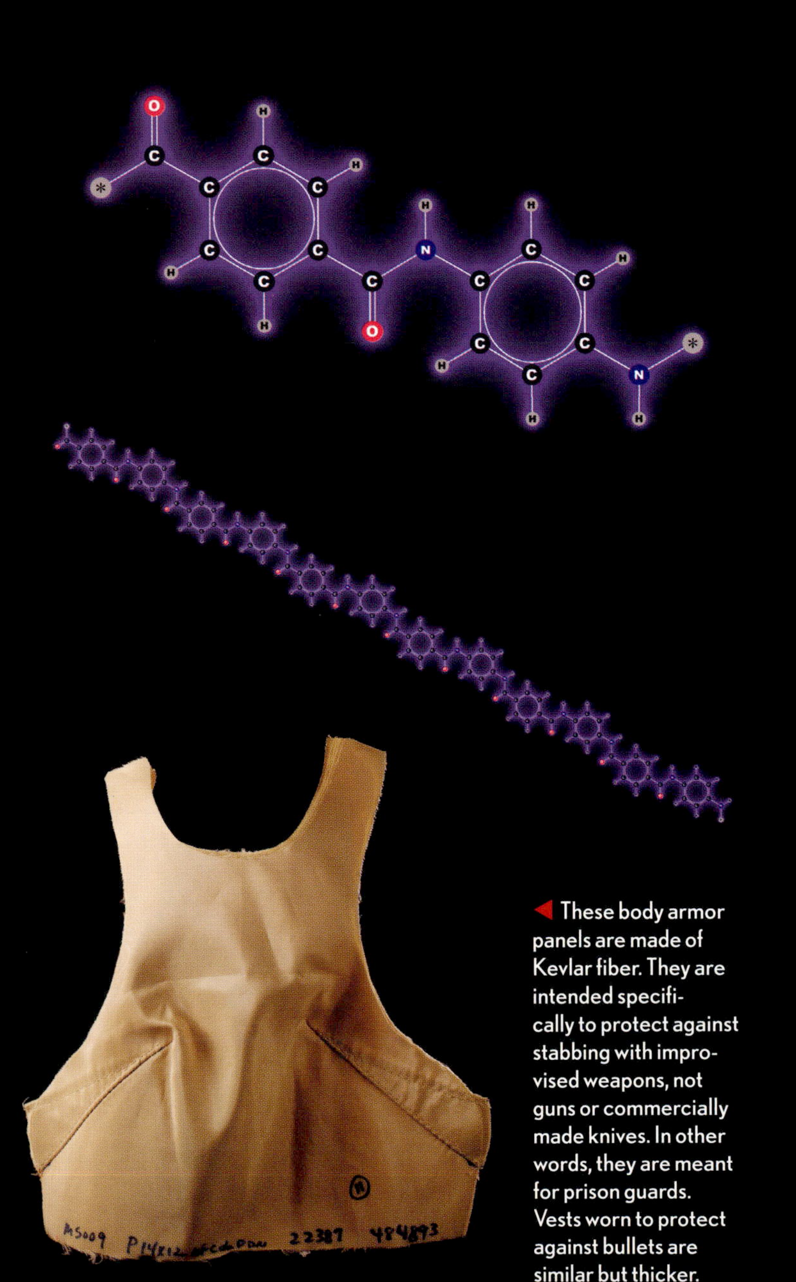

◀ These body armor panels are made of Kevlar fiber. They are intended specifically to protect against stabbing with improvised weapons, not guns or commercially made knives. In other words, they are meant for prison guards. Vests worn to protect against bullets are similar but thicker.

◀ I wrote a column for *Popular Science* magazine once about this material: it's bomb-proof wallpaper! It actually worked pretty well in the test I set up with a wrecking ball. (Not having a bomb handy, I had to improvise.) The strength comes from Kevlar fiber embedded in a thick, rubbery sheet. Together, these materials give and stretch to absorb the energy of a blast.

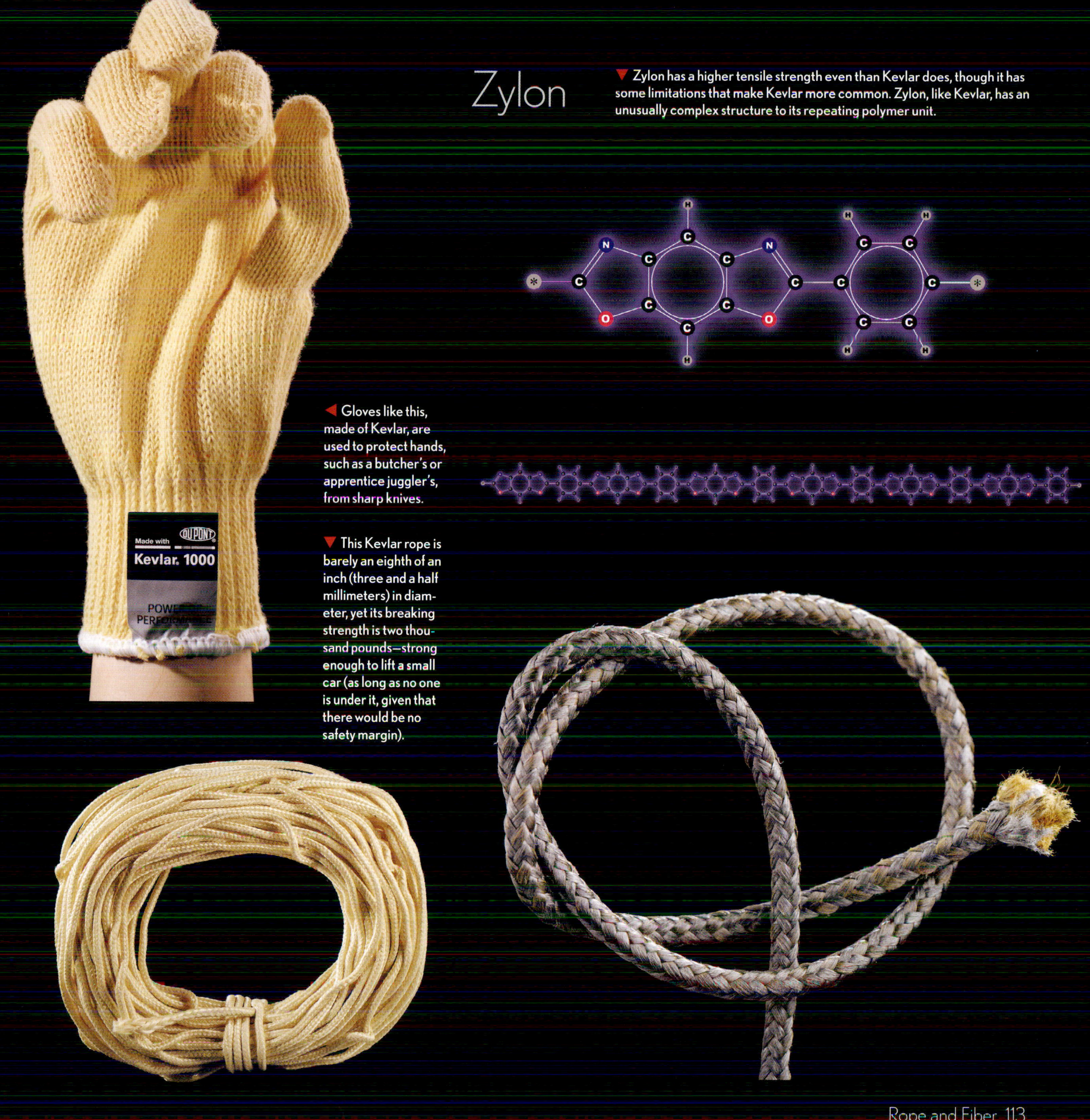

Zylon

▼ Zylon has a higher tensile strength even than Kevlar does, though it has some limitations that make Kevlar more common. Zylon, like Kevlar, has an unusually complex structure to its repeating polymer unit.

◀ Gloves like this, made of Kevlar, are used to protect hands, such as a butcher's or apprentice juggler's, from sharp knives.

▼ This Kevlar rope is barely an eighth of an inch (three and a half millimeters) in diameter, yet its breaking strength is two thousand pounds—strong enough to lift a small car (as long as no one is under it, given that there would be no safety margin).

Polypropylene

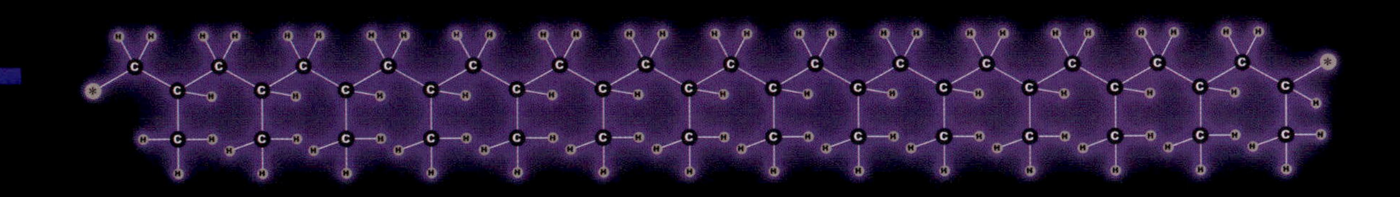

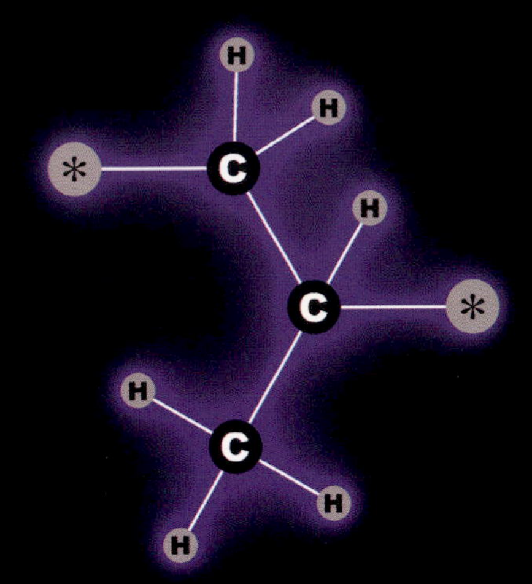

▼ This is a very common type of polypropylene rope that I dislike. I don't think this is the fault of polypropylene itself but rather that the fibers in this particular type of rope are quite thick, making it very rough and hard on the hands. It's like a bunch of individual monofilament fishing lines all twisted together. Even thinking about it makes my hands hurt from the memory of all the times I've tried to tie knots in it. I have provided here an example of a very bad knot.

▲ This is a big bag! It holds about one cubic yard (about the same as one cubic meter) and has a weight capacity of 2,750 pounds (1,250 kilograms). The loops are designed to be picked up with chains or forklift tines, and there's a spout at the bottom that can be untied to let the contents (for example, sand) out. The material is polypropylene.

Polyester

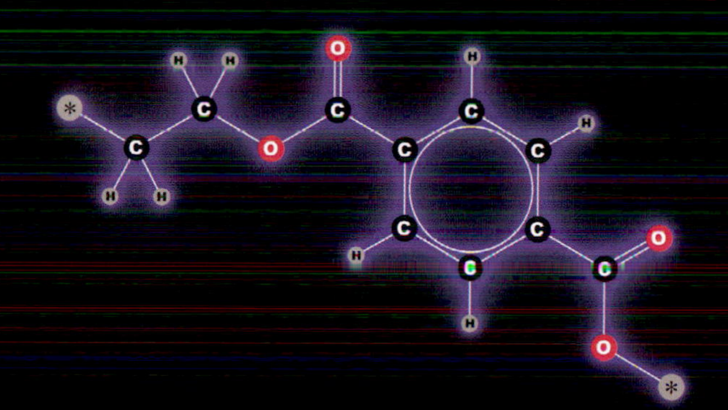

This six-inch- (fifteen-centimeter-) wide belt is made for pulling heavy things, such as trucks or tractors, out of places they are not supposed to be. It has an ultimate breaking strength of sixty thousand pounds (thirty tons or twenty-seven thousand kilograms). It's made of polyester, which allows it to stretch quite a lot before breaking—unlike the commonly used alternative, a steel chain, which has virtually no give before snapping. Between a polyester strap and a steel chain with equal ultimate breaking strength, the polyester strap is able to absorb much more energy without breaking because of its ability to stretch. However, this also makes it significantly more dangerous: by the time a strap like this does break, it has stored up a lot of energy, which is released through a violent recoil of the strap. That's why you must never, ever stand directly in line with a rope under strong tension. A steel chain, by contrast, will snap back only a modest distance if it breaks. Another difference: a steel chain is a cold and hard thing, while this polyester strap is luxuriously soft. It feels like fine silk thread, only cheaper and more likely to be wrapped around a truck axle than an elegant human neck. Hopefully.

Polyglycolide and Polydioxanone

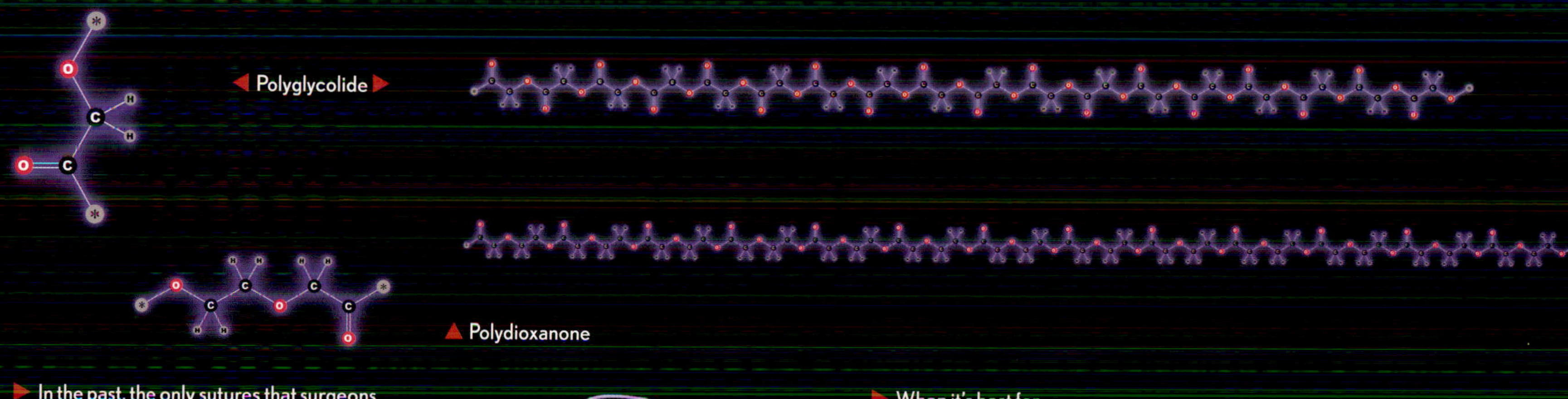

Polyglycolide

Polydioxanone

In the past, the only sutures that surgeons could use that would be absorbed naturally by the body over a period of time were made of catgut. These two modern synthetic sutures are made of polyglycolic acid and polydioxanone, which are easily absorbed by the body but do not have the disadvantages associated with using a natural product (including less predictable physical characteristics and greater risk of contamination).

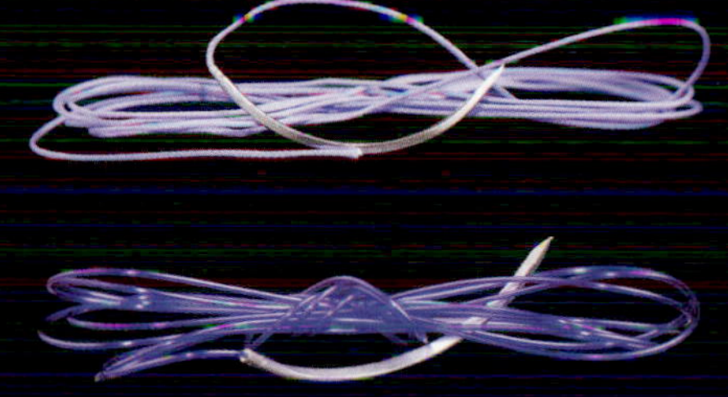

When it's best for a surgical suture *not* to be absorbed by the body, nylon or polypropylene are used.

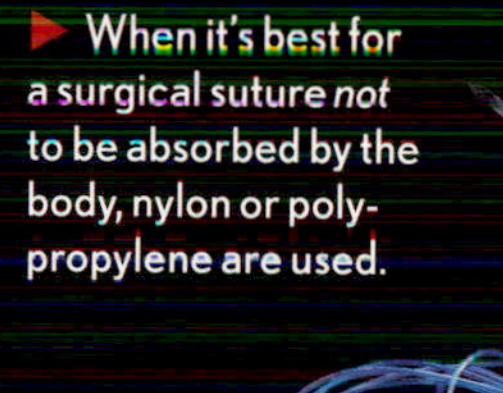

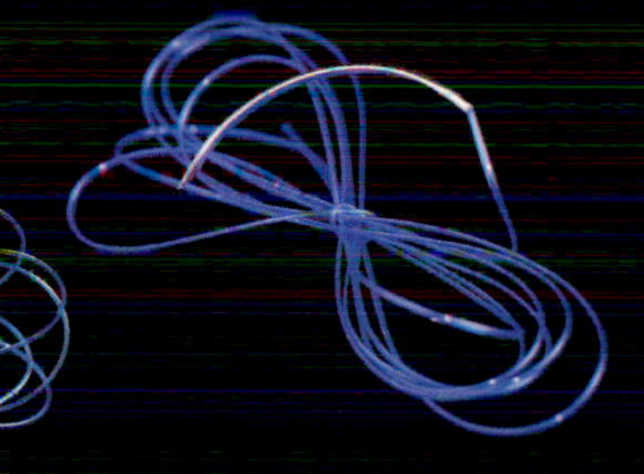

Plant Fibers Are Made of Sugar

THE WORLD OF natural fibers is rich and diverse. Practically anything hairy, from coconuts to camels, is used to create rope, yarn, thread, cloth, or batting. Socks made of dog hair might seem odd, but how are they really any stranger than socks made from sheep hair (wool) or goat hair (mohair)? Even human hair has been used to weave bracelets and necklaces.

Fibers made by plants are quite chemically simple, similar in many ways to synthetic fibers. The great majority of plant fibers are made mainly of cellulose, whose individual repeating units are glucose sugar molecules.

This accounts for the fact that some microorganisms (and animals that host these microorganisms in their guts) can eat cellulose fibers and live off the energy from their sugars (in other words, they can eat grass). Other animals, like us, don't have the right enzymes to digest cellulose, so in order for us to use the energy content in the fiber, we have to feed it to other animals and then eat those animals or their milk (this is called cattle ranching).

▼ Many plant fibers also include some percentage of lignin, whose repeating units are a mixture of three types of alcohol molecule: sinapyl alcohol, coniferyl alcohol, and paracoumaryl alcohol. (See page 38 for more about the chemical definition of an alcohol.)

▶ Sinapyl alcohol

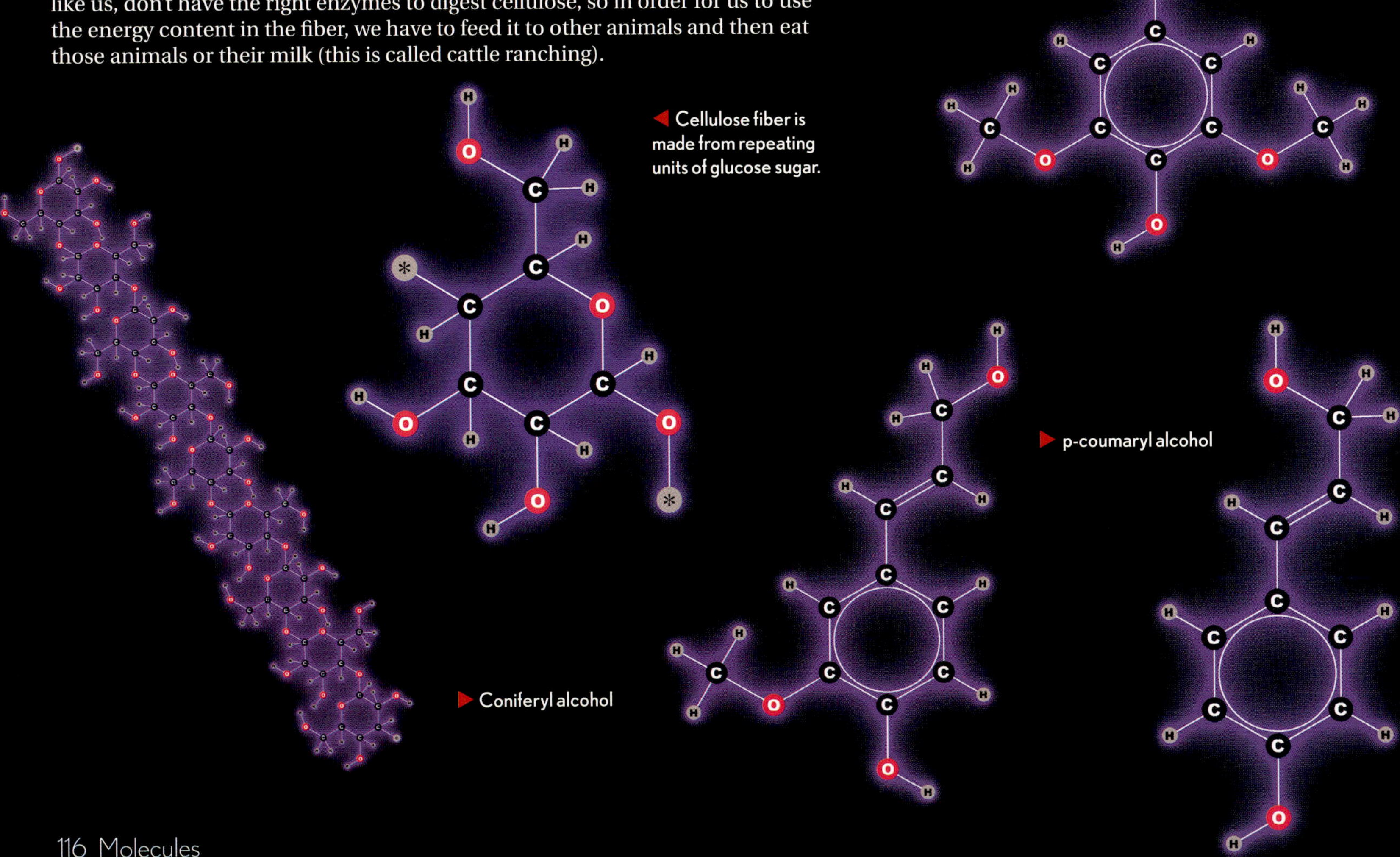

◀ Cellulose fiber is made from repeating units of glucose sugar.

▶ Coniferyl alcohol

▶ p-coumaryl alcohol

Wood is about 70 percent cellulose and 30 percent lignin. This sort of "wood wool" was once used quite commonly as packing material, but I haven't seen any examples for quite a while. This material came from my basement and is at least forty years old: I inherited it from my parents. Wood is a very fibrous material but is rarely used to make rope or thread. Instead, wood's fibers are used to make paper, cardboard, books, and, of course, natural and engineered wood objects such as tables, chairs, bookshelves, and structural beams.

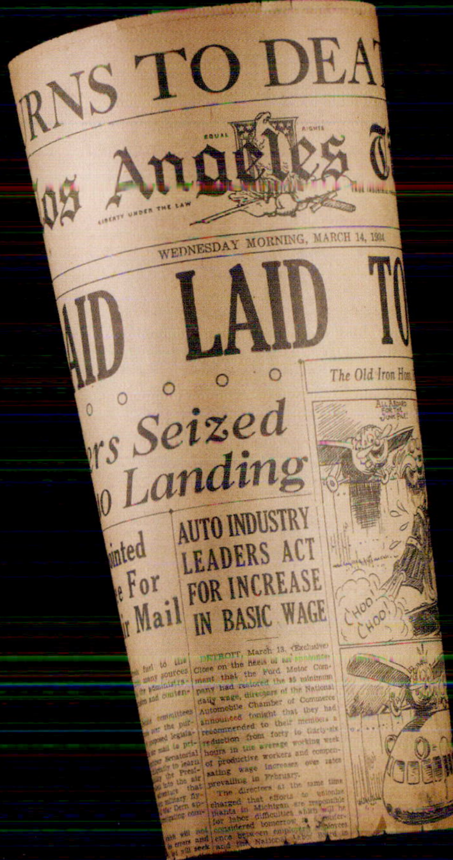
RNS TO DEAT
os Angeles T
WEDNESDAY MORNING, MARCH 14, 1934
AID LAID TO
rs Seized
o Landing
AUTO INDUSTRY LEADERS ACT FOR INCREASE IN BASIC WAGE

Cheap paper made from wood fiber, common in newspapers or cheap paperback books, contains a significant amount of lignin, which releases acid over time that yellows and ultimately destroys the paper. Older and more expensive cotton paper does not have this problem because cotton naturally contains almost no lignin.

This paper, handmade in India, is pure cotton, which means nearly pure cellulose. Cotton is desirable for long-lasting archival paper because, unlike wood fiber, cotton fibers naturally contain very little lignin, the substance that causes cheap paper to yellow over time.

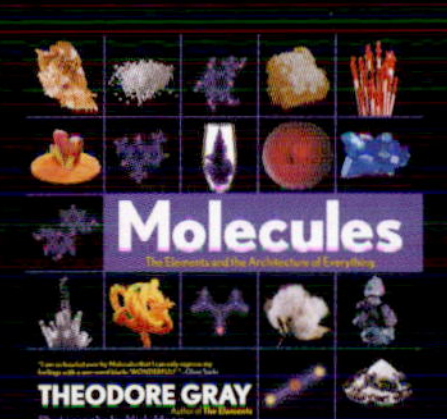

The print edition of this book is printed on paper made of wood fiber with its lignin content removed. This is the most common form of paper found in books meant to be reasonably permanent and in other applications where the paper needs to be a bright, bleached white. This paper is known as acid-free paper, but to be truly of archival quality it would need to have additional buffers and neutralizers added to protect it from acid it may pick up from the atmosphere.

Plant Fibers Are Made of Sugar

▼ Sisal fiber comes from a variety of agave, the same type of plant that tequila is distilled from. While this fiber has many uses, its signature use seems to be cat scratching posts. I don't think cats properly appreciate how much effort is put into finding just the right materials to please them.

▲ Coconut fibers, after they have been separated out from a coconut husk, are called coir.

◀ When people outside the tropics buy a coconut at the store, they are getting just the seed part, the hard shell with liquid interior. But when a coconut falls on your head fresh from the tree, that seed is surrounded by a thick fibrous husk, which is what yields the fibers used to make ropes, mats, and seed-starting medium.

▼ Rope made from coconut fiber isn't really very good rope, but apparently parrots like it as much as cats like rope made of sisal. Pet stores are where you can usually find this fiber for sale.

▶ This agave plant, species *agave sisalana*, is the source of sisal fibers used to make ropes, paper, and almost all the world's cat scratching posts. Only a few percent of each leaf is fiber; the rest is wasted when the plant is processed.

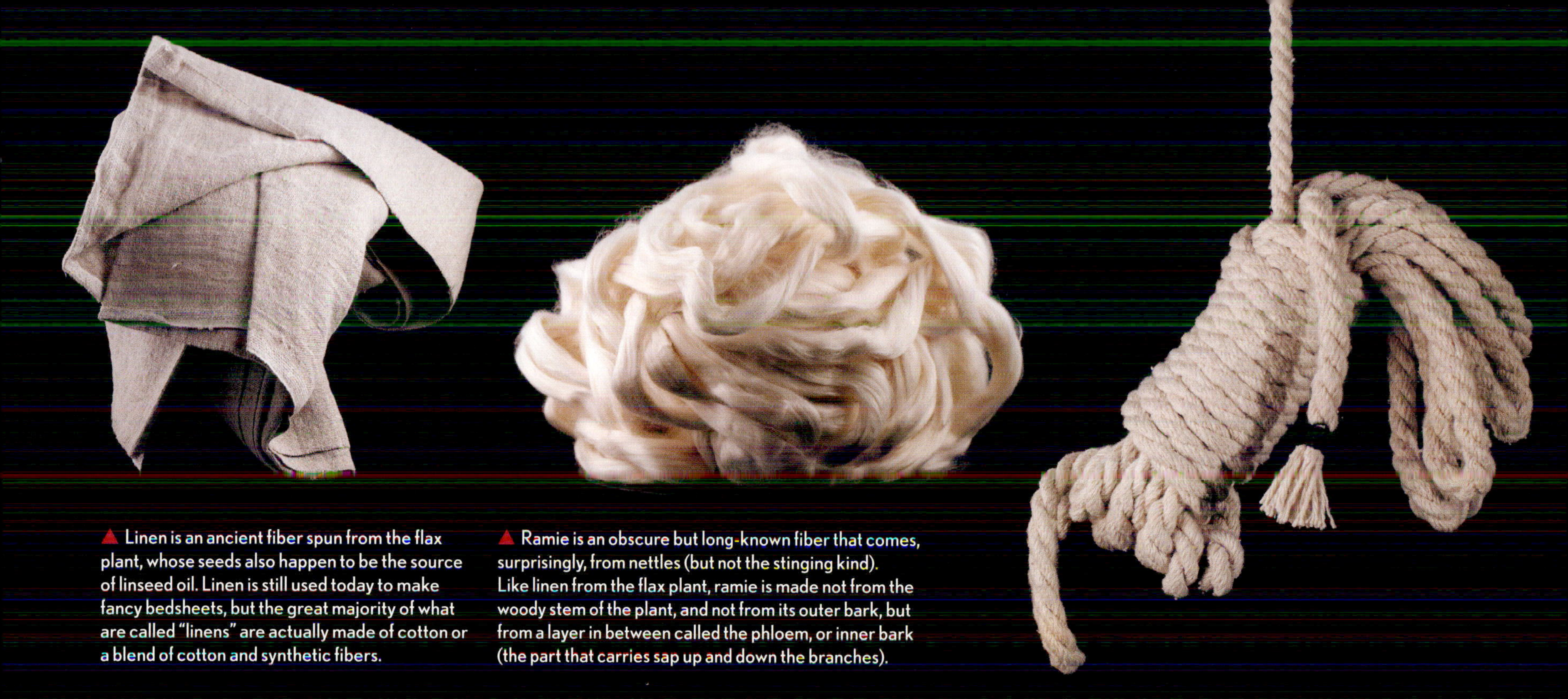

▲ Linen is an ancient fiber spun from the flax plant, whose seeds also happen to be the source of linseed oil. Linen is still used today to make fancy bedsheets, but the great majority of what are called "linens" are actually made of cotton or a blend of cotton and synthetic fibers.

▲ Ramie is an obscure but long-known fiber that comes, surprisingly, from nettles (but not the stinging kind). Like linen from the flax plant, ramie is made not from the woody stem of the plant, and not from its outer bark, but from a layer in between called the phloem, or inner bark (the part that carries sap up and down the branches).

▼ Hemp used to be the go-to fiber for a wide range of applications, from clothing to ship's ropes. Its cultivation was a huge industry worldwide. Then, concerns about the fact that some varieties of hemp are better known as marijuana led to strong discouragement or complete banning of the cultivation and sale of all kinds of hemp, even the kind that contains no pyschoactive component. Hemp is making a comeback, as the ecological benefits of hemp fiber have been recognized.

▶ Fiber sold as bamboo is in an interesting classificational gray area. It is possible to make rope and thread directly from the soft interior of bamboo stems by mechanical processing of the material, but it seems that the majority of fiber sold as "bamboo" is actually rayon made from bamboo. The source may be the bamboo plant, but rayon is cellulose that has been chemically reprocessed to the point that it really doesn't matter anymore where the cellulose came from. If a fiber shares none of the physical, structural properties of the original bamboo fiber, in what sense is it bamboo?
But not to worry, *this* bamboo rope comes from a specialist supplier who assures me that it is genuine, raw bamboo fiber, not reconstituted rayon.

▶ Rayon sounds like a synthetic fiber, right? In some ways it is, but in other ways it isn't. To make rayon, plant cellulose from any of a variety of sources is purified, dissolved, and extruded into new fibers. Chemically, rayon is completely natural plant cellulose, very similar to cotton (which is nearly pure cellulose). Physically, it is in a completely artificial, man-made form.

▼ Jute is the second most widely used fiber after cotton. You might know it best as the fiber used to make burlap sacks, but it's also used to make twine for baling hay and a thousand other things.

Plant Fibers Are Made of Sugar

▶ This is the kind of rope used to torment students in middle school who are made to climb it in gym class. It's prickly, smells funny, and climbing it is not fun at all. (None of this is directly the fault of the manila fiber, other than the prickly and smelly parts.) Manila fiber comes from the abacá plant, a close relative of the banana.

◀ Papyrus was in use in Egypt at least five thousand to six thousand years ago. It is made of cellulose from the inner pith of the papyrus reed, which grows in shallow water. It works pretty well in an area with a dry climate such as Egypt, but in Europe it doesn't last very long and was thus phased out in favor of parchment, which is made of animal collagen protein (in the form of skin) rather than plant cellulose. Later, there was a trend back to cellulose in the form of cotton and wood fiber paper.

▲ The term "cotton candy" is far more accurate than you might think. Not only does cotton candy *look* a whole lot like cotton, it is chemically very closely related. The cellulose polymer molecules in cotton are long chains of glucose sugar molecules linked together (see page 116). Cotton candy is made of sucrose (table sugar), which consists of two sugar molecules linked together (one glucose and one very similar fructose). The only real difference is that the links in cellulose are made at a different point on the molecule, so the enzymes in our digestive system are not able to break it.

▶ If this beak were from a bird, it would be made of keratin protein just like bird feathers and the hair of mammals are (see next section). But this one is the beak of a 120-pound (50-kilogram) giant Humboldt squid and is therefore made of chitin, a much simpler chemical. Chitin is a polymer whose repeating unit, n-acetylglucosamine, is a derivative of glucose (sugar). Chemically, this makes chitin very similar to plant cellulose and very different from animal keratin.

Animals Make Complicated Fibers

NO FIBERS IN regular commercial use come from mollusks, crustaceans, spiders, or fungi, but we do use fibers made by insects and mammals. Their fibers are far more complex, chemically speaking, than those made by plants, and they possess properties impossible to match using simpler chemical structures.

Animal fibers are proteins—long molecules made by linking together units consisting of amino acids. Each of the twenty-odd biologically significant amino acids has an identical linking part that connects it with its neighbors in the protein chain, plus a variable "side chain" that gives it unique properties.

Amino acid side chains differ in how large they are and in whether their loose ends have a positive or negative charge or are attracted to or repelled by water. This array of options allows proteins to play an astonishing range of roles, from enzymes that catalyze reactions in the body to the structural components of the body itself.

These options also allow protein fibers to have a wide range of interesting properties. For example, a protein can combine segments of its chain that are attracted to water and others that are repelled by water in such a way that the protein curls up when dry but stretches out when wet, or vice versa. (See page 67 for an example of such a folding mechanism.)

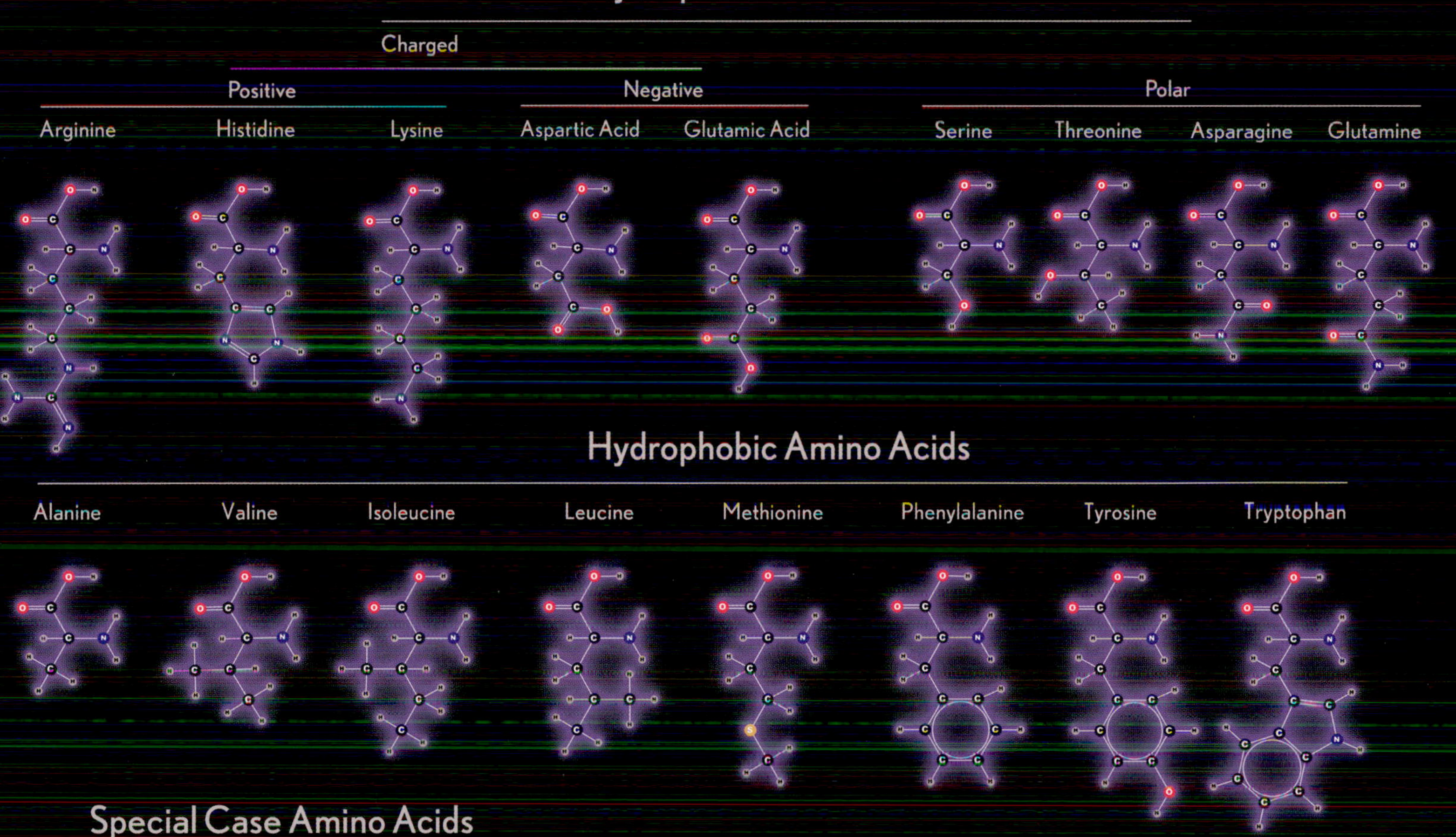

▲ Animals make use of twenty-one different amino acids, each with its own unique characteristics. Joined together into protein chains that may be thousands of amino acid units long, the potential for creating megamolecules with interesting properties is almost unlimited.

Protein Fibers that Occur Outside Animals

WARM-BLOODED ANIMALS extrude one particular kind of protein more than any other: keratin. This complex protein contains a relatively high percentage of cystine, an amino acid made from two sulfur-containing cysteine molecules that link to each other via a sulfur–sulfur bond. These bonds are exactly analogous to the sulfur links that give vulcanized rubber its strength. Just as with rubber, the more sulfur bonds, the more rigid the protein.

Cystine content and sulfur bonding account for the different degrees of stiffness you find in keratin, from your sweetheart's soft curls to the rhinoceros horn that throws you ten feet in the air (followed by your certain death under the hooves of the beast, which are also made of keratin).

▶ Keratin forms complex super-helical structures twisted always in a left-handed direction. (All protein molecules on Earth are left-handed. We are a left-handed planet. By the way, there are two good ways to tell if a life form is alien: if its molecules are predominantly right handed, and if the distribution of isotopes of the elements in its body are significantly different from what is common on our planet. The first proves that it evolved independently from us [on our planet or a different one], while the second proves that, wherever it evolved, this particular specimen grew up on a different planet. Both of these differences would be virtually impossible to disguise, should an alien wish to walk among us undetected.)

◀ Horse hair blankets are notoriously uncomfortable. You can see why from the coarseness of these bundles of horse hair, contrasted with much finer human hair. Both are available from commercial sources. Horse hair is used for crafts such as weaving or braiding and for the bows of stringed instruments such as violins. Human hair is sold for wigs and hair extensions.

▲ Bracelets and necklaces made of human hair (often from a dead loved one and sometimes holding a small image or engraved name of the person in question) were quite popular in Victorian England.

▶ Claws and fingernails are made of the same protein as hair. This particular claw came from a badger. (Bear claw necklaces are a more potent symbol to many cultures but are much harder to get; most that you see for sale are imitation.)

◀ The keratin protein making up this tough rhinoceros horn is very high in the amino acid cystine, which forms a large number of sulfur–sulfur bonds that make the protein hard. Unlike most of the things you see here, this horn is not in my collection. It is locked in a vault in an undisclosed location at The Field Museum in Chicago. Rhinoceros horns are so valuable for use in Chinese medicine and as a purported aphrodisiac, and so illegal to collect from the wild, that several high-profile thefts have caused museums to secret away nearly all of them that had been on public display. Kind thanks are due to The Field Museum for allowing us to photograph this rare object.

▲ The outer sheaths of bird beaks, like this one from a black hornbill, are made of keratin protein, just like hair and nails. Inside is a hollow, bony structure supporting the keratin.

▲ Several species of sea sponges are the origin of the word and concept of "sponges" used for cleaning and bathing purposes. These days, almost all sponges are synthetic, but you can still buy natural sea sponges; the part you get is the skeletons of these odd, clustering creatures. Sponges have no brain, nervous system, digestive system, or any other system. They are just a community of cells that grow together with a skeleton made of collagen protein. So it's a bit hard to say whether this item belongs in the category of collagen found inside or outside of animals. It's difficult to define what is the inside or outside of a sea sponge.

▶ It may *look* like a sea sponge, but the popular "loofah" bath scrub is not from an animal at all; it's from a plant (the luffa gourd to be more specific). Normally, you see loofahs cut into sections, but this complete one lets you see the shape of the original plant. The fiber is cellulose and lignin.

▲ This peculiar material is called byssus, and it's a counterexample, sort of, to the claim that no commercial fibers come from mollusks. Byssus is the fiber made by clams and mussels to glue themselves to rocks underwater. This keratin-based material is from a common type of clam that produces fibers a couple of inches (five centimeters) long, but some varieties of byssus are up to six inches (twenty centimeters) long. Some spectacular fabrics have been made from this fiber, but currently it appears that only one artist, who lives in Sardinia, is making anything out of byssus fiber, so it's fair to say that it's not in *regular* commercial use.

So Many Hairs!

IT'S REALLY QUITE remarkable how many different kinds of animal hair are readily available from commercial sources (by which I mean eBay). Each kind of hair has unique properties, varying in stiffness, tendency to retain static electricity, surface roughness, color, and cool-sounding-ness of its origin (which may be of paramount concern in fashion applications).

▲ Applying gold leaf is an ancient and delicate art. The leaf is so incredibly thin that touching it with fingers destroys it instantly. The only way to pick it up is to use a tiny bit of static electricity on the end of a brush, and the correct kind of hair to use for such a brush is squirrel hair. It is unclear whether gray, red, blue, or brown squirrel hair is best, and all varieties are available in the form of these "gilder's tips."

▲ This goat hair brush is used to apply cosmetics.

▼ Sables are ferret-like creatures (actually martens) about half the weight of a cat. Their fur is considered a great luxury, like mink, and as with mink, you can get brushes made from it.

▶ Blue squirrel hair seems particularly common in specialized paint brushes.

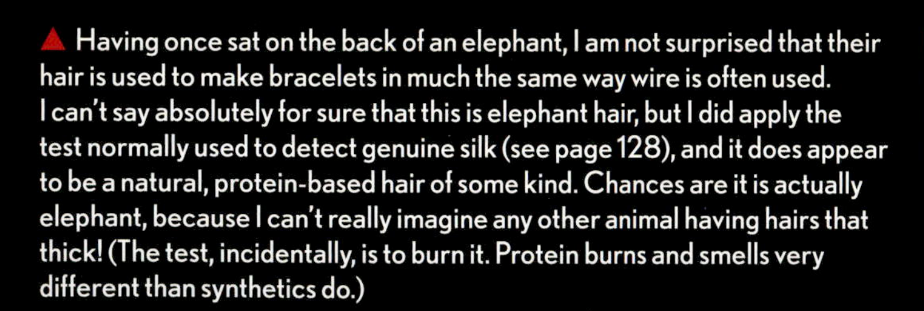

▲ Having once sat on the back of an elephant, I am not surprised that their hair is used to make bracelets in much the same way wire is often used. I can't say absolutely for sure that this is elephant hair, but I did apply the test normally used to detect genuine silk (see page 128), and it does appear to be a natural, protein-based hair of some kind. Chances are it is actually elephant, because I can't really imagine any other animal having hairs that thick! (The test, incidentally, is to burn it. Protein burns and smells very different than synthetics do.)

▶ OK, this one caught me off guard. The fact that it is possible to buy giraffe hair bracelets strikes me as both marvelous and disturbing. It's a marvelous sign of just how connected our world is; sitting in my living room, I can send an electronic signal to someone in South Africa requesting them to please put some giraffe hair on a flying machine, and a few days later I have it. But can we really sustain a society in which such a thing is possible? I don't mean just from an ecological point of view, I mean simply from the point of view of how *complicated* the whole place has gotten.

Keratin Extruded by Warm, Fuzzy Animals

THE MOST WIDELY used animal fibers come from soft and warm animals such as sheep and fluffy birds. This isn't surprising, as we, too, use these fibers to keep warm and to give ourselves soft things to wear as well as soft places to sit, walk, and sleep.

▼ Sheep hair, more specifically the type of hair known as wool, is very widely used. More than a million tons are produced annually. This hair is from a Shetland sheep raised in Montana. I should note that calling this material "sheep hair" would upset the people who grow the sheep, because in their terminology there is a difference between the sheep's "hair" (which is too straight and smooth to hold together when spun into yarn) and the "wool" that grows underneath and is protected by the hair. But speaking more generically, wool is a type of hair that happens to grow with a lot of kinks in it and with a rough surface that allows it to cling to itself. Like all other kinds of hair, wool gets its properties from the details of the amino acid sequences in the keratin proteins it is made of.

▲ Mohair is the coat of the angora goat (not to be confused with angora wool, which is the fur of the angora rabbit). This form of goat hair is used for sweaters, swanky coats, and, more amusingly, for doll wigs. I'm not sure why it isn't also used for human wigs.

▲ Yes, there really are dog hair socks. I got these at a dog show. They are made from the hair of Nova Scotia Duck Tolling Retrievers. Dogs of this breed have a combination of red-orange hair with white tufts on their chest; the yarn with which these socks were knitted has not been dyed. This is exactly what the dogs look like.

► Camel hair (actually the camel's undercoat, after coarse hairs are removed) is surprisingly soft and widely used to make coats. "Camel hair" paintbrushes, on the other hand, are often actually made from less expensive sources such as squirrel hair.

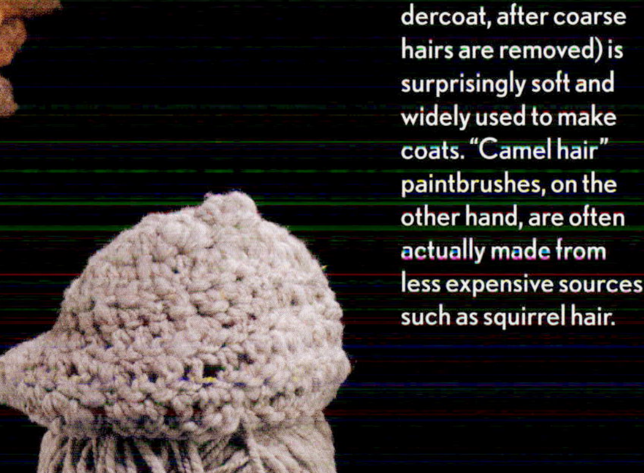

► Most wool now comes from Australia, New Zealand, and China, but communities of local sheep farmers thrive all over the world. This wool came from just a few miles down the road from my house in Central Illinois, which is about as well known for its sheep industry as for its wineries. My girlfriend's mom was going to knit it into a sheep but only got as far as the rear end. So basically the situation we are dealing with here is that we have photographed a knitted sheep's butt and put it in a book.

Keratin Extruded by Warm, Fuzzy Animals

▶ Bird feathers (like these duck feathers from inside a pillow I raided) are made of protein chains similar to, but stiffer than, the keratin protein that makes up human and other animal hairs. Feather proteins are more closely related to the keratin in our fingernails.

▲ There seems to be some doubt as to whether it's legal to import raw eiderdown into the United States, so my sample was supplied in the form of a cute little silk pillow.

▼ Ostrich feathers are widely used even today for dusting things. Synthetic alternatives are of course available—and cheaper—but ostrich feathers are said to work better because the microstructure on the surface of the feather filaments trap dust rather than just move it around. One thing nature is very good at is creating incredibly complex microscopic structures. That's because nature's machines are the size of molecules, while our crude versions are the size of rooms.

▲ A warm comforter made of this particular eiderdown costs about $15,000, making this material perhaps the most expensive form of keratin money can (legally) buy. Why would anyone pay so much for down that a duck grows on its belly? Down is to feathers as wool is to hair. Both down and wool are the soft undercoat that provides warmth and is protected by a longer, stiffer, more waterproof outer coat. Down is superior to feathers in both warmth and softness, so the best, most expensive coats and blankets are filled with pure down, while cheaper ones are filled with feathers or a mixture. Not all down is created equal: down from birds that live in colder areas is thicker and warmer and thus most desirable. This particular down, eiderdown, is collected almost exclusively in Iceland from the nests of the Eider duck in a process that is said to be harmless to the birds and eggs. Each nest yields about as much eiderdown as you see here: twenty grams (two-thirds of an ounce). The total annual production could fit into one small truck.

Silk

THE UNDISPUTED KING of natural fibers comes not from a cute mammal but from one of the lowliest of creatures: worms, more specifically silkworms, which aren't really worms at all, but the caterpillars of the silkmoth. Known since antiquity, silk is silky soft, smooth, and incredibly strong. It's also expensive and must be cleaned carefully, making it a fussy, luxury fiber compared with the workhorses of cotton, wool, and synthetics.

Like hair, silk is a protein, but a slightly different one: fibroin.

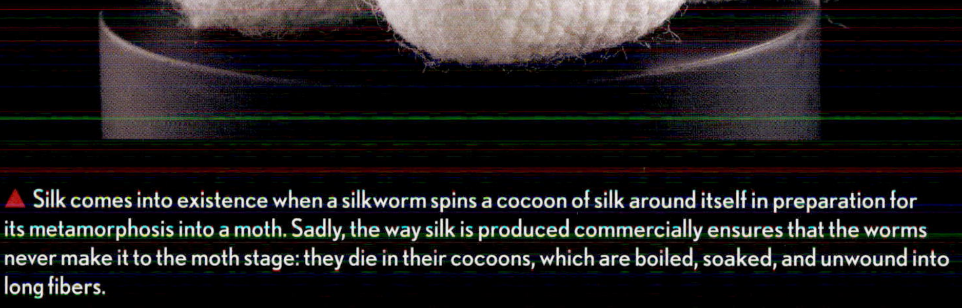

▲ Silk comes into existence when a silkworm spins a cocoon of silk around itself in preparation for its metamorphosis into a moth. Sadly, the way silk is produced commercially ensures that the worms never make it to the moth stage: they die in their cocoons, which are boiled, soaked, and unwound into long fibers.

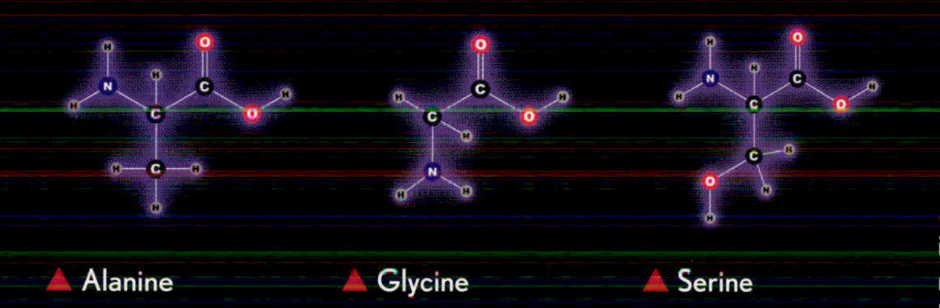

▲ Alanine ▲ Glycine ▲ Serine

The chemical structure of fibroin is relatively simple, consisting of a repeating structure of just three different amino acids. But its physical structure is complex, with the protein backbone folded into loops and sheets that give silk its strength and sheen.

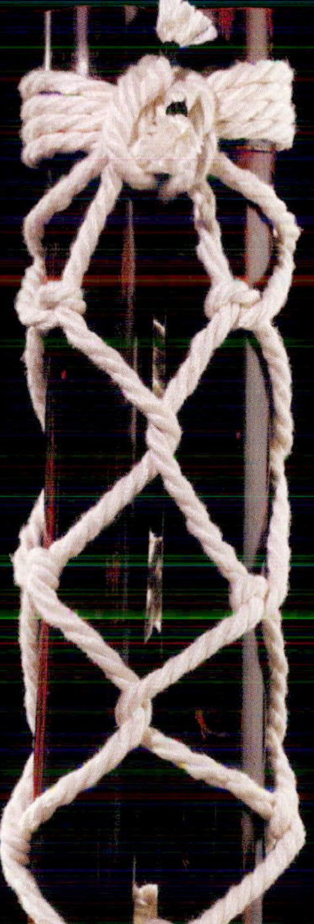

◀ Silk rope is kind of crazy: it's far more expensive than can possibly be justified for all but one commercial application.

▼ Silk medical sutures are strong but have been largely replaced by superior synthetic alternatives.

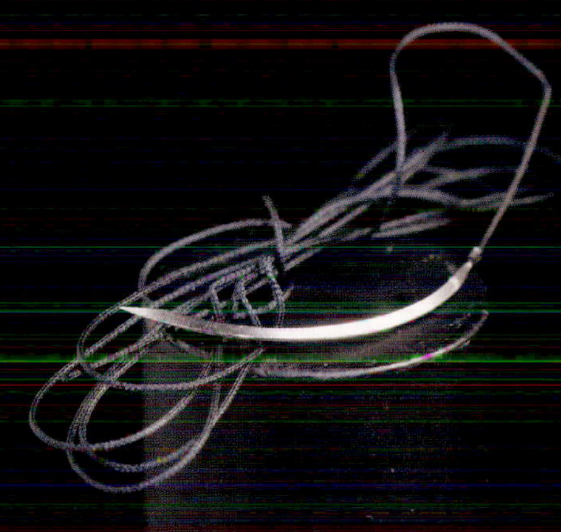

▼ Raw silk fiber before it is spun into thread is a beautiful, lustrous, and shiny thing.

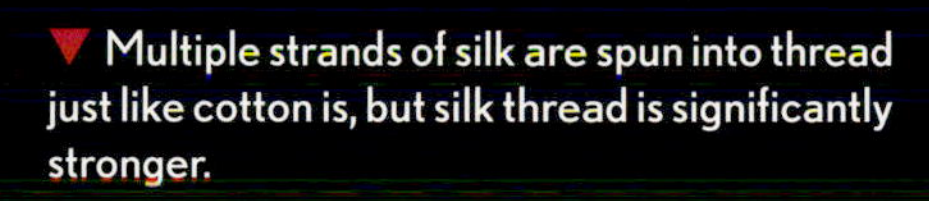

▼ Multiple strands of silk are spun into thread just like cotton is, but silk thread is significantly stronger.

▼ Because of its great strength and light weight, parachutes were made from silk until advanced nylon fabrics replaced it. This is a scrap of fabric from a World War II silk parachute.

▼ Rough, hand-woven silk cloth is anything but rough: even in this form it exhibits silk's silky softness.

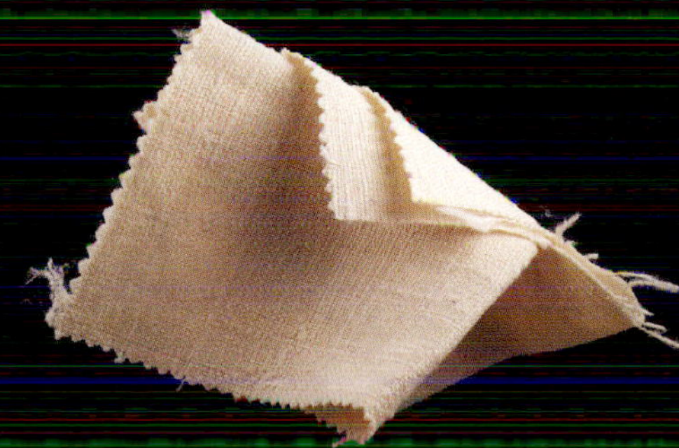

Tested by Fire

THERE'S REALLY ONLY one sure way to test whether an item is genuine silk outside of a laboratory: burn a small sample of it. When a natural protein fiber—silk, hair, or leather—burns, it will melt a little bit, but for the most part it turns into a black charred mass.

Most synthetic fibers, such as nylon, behave very differently: they melt and retract into a ball from which smaller flaming balls of molten plastic drip down to the ground, leaving nothing behind when they have burned completely. Once you've seen the two behaviors side by side, there's no mistaking which one is the synthetic.

Plant fibers such as cotton or wood burn without any sign of melting: they just turn into ash slowly as they burn. And, amusingly, even steel wool fibers burn easily when they are fine enough.

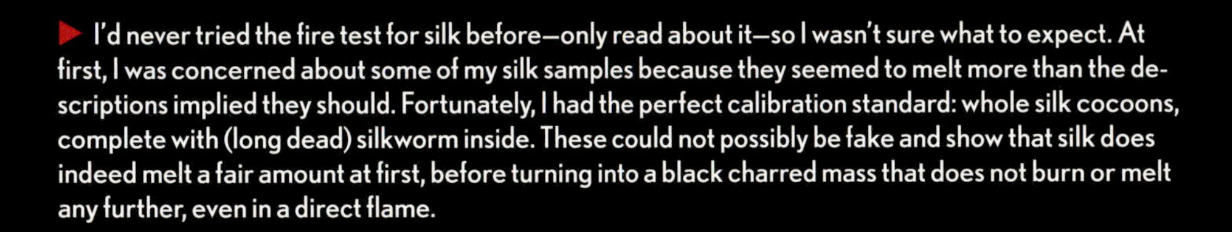

▶ I'd never tried the fire test for silk before—only read about it—so I wasn't sure what to expect. At first, I was concerned about some of my silk samples because they seemed to melt more than the descriptions implied they should. Fortunately, I had the perfect calibration standard: whole silk cocoons, complete with (long dead) silkworm inside. These could not possibly be fake and show that silk does indeed melt a fair amount at first, before turning into a black charred mass that does not burn or melt any further, even in a direct flame.

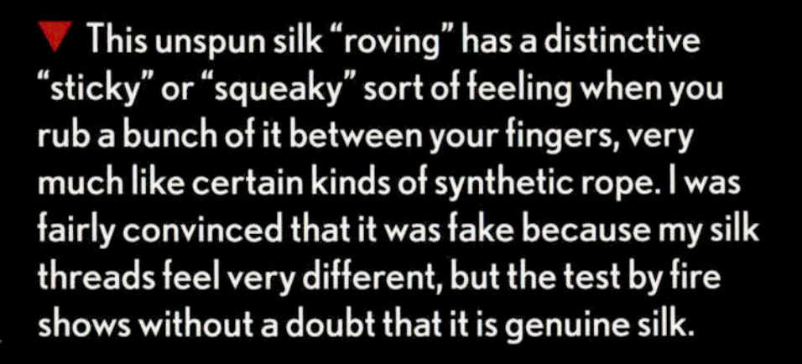

▼ This unspun silk "roving" has a distinctive "sticky" or "squeaky" sort of feeling when you rub a bunch of it between your fingers, very much like certain kinds of synthetic rope. I was fairly convinced that it was fake because my silk threads feel very different, but the test by fire shows without a doubt that it is genuine silk.

▼ This silk thread flows in the hands without any of the squeaky feel of silk roving, but under the flame it behaves exactly the same way, melting at first and then turning into a solid char. It too is genuine silk.

▼ Finely spun nylon rope looks and feels a lot like silk, but when you burn it, all doubt is removed. It immediately melts and pulls back into balls of hot liquid, which drip down while still on fire. The sound these flaming balls make as they fall through the air is quite amusing, and diagnostic of synthetic fiber. (The flaming balls are also rather dangerous if you're doing this over a flammable surface, for example a synthetic fiber carpet.)

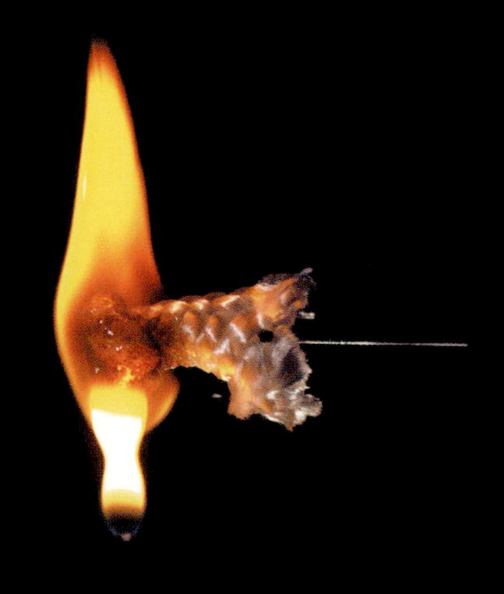

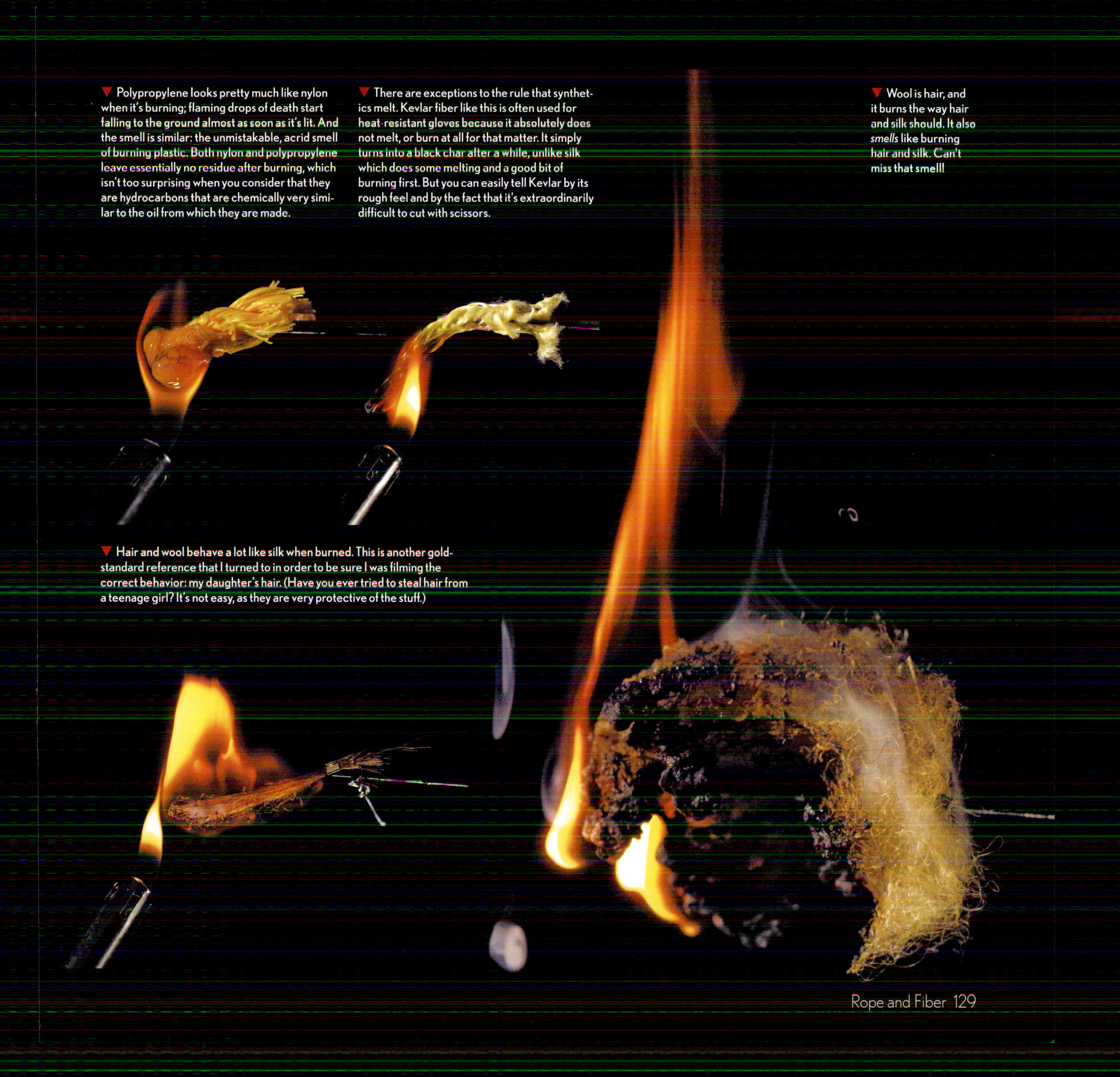

Polypropylene looks pretty much like nylon when it's burning; flaming drops of death start falling to the ground almost as soon as it's lit. And the smell is similar: the unmistakable, acrid smell of burning plastic. Both nylon and polypropylene leave essentially no residue after burning, which isn't too surprising when you consider that they are hydrocarbons that are chemically very similar to the oil from which they are made.

There are exceptions to the rule that synthetics melt. Kevlar fiber like this is often used for heat-resistant gloves because it absolutely does not melt, or burn at all for that matter. It simply turns into a black char after a while, unlike silk which does some melting and a good bit of burning first. But you can easily tell Kevlar by its rough feel and by the fact that it's extraordinarily difficult to cut with scissors.

Wool is hair, and it burns the way hair and silk should. It also *smells* like burning hair and silk. Can't miss that smell!

Hair and wool behave a lot like silk when burned. This is another gold-standard reference that I turned to in order to be sure I was filming the correct behavior: my daughter's hair. (Have you ever tried to steal hair from a teenage girl? It's not easy, as they are very protective of the stuff.)

Tested by Fire

▲ Buyer beware! This material was sold as genuine suede leather, but it burns unmistakably like a synthetic polymer. It is utterly fake, made probably of some kind of polyurethane plastic.

▲ Real leather strap looks surprisingly similar to the fake stuff but is considerably stronger and difficult to burn at all.

▶ Cotton burns cleanly and beautifully, leaving almost no ash.

▼ Duck feathers act a lot like hair or silk when on fire: a bit of melting, no dripping, and they leave a black char behind, unlike most synthetics.

▼ I created this reference standard for burning leather by cutting a small strip off a sheepskin that I know is real because it's got wool growing on it. Real leather burns a lot like hair, leaving a black char.

◀ The flaming balls dropping from this burning "suede" are a dead giveaway: it is synthetic.

▶ All plant fibers, including hemp and coconut, burn a lot like wood. (The picture shows hemp rope.) They, like cotton, are mostly plant cellulose. The main difference is that cotton is pure cellulose, hemp fibers are mostly cellulose and lignin, while wood also has rosins and oils that sometimes bubble up and cause irregular flare-ups.

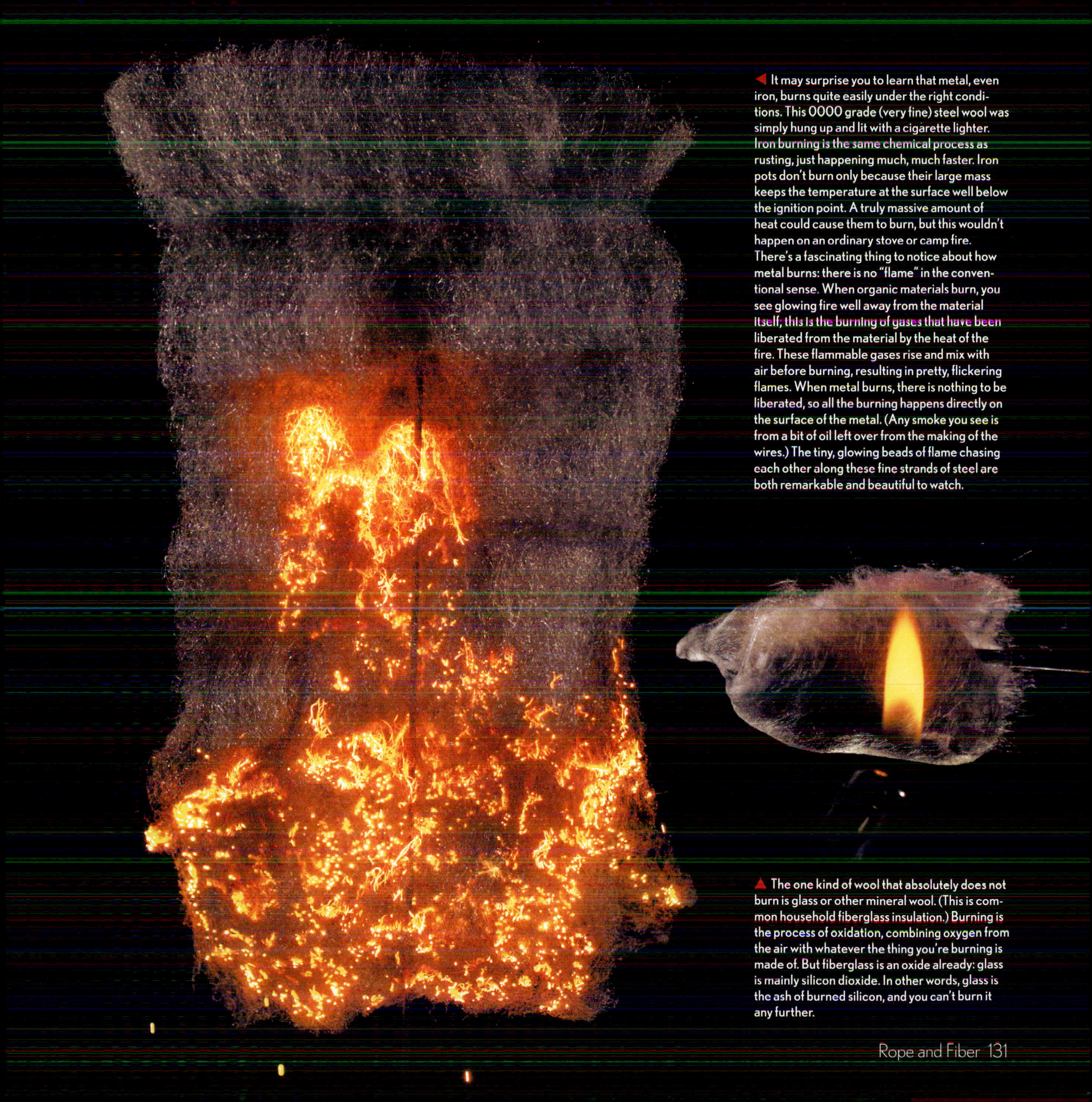

◀ It may surprise you to learn that metal, even iron, burns quite easily under the right conditions. This 0000 grade (very fine) steel wool was simply hung up and lit with a cigarette lighter. Iron burning is the same chemical process as rusting, just happening much, much faster. Iron pots don't burn only because their large mass keeps the temperature at the surface well below the ignition point. A truly massive amount of heat could cause them to burn, but this wouldn't happen on an ordinary stove or camp fire. There's a fascinating thing to notice about how metal burns: there is no "flame" in the conventional sense. When organic materials burn, you see glowing fire well away from the material itself; this is the burning of gases that have been liberated from the material by the heat of the fire. These flammable gases rise and mix with air before burning, resulting in pretty, flickering flames. When metal burns, there is nothing to be liberated, so all the burning happens directly on the surface of the metal. (Any smoke you see is from a bit of oil left over from the making of the wires.) The tiny, glowing beads of flame chasing each other along these fine strands of steel are both remarkable and beautiful to watch.

▲ The one kind of wool that absolutely does not burn is glass or other mineral wool. (This is common household fiberglass insulation.) Burning is the process of oxidation, combining oxygen from the air with whatever the thing you're burning is made of. But fiberglass is an oxide already: glass is mainly silicon dioxide. In other words, glass is the ash of burned silicon, and you can't burn it any further.

Protein Fibers that Occur Inside Animals

◀ Leather can be cut into strips, which in turn can be twisted, woven, or braided just like any other fiber. This braided leather whip is a fairly menacing example of collagen fiber.

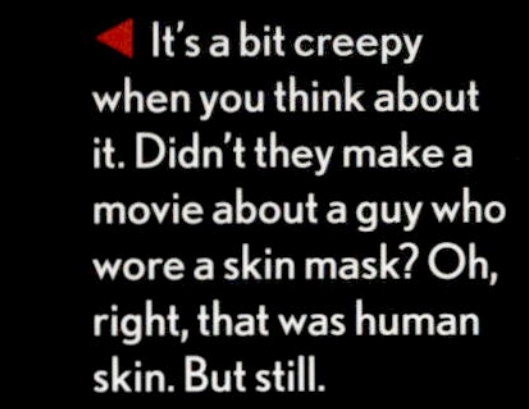

◀ It's a bit creepy when you think about it. Didn't they make a movie about a guy who wore a skin mask? Oh, right, that was human skin. But still.

▼ Sinew from the tendons that run along the backbone of whitetail deer are used to strengthen bows made in the primitive style.

GATHERING KERATIN PROTEIN from an animal generally doesn't kill it unless you decide to take the whole skin along with the hair. But animals also make a different kind of fibrous protein: collagen. This is the protein that makes up skin, ligaments, sinews, and other connective tissue. By far the most common example of the use of collagen is leather, which is made into coats, shoes, bags, belts, and a thousand other things.

Somewhat more exotic is the use of animal sinew as a fiber. Sinew fiber manufacture persists primarily as a hobby, now that much better synthetic alternatives are available. Catgut, another form of collagen connective tissue, is still used in a few ways.

▲ Leather is an incredibly versatile material. I wasn't able to find a cow made of cow leather or a horse made of horse leather, but here is a horse made of cow leather.

▼ Collagen is a protein, like keratin, but with a different amino acid sequence and therefore a different overall physical structure.

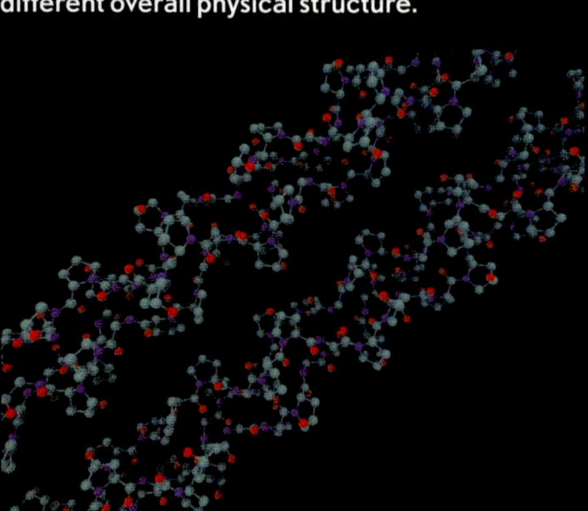

▶ Catgut does not come from cats! Calm down! It comes from the intestines (gut) of sheep, goats, cattle, pigs, horses, donkeys, and so on. Not cats. Even the word doesn't come from cats. The gut part comes from the fact that it is made of gut, but the cat part likely derives from the word *kit*, an old word for fiddle. Catgut has been and still is sometimes used for stringed instruments. These catgut strings are meant for a Persian instrument known as a Tar.

▲ Catgut finds application even today for sutures used to sew up the insides of animals you wish to keep alive (as opposed to the ones you killed to get the catgut). Its advantage is that it is slowly absorbed by the body, so the sutures don't need to be removed later.

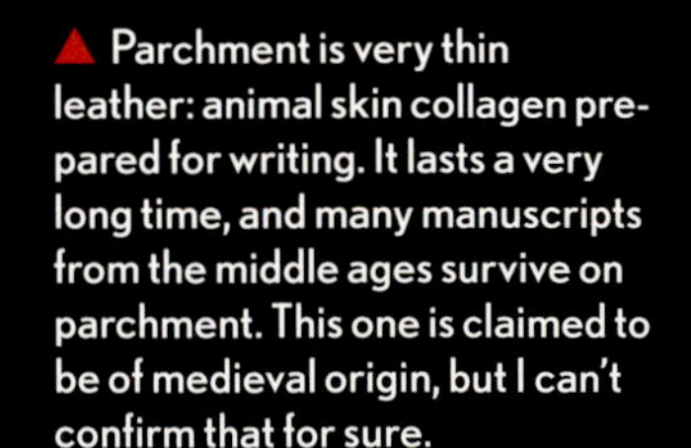

▲ Parchment is very thin leather: animal skin collagen prepared for writing. It lasts a very long time, and many manuscripts from the middle ages survive on parchment. This one is claimed to be of medieval origin, but I can't confirm that for sure.

# of a Rock

MOST FIBERS are organic compounds, but there are important inorganic examples as well: steel wire and rope, of course, as well as carbon fiber and silica (glass) fiber. One of the most beautiful natural fibers, asbestos, has a bad name these days (see page 226) but was once a wonder material for its light weight, fire resistance, and insulating properties.

Unlike all the fibers we've talked about so far, inorganic fibers, as a general rule, are not made of long, thin molecules—or necessarily of individual molecules at all. Metal fibers, for example, are simply long, thin pieces of metal alloys. They are not joined into discrete molecules, with no preferred orientation of the atoms. Glass and rock fibers are similarly just long, thin physical forms of simple molecules of a few kinds of atoms linked in three-dimensional matrices, not linear chains.

The properties of inorganic fibers are not nearly as diverse as those of organic fibers, but they play an important role as the only kind of fiber that can withstand truly high temperatures. And they can last essentially forever under all but the most extreme conditions, much like the rocks some of them are made from.

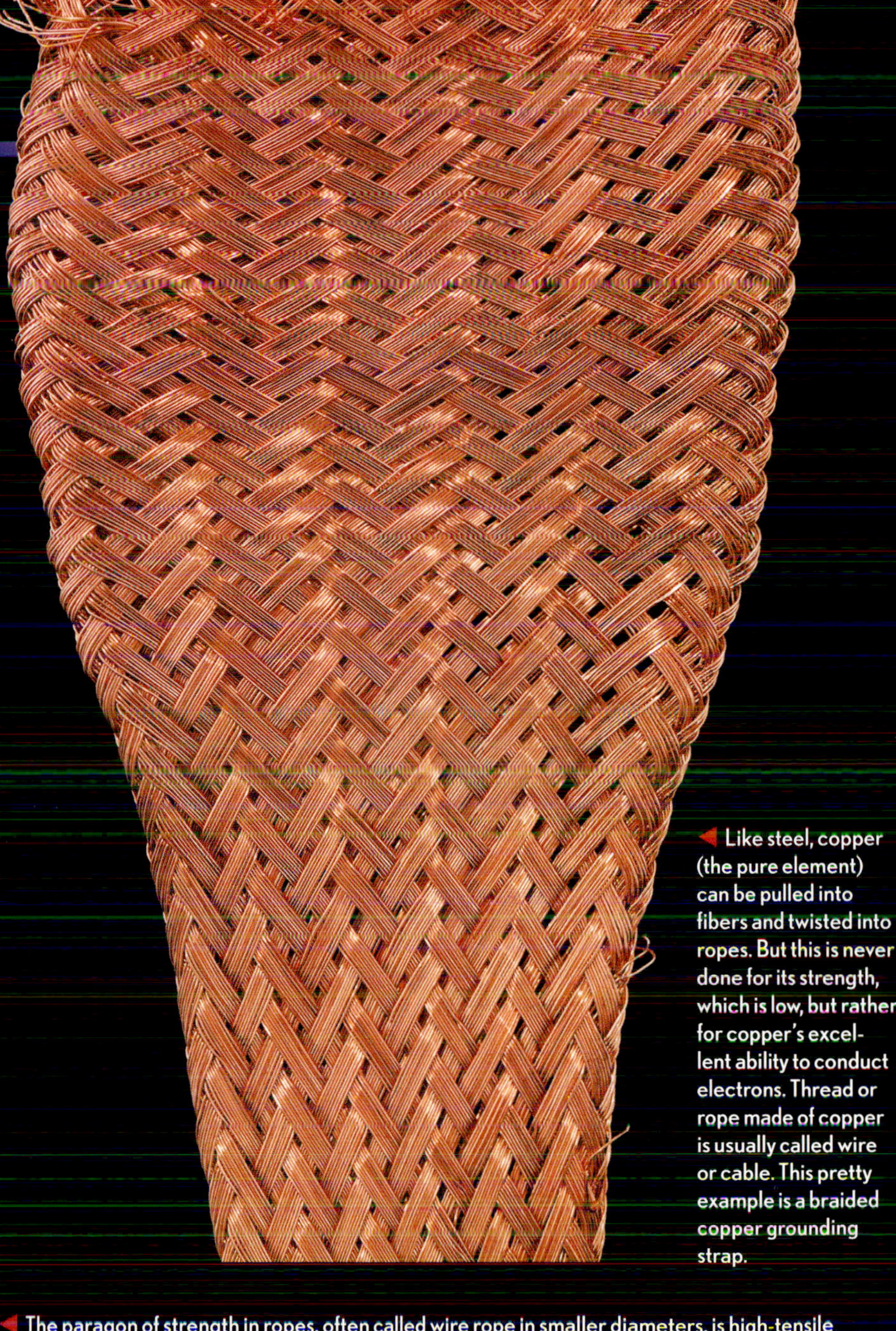

Like steel, copper (the pure element) can be pulled into fibers and twisted into ropes. But this is never done for its strength, which is low, but rather for copper's excellent ability to conduct electrons. Thread or rope made of copper is usually called wire or cable. This pretty example is a braided copper grounding strap.

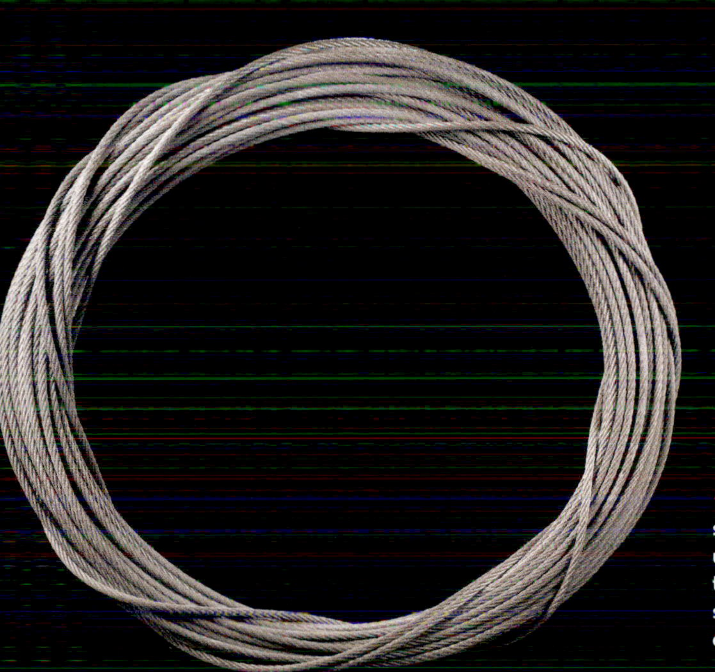

The paragon of strength in ropes, often called wire rope in smaller diameters, is high-tensile steel (made primarily of iron plus a bit of carbon). Some synthetic organic fibers, such as Kevlar and ultra-high-molecular-weight polyethylene, and natural fibers, such as silk, are significantly stronger than steel on a pound-for-pound basis, but none of these materials combine the stiffness, durability, strength, and low cost of a good steel cable. For use in construction cranes, building elevators, and elevated cable cars, steel is the fiber of choice.

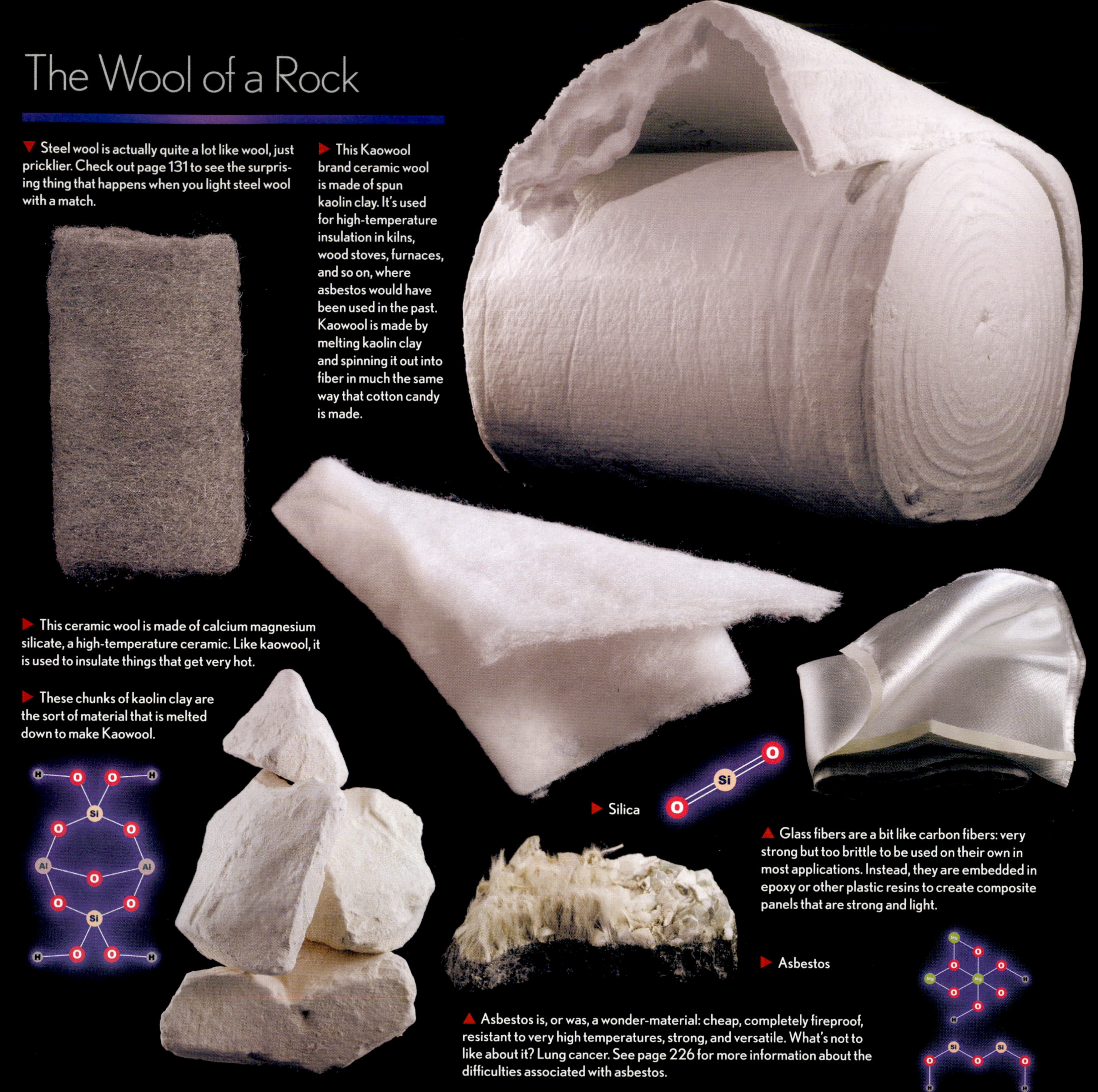

The Wool of a Rock

▼ Steel wool is actually quite a lot like wool, just pricklier. Check out page 131 to see the surprising thing that happens when you light steel wool with a match.

▶ This Kaowool brand ceramic wool is made of spun kaolin clay. It's used for high-temperature insulation in kilns, wood stoves, furnaces, and so on, where asbestos would have been used in the past. Kaowool is made by melting kaolin clay and spinning it out into fiber in much the same way that cotton candy is made.

▶ This ceramic wool is made of calcium magnesium silicate, a high-temperature ceramic. Like kaowool, it is used to insulate things that get very hot.

▶ These chunks of kaolin clay are the sort of material that is melted down to make Kaowool.

▶ Silica

▲ Glass fibers are a bit like carbon fibers: very strong but too brittle to be used on their own in most applications. Instead, they are embedded in epoxy or other plastic resins to create composite panels that are strong and light.

▶ Asbestos

▲ Asbestos is, or was, a wonder-material: cheap, completely fireproof, resistant to very high temperatures, strong, and versatile. What's not to like about it? Lung cancer. See page 226 for more information about the difficulties associated with asbestos.

▲ Heat-resistant work gloves today are typically made of Kevlar or textured fiberglass, with wool and cotton insulation inside. But older furnace gloves like these were always made of asbestos.

▲ Zetex is a brand name for textured fiberglass fabric used to make high-temperature gloves. It's more heat resistant than Kevlar and doesn't kill you like asbestos does.

► This is lovely Miraflax brand fiberglass made by Dow Corning years ago. I insulated a whole house with it and loved the way it didn't make me itch. It feels super-soft. To be honest, though, it does actually cause some itching eventually—but nothing like regular fiberglass. For some reason, they stopped making it, and if anyone knows why, I'd love to know.

▲ This type of insulation is sold alongside fiberglass insulation in home improvement centers, and it can be installed in more or less the same way. However, it is made from spun basalt rock, not from glass. It's denser and provides better sound insulation than fiberglass, but is otherwise surprisingly similar, considering that it's made of bedrock.

▼ This wool is spun from molten basalt and chalk and is used as a growing medium for sprouting seeds.

▲ Vast quantities of soda-lime glass are spun into fiberglass insulation for use in homes, appliances, commercial buildings, and so forth. Glass fiber is a nearly ideal material in many ways: cheap, effective, nonflammable, very long-lasting, and easy to install. The only real downside is that it's incredibly irritating to skin. You may wonder whether breathing fiberglass could cause lung disease for the same reasons that asbestos does, but in practice it does not. This isn't because glass fibers are any less sharp than asbestos fibers, but rather because the chemical environment of the lungs dissolves glass fibers relatively quickly, so they don't hang around for many years, as asbestos fibers do.

▲ Ordinary fiberglass is made of ordinary glass, but this superior material is made of borosilicate (Pyrex) heat-resistant glass, for use not as insulation but as a filter material in chemical apparatuses.

The Wool of a Rock

▶ Carbon fiber consists almost entirely of carbon atoms arranged in a hexagonal grid, like graphite. Unlike graphite's flat sheets, the hexagons are arranged in long fibers. Carbon fibers are tremendously strong but brittle and, thus, are often protected by being embedded in a plastic matrix. The exceptionally light, strong, stiff carbon fiber composite components found in airplanes, sports equipment, and fancy photographic tripods are made this way.

▼ Carbon fibers are often used to reinforce and stiffen organic resins such as epoxy or polystyrene. So it's not always necessary for them to be particularly long. These carbon fibers started out long but were intentionally chopped into pieces about one-quarter inch (half a centimeter) long for use as strengthening filler. Glass fibers are often chopped up for the same reason: to make fiberglass-reinforced panels for boats and sports cars.

▶ Long, unchopped carbon fibers embedded in epoxy resin form very lightweight, strong, and stiff structures, such as the frame of this expensive bicycle owned by the photographer for this book.

Is the Electrostatic Force Really What Holds This All Together?

EVERY ONCE IN A WHILE, most often in airplanes, I worry about the fact that the contraption I'm in is held together by static electricity. The force that keeps all substances together—all metals, all ropes, all chains, all airplanes, everything—is the same force that makes a balloon stick to the wall after you rub it on your shirt. Balloons don't stick to the wall very well.

The objects in our macroscopic world have vast numbers of positive and negative charges in them (all their protons and electrons respectively), but almost all those positives and negatives are perfectly matched and exactly cancel each other out. Even what would be considered a very large static electric charge contains a tiny number of electrons, compared to the total number in the atoms of whatever is holding that charge (for example, a balloon).

If you could separate all the protons and electrons in an object, the force between them would be mind-boggling.

For example, consider one gram of iron. You could use it to make about one centimeter of 5/32-inch steel aircraft cable, which would have a breaking strength of about three thousand pounds—strong enough to hold up a cube of iron about twenty inches (fifty centimeters) on edge, or a small car.

But if you could separate all the protons in that piece of iron from all its electrons and put the protons on one side and the electrons on the other side of a one-centimeter gap, the attractive force between them would be strong enough to hold up a cube of iron about eight *miles* (thirteen kilometers) on edge, or a good-sized mountain.

The electrostatic force is a whale of a strong force. Even a tiny fraction of it is more than enough to hold together the skin of an airplane.

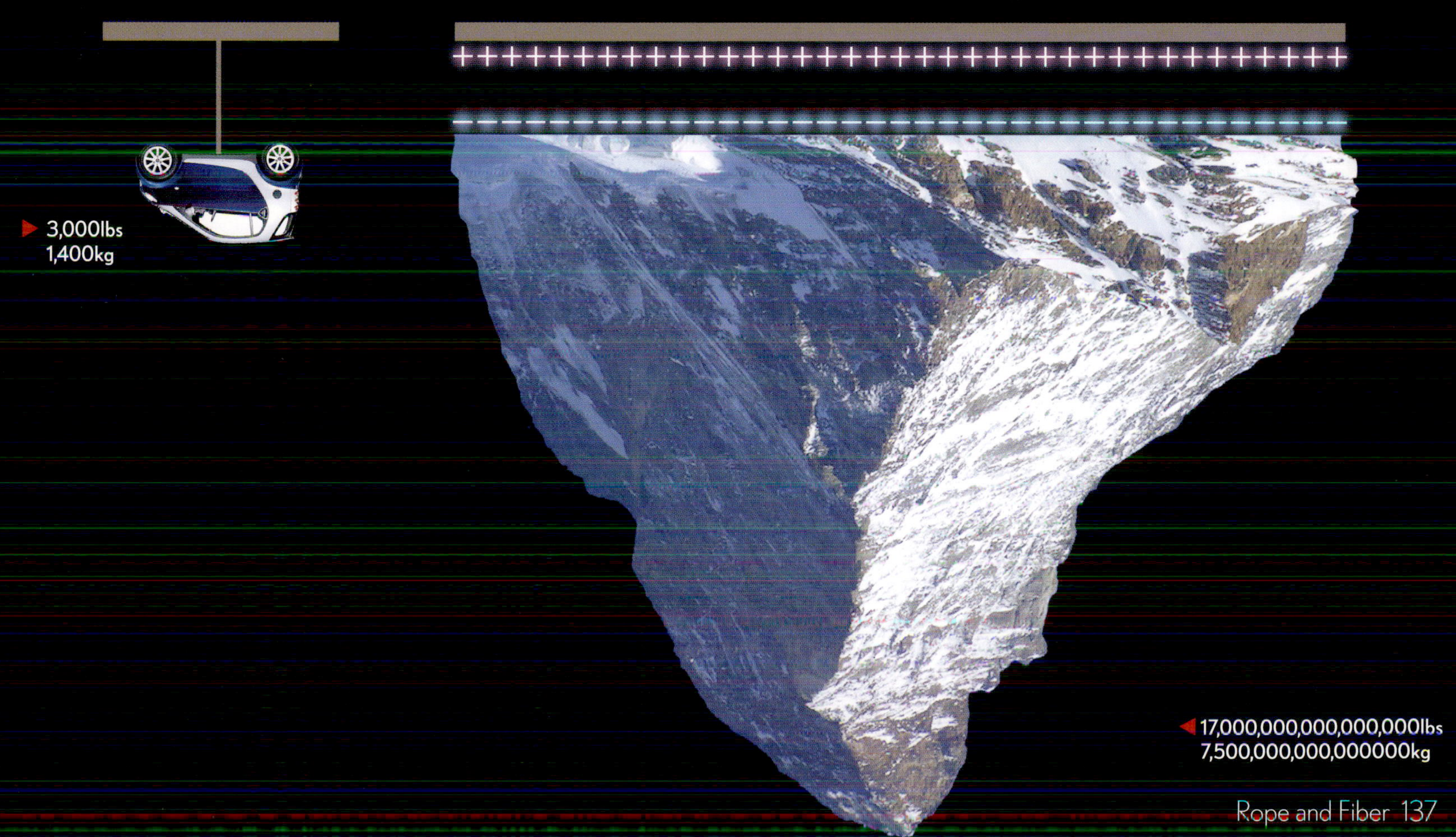

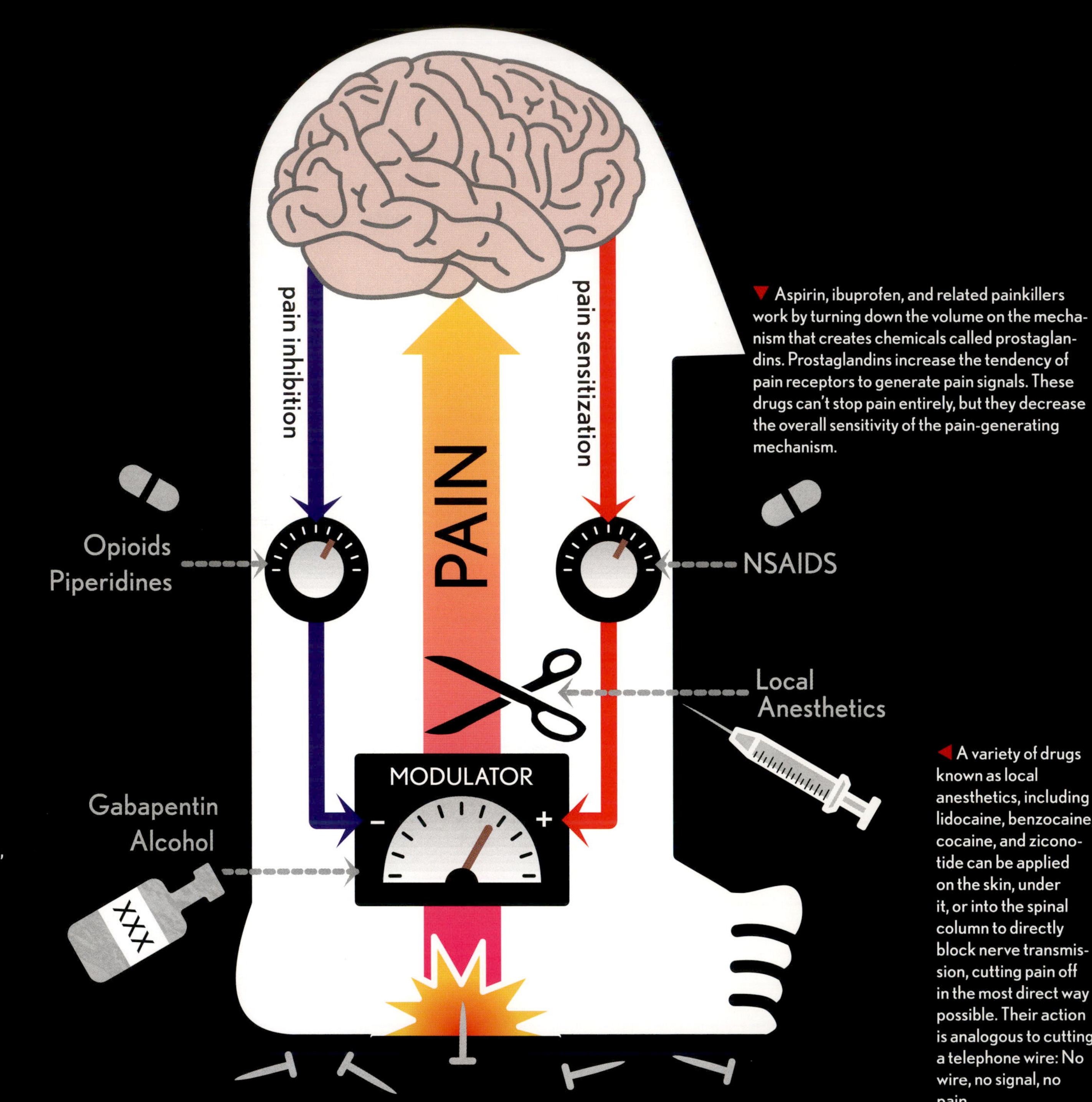

▼ Aspirin, ibuprofen, and related painkillers work by turning down the volume on the mechanism that creates chemicals called prostaglandins. Prostaglandins increase the tendency of pain receptors to generate pain signals. These drugs can't stop pain entirely, but they decrease the overall sensitivity of the pain-generating mechanism.

▶ Opioids like cocaine can be used as topical anesthetics, but when diffused throughout the brain (after eating or injection into the bloodstream), they work in a very different way, turning up the volume on the dopamine mechanism, which signals the pain mechanism to turn down its volume.

◀ A variety of drugs known as local anesthetics, including lidocaine, benzocaine, cocaine, and ziconotide can be applied on the skin, under it, or into the spinal column to directly block nerve transmission, cutting pain off in the most direct way possible. Their action is analogous to cutting a telephone wire: No wire, no signal, no pain.

▲ Some painkillers, for example gabapentin, as well as alcohol, turn down the volume on the pain-generating mechanism directly, increasing the threshold above which pain receptors will generate their unpleasant signals.

Chapter 8

Pain and Pleasure

PAIN IS SOMETHING you don't think about much unless you have some, in which case it's the *only* thing you think about. Making the pain go away drives us to do everything from jerking our hands back from a hot stove to running billion-dollar drug research programs looking for better painkillers, (which is to say looking for better molecules).

Pain is just information. It's like a blinking light you see in the distance; the faster it blinks, the stronger the pain. But this light has no inherent power and is of no significance unless perceived by the brain. A simple sheet of paper placed anywhere between you and the blinking light will block it and stop the pain, no matter how powerful it seems. The knowledge that pain has no power—no reality in the world other than that which your mind gives it—doesn't help you when you're suffering, but it does mean that drugs don't have to be big and muscular to stop it. They just have to be clever.

Many painkillers in use today are either purified natural plant extracts, identical synthetic copies of those extracts, or synthetic compounds that are close chemical relatives of their natural inspirations.

It's not that plants are trying to help us; it's more the opposite. Quite a few of the effective medicines derived from plants are used by those plants as defensive poisons, and it's this very fact that makes them effective as medicines. A substance that blocks the action of nerves can either kill you, if it blocks the nerves running your heart, or stop the pain of an operation, if it blocks the nerves between your brain and the site of the incision. That's why field researchers looking for drug candidates get excited when they find a new, especially poisonous, plant, insect, frog, bacteria, or fungus.

◀ Einstein famously said that everything should be made as simple as possible, but not simpler. He probably wouldn't have liked this diagram. The real details of how pain transmission and regulation work are very, very complicated: [illegible]t you see here is a gross oversimplification, so please don't take the diagram too literally and write me angry letters about how I've totally misrepresented the enkephalin binding mechanism, or something.

The Bark of the Willow

PAINKILLERS RANGE from drugs so weak that banging your head on the wall would do more good, to those so strong their main use is immobilizing elephants.

Although not the strongest or the oldest, the most widely used painkillers by far were inspired by the bark of the willow tree. Every schoolchild in the United States learns how Native Americans chewed willow bark to relieve pain because it contains aspirin. Willow bark has indeed been used for at least three thousand years, but it doesn't actually contain aspirin. Instead, it contains salicin, a compound that is similar to the active ingredient in modern aspirin but somewhat more toxic and not any better as a painkiller.

This points to an important fact about drugs: if you happen to find one in nature, it's worth trying out chemical variations of it because there's a fair chance you'll find something even better. In this case, an artificial variation, acetyl salicylic acid, turned out to be the best choice, and it became modern aspirin.

Today, synthetic variations of aspirin, some similar, some quite different, are very common. These drugs are called NSAIDs, which stands for nonsteroidal anti-inflammatory drugs (because they reduce inflammation and they are not steroids). They include the four ubiquitous nonprescription painkillers found in pharmacies around the world: aspirin, acetaminophen (called paracetamol in the United Kingdom), ibuprofen, and naproxen sodium.

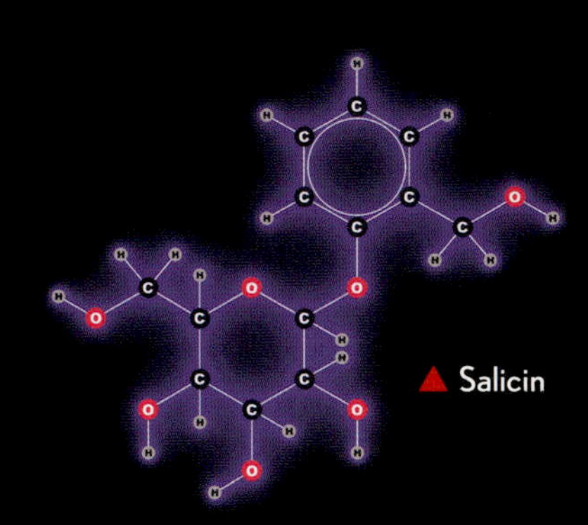

▲ Salicin

▲ Shredded willow bark has been used as a treatment for pain all over the world for thousands of years. The main active ingredient is salicin, but the plant also contains polyphenols and flavonoids, which might actually be beneficial as well.

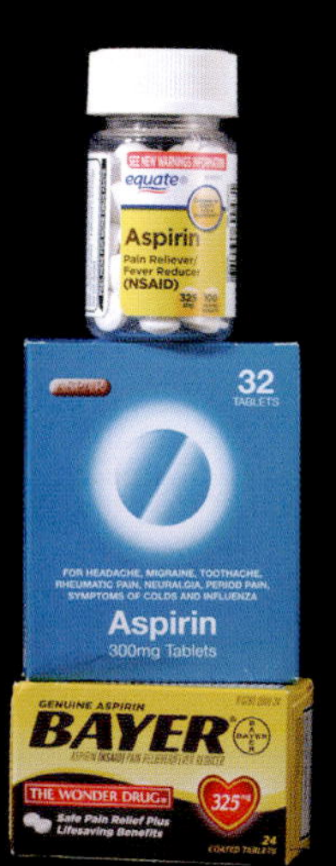

◀ Acetylsalicylic acid (aspirin) is sold under dozens of brand names, but only Bayer is the original. Aspirin was first marketed commercially by the Bayer company, which got its start in 1863 making the synthetic dye fuchsine (see page 202).

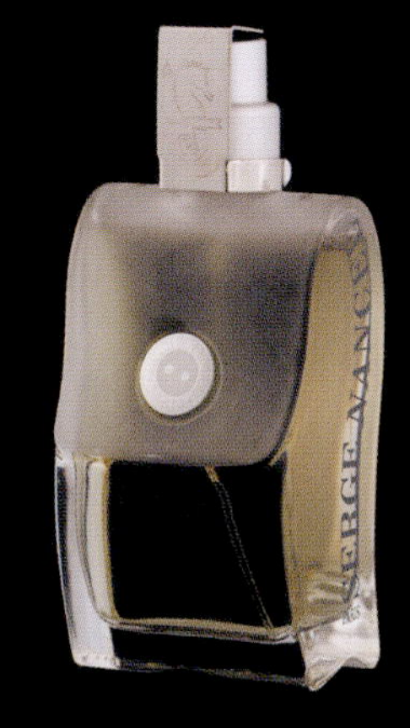

◀ Castoreum, extracted from scent glands in the rear ends of beavers and used by them to mark territory, contains salicin, the same painkilling ingredient found in willow bark. While there is some record of castoreum being used as a painkiller, its main use today is in cologne. For more about things people like the smell of, or don't, see Chapter 11.

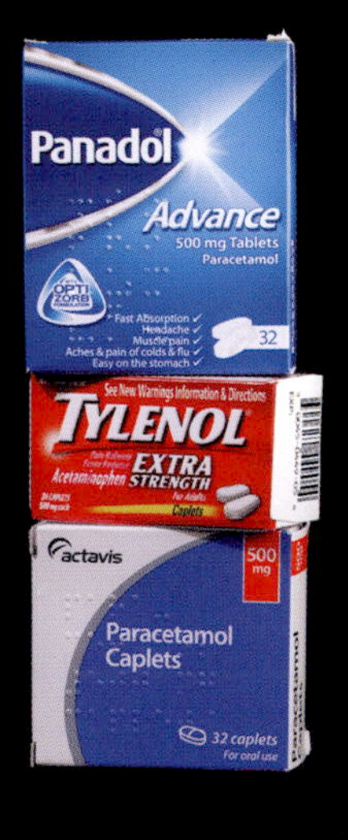

◀ Both generic names, acetaminophen (used in the United States) and paracetamol (used elsewhere), are shortenings of the full chemical name para-acetylaminophenol. It's just a matter of choosing which letters to leave out. This chemical, like aspirin, is sold under dozens of brand names, for example Tylenol in the United States and Panadol in the United Kingdom.

▼ Acetylsalicylic acid

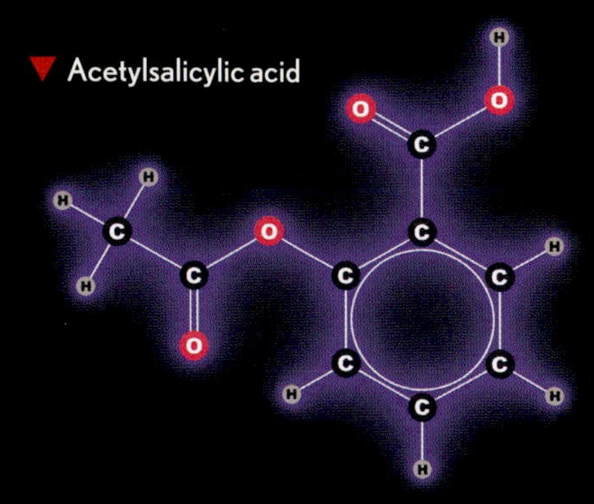

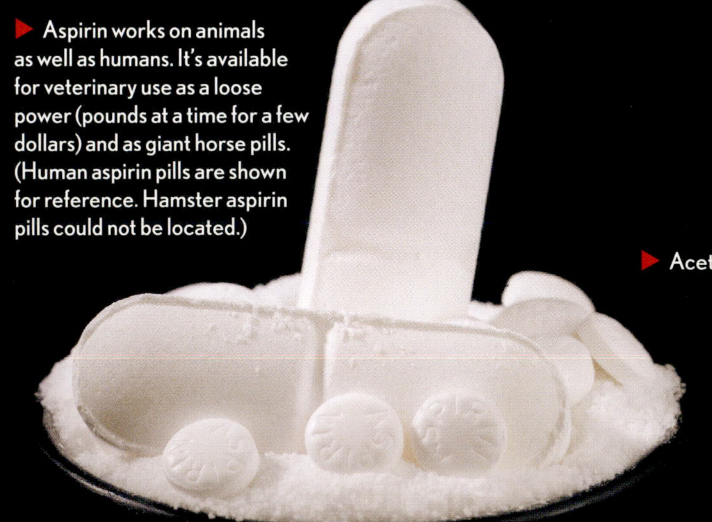

▶ Aspirin works on animals as well as humans. It's available for veterinary use as a loose power (pounds at a time for a few dollars) and as giant horse pills. (Human aspirin pills are shown for reference. Hamster aspirin pills could not be located.)

▶ Acetaminophen

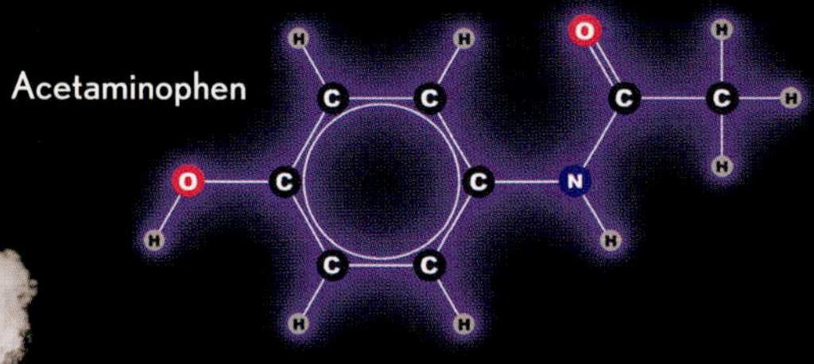

▼ Common painkillers are available in a dizzying variety of combinations, including some with caffeine or antihistamines to give a combined effect.

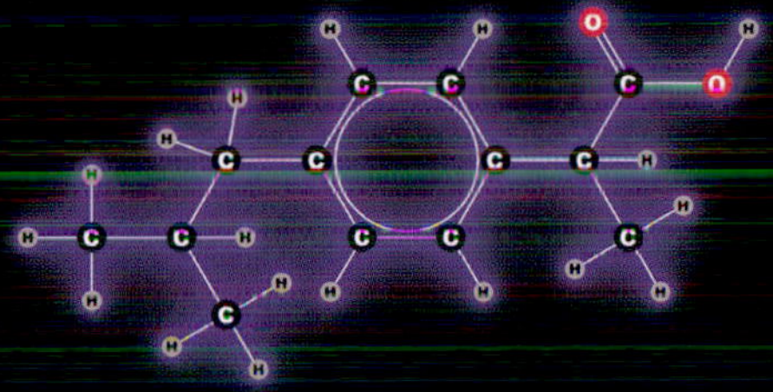

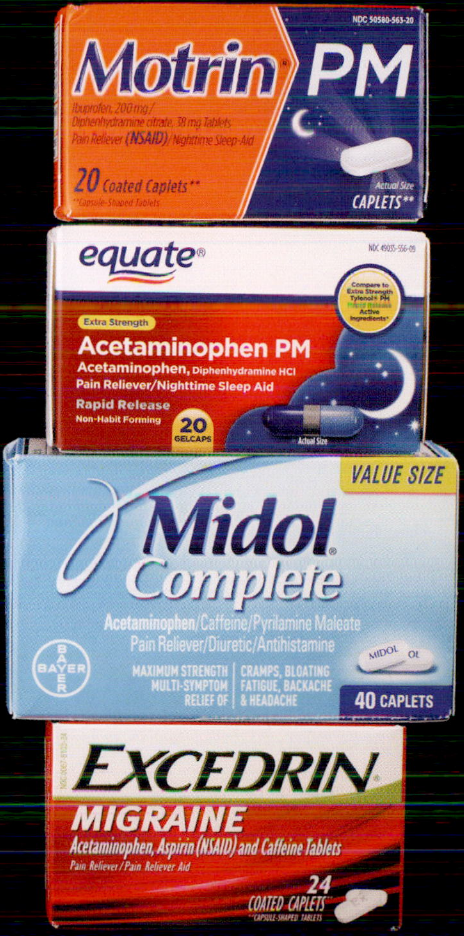

▶ Motrin PM: Ibuprofen, Diphenhydramine

▶ Equate PM: Acetaminophen, Diphenhydramine

▶ Midol Complete: Acetaminophen, Caffeine, Pyrilamine Maleate

▶ Excedrin Migraine: Acetaminophen, Caffeine

▲ Ibuprofen (Painkiller)

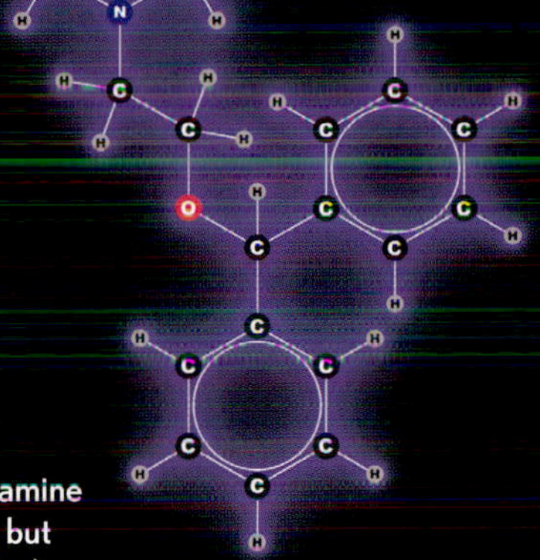

▶ Diphenhydramine (Antihistamine, but used as a sedative)

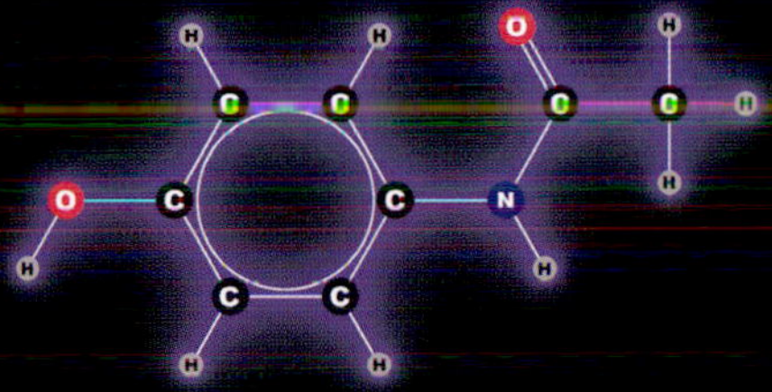

◀ Acetaminophen (Painkiller)

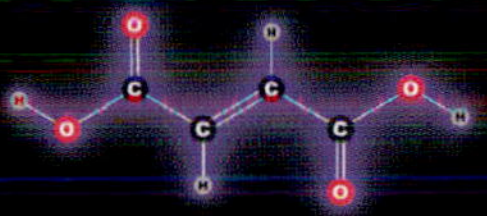

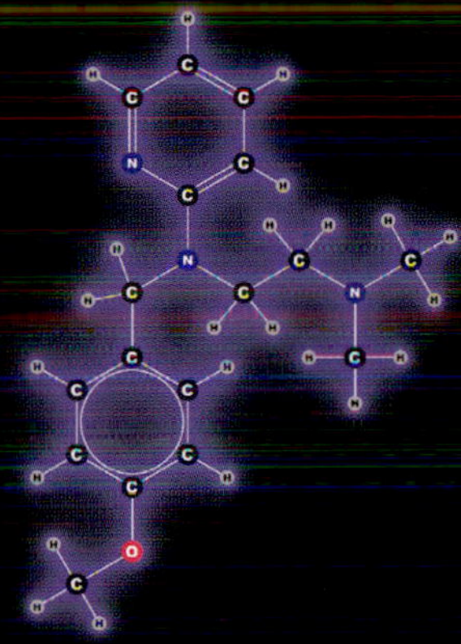

▼ Caffeine (Stimulant, but seems to make painkillers work better)

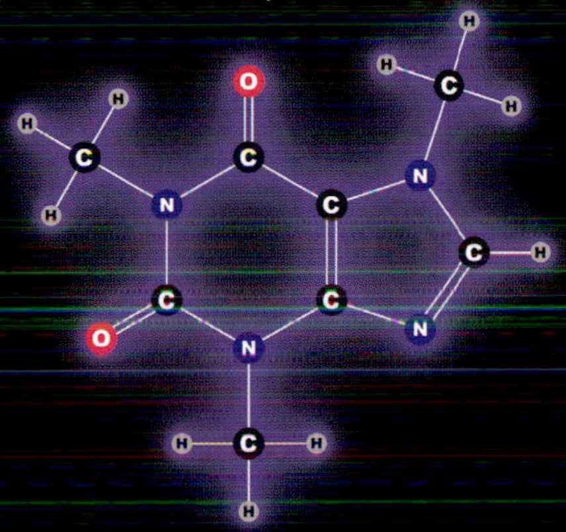

▲ Pyrilamine maleate (Antihistamine, but used as a sedative)

◀ Naproxen sodium is a newer, over-the-counter painkiller that shares the same acid structure as aspirin, but instead of a single benzene ring, it boasts an elegant double ring.

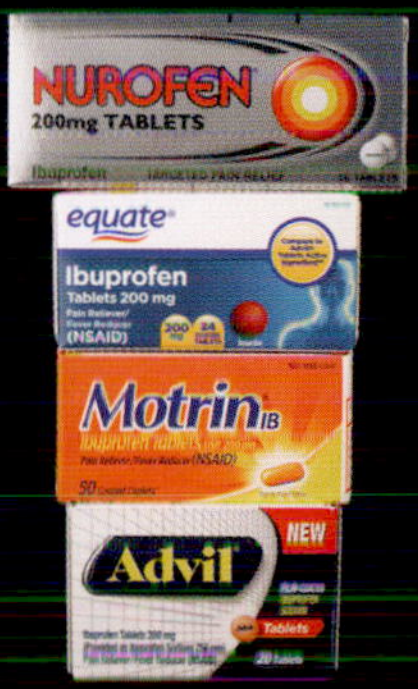

◀ Ibuprofen is a weak organic acid, like aspirin, and it shares the same six-member benzine ring. But in many situations, it's more effective than aspirin as a painkiller and inflammation reducer.

▼ Naproxen (Painkiller)

▶ Ibuprofen (Painkiller)

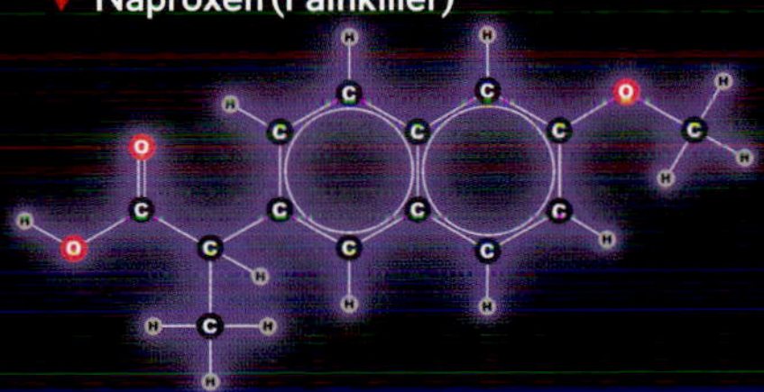

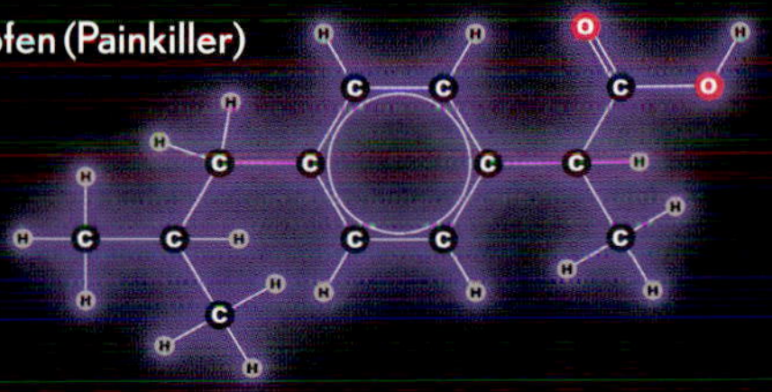

▲ Caffeine seems like an odd thing to give for headaches, but it appears to increase the effectiveness of other painkillers, at least for some people. The mechanism by which this works is not known.

Opium and Its Cousins

SURPRISINGLY, ONE OF the most powerful of all painkillers, one so effective it is in common use today in hospitals around the world, is also by far the oldest, predating the use of willow bark by thousands of years.

Opium, extracted from the flower known as the opium poppy, contains three very similar compounds: morphine, codeine, and thebaine. That two of these chemicals are modern medicines still in common use today shows just how marvelous a plant this poppy is. For many thousands of years, we had no antibiotics or vaccines, but at least we had a really, really good painkiller.

A wide array of close chemical relatives of opium are now in use, some of them thousands of times more potent than opium's main component, morphine. (For comparison, common aspirin is several hundred times *less* potent than morphine as a painkiller.) Each variation has its particular advantages: some act for long periods, staying in the body for days. Others are specifically useful because they clear out quickly.

Opium and its synthetic derivatives create chemical dependence (addiction). This, combined with the fact that they eliminate both physical and emotional pain, makes them dangerous trapdoors. Both legal and illegal forms of them are frequently abused, and doctors are very reluctant to prescribe them even to people in serious pain. Unfortunately, when the prescriptions stop, people who have become addicted to legal drugs may turn to street heroin—a dangerous and often contaminated synthetic variant of morphine.

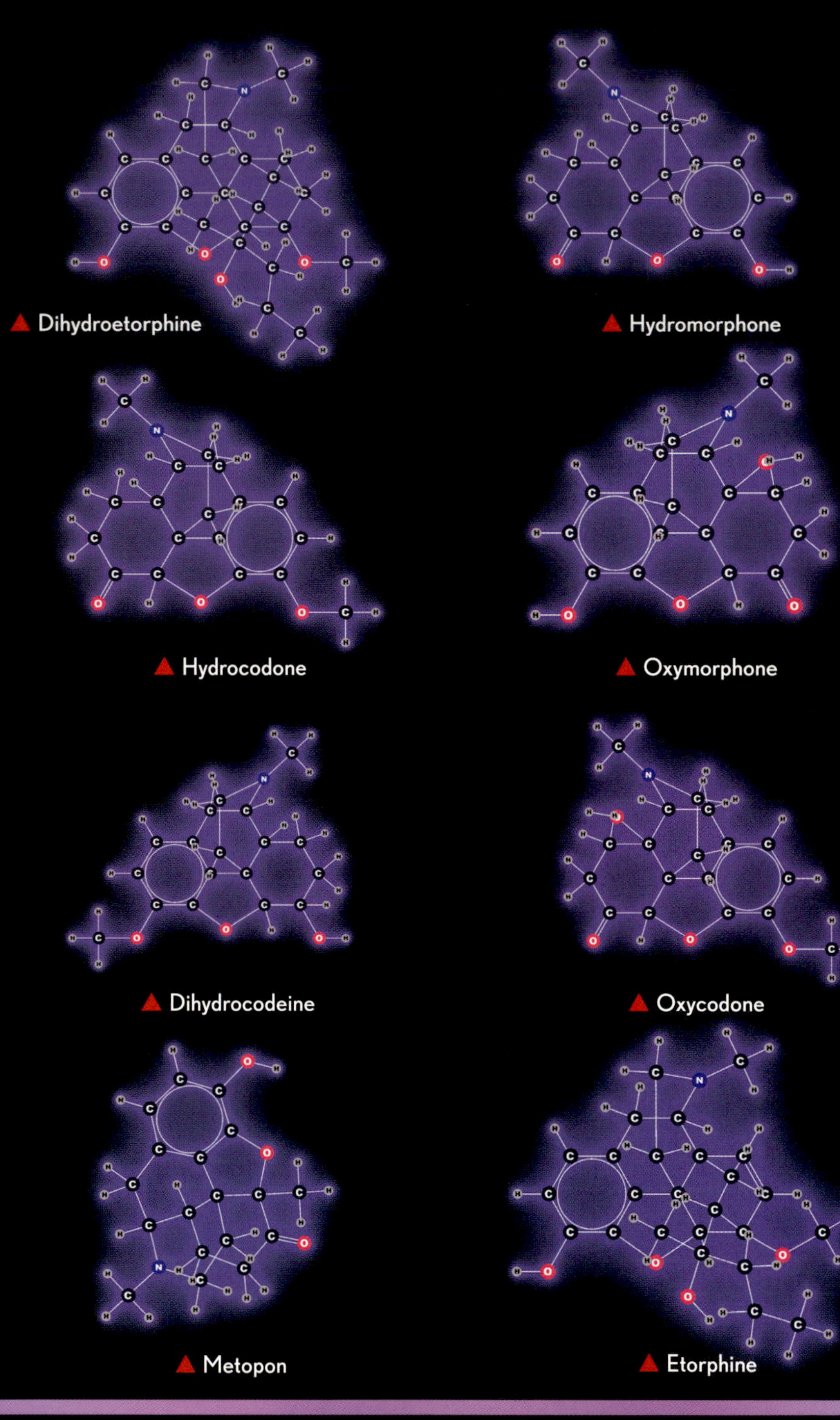

▲ Dihydroetorphine

▲ Hydromorphone

▲ Hydrocodone

▲ Oxymorphone

▲ Dihydrocodeine

▲ Oxycodone

▲ Metopon

▲ Etorphine

▶ All of the compounds pictured on this page, some natural and some synthetic, share the same four-ring structure as morphine, and all are powerful painkillers. These three, morphine, codeine, and thebaine, together make up the natural resin of the opium poppy.

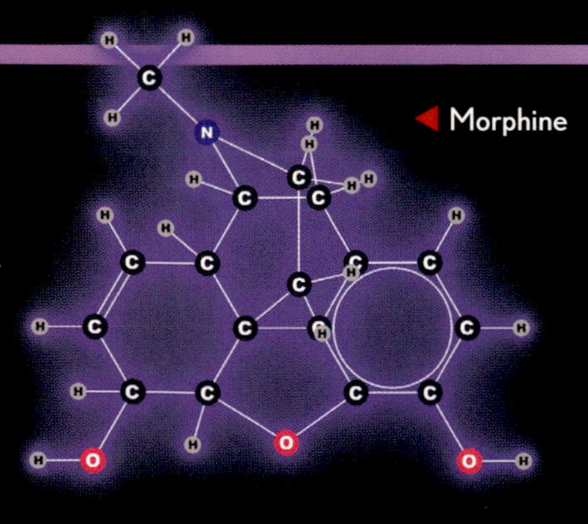

◀ Morphine

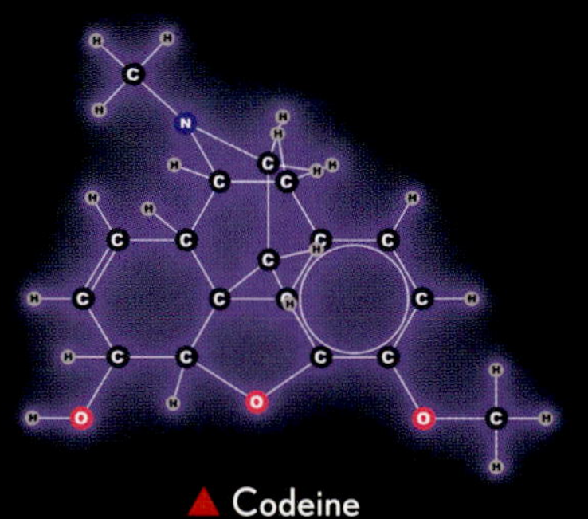

▲ Codeine

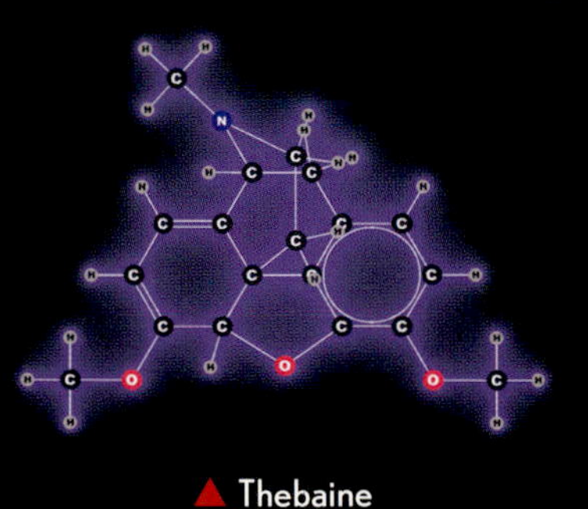

▲ Thebaine

◀ Resin from these poppies contains very high concentrations of morphine, codeine, and thebaine.

Opium and Its Cousins

▶ Opium was traded in China and throughout the Orient for thousands of years. Precision "violin scales" like this one were used to weigh out tiny amounts of the valuable substance.

▶ If you had this four-inch box entirely full of opium, that would be . . . a lot. It is more likely that a box like this, though informally referred to as an opium box in the antique trade, would have been used for tobacco in one form or another.

▶ Vicodin is one of several brands of painkillers that combine acetaminophen with the synthetic opiate hydrocodone, which makes it addictive, seductive, and available by prescription only. Pills like this are widely traded on the black market for purposes of abuse as well as to treat pain in people who can't get them legitimately. (The red speckle is intentionally added by the manufacturer to clearly and unmistakably mark these pills as containing hydrocodone.)

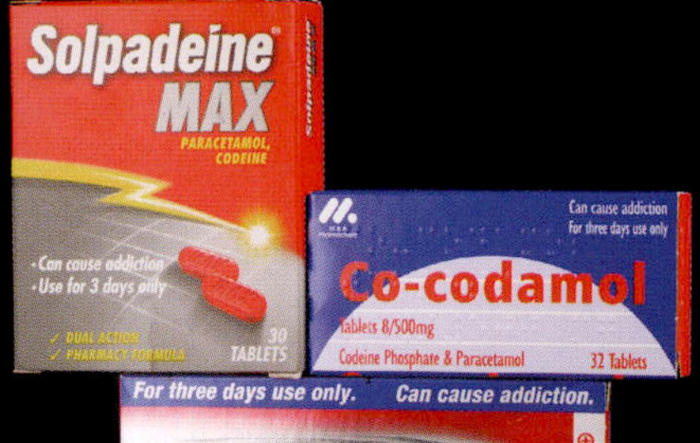

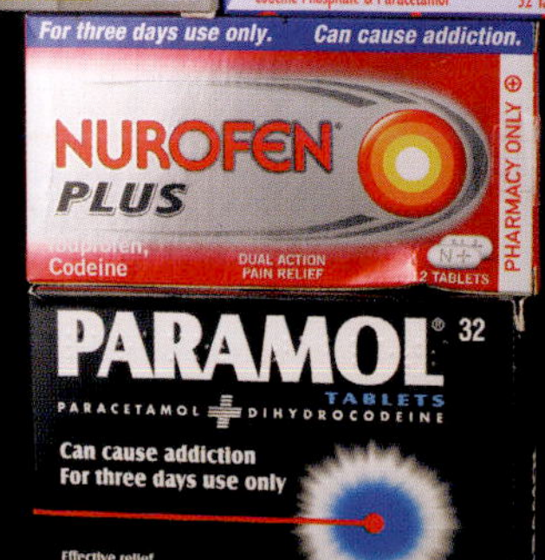

◀ In the United States, codeine is available only with a doctor's prescription, but other countries have different rules. In England, for example, formulations including codeine are available with the consent of a pharmacist. So as long as you don't want an unreasonably large amount, you can just walk into a drugstore and buy it.

▶ Morphine was, and is, a critical comfort on the battlefield. This morphine self-injection syringe is from World War II. The instructions show how to use a small pin inserted into the injection needle to puncture an inner foil seal inside the tube.

▲ Some opium boxes from the turn of the century were made of a single coin and might have actually fooled someone. This one is more of a novelty and is said to be from 1906.

In chemical structure, these compounds may seem almost indistinguishable from morphine and its cousins. But these are all actually antiopioids. They counteract the effects of opium and its derivatives by blocking the same chemical pathways in the body that opium affects. This means they can be used as antidotes for morphine overdose and to help overcome opium or heroin addiction.

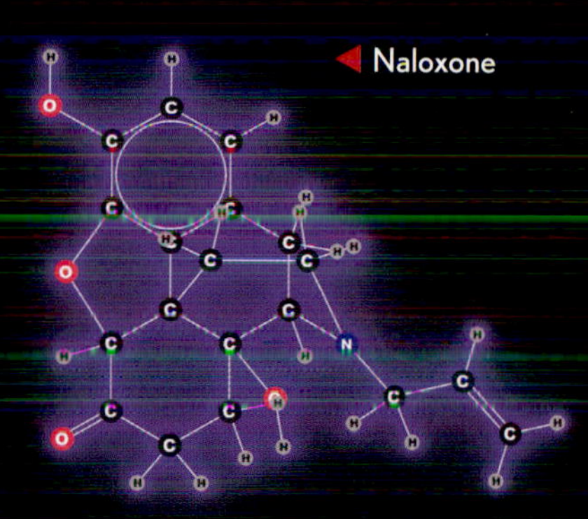

Naloxone

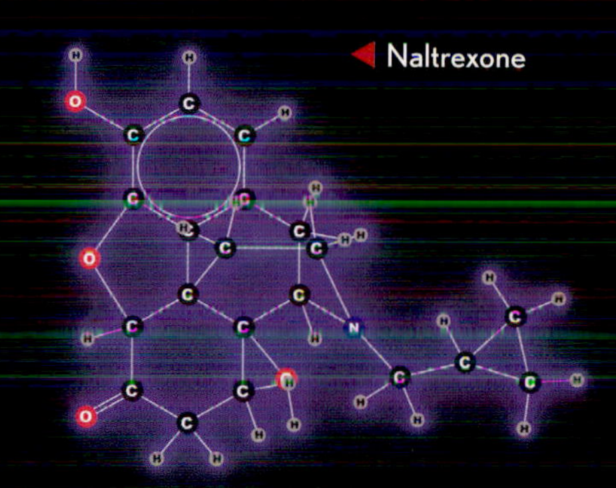

Naltrexone

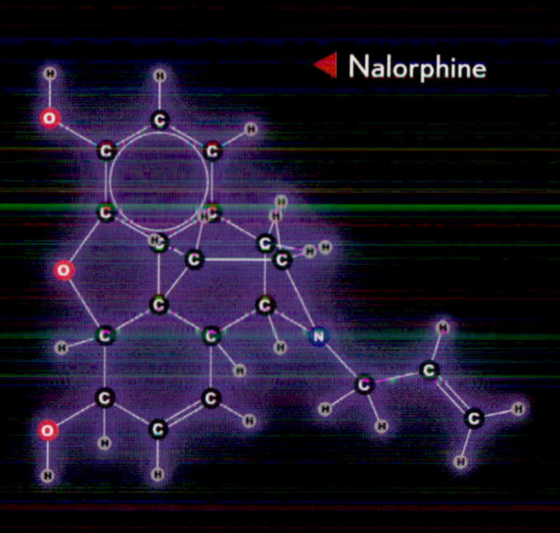

Nalorphine

Ingesting a tiny morphine pill can provide powerful pain relief. It can also easily lead to addiction.

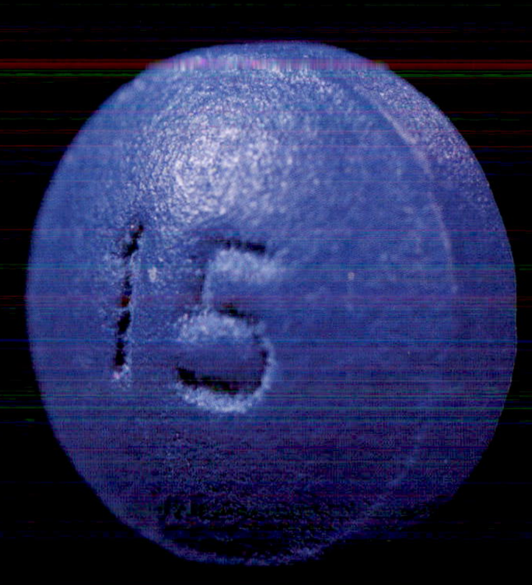

Heroin is the black sheep of the opium family. It exists primarily as an illegal drug, having very few medical advantages over other opioids. (But when it is used in medicine, it's called diamorphine.)

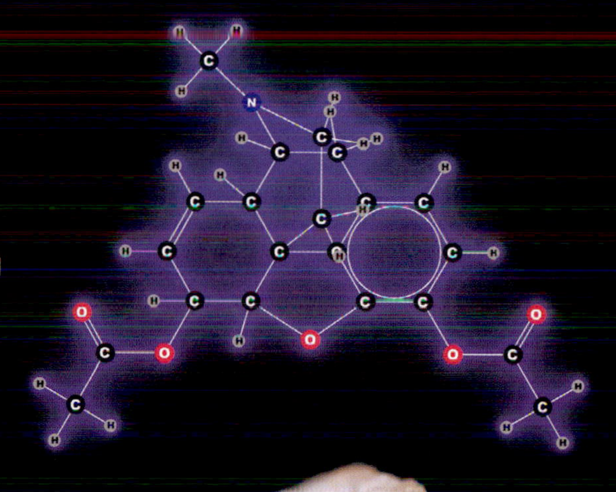

Naloxone is given by injection, and despite being very opium-like, it's not highly regulated because it has no potential for abuse. It's used to treat opiate overdoses because it reverses many of the effects of these drugs.

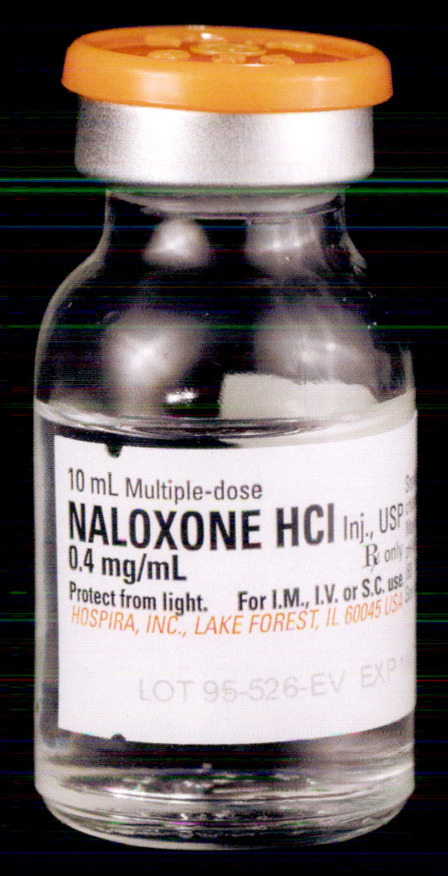

The pure chemical in heroin, diacetylmorphine, is supposed to be white! This example of illegal heroin shows the streaks of impurities running through it, which may include unpredictable amounts of psychoactive substances even more powerful than heroin. This kind of street drug is very, very dangerous.

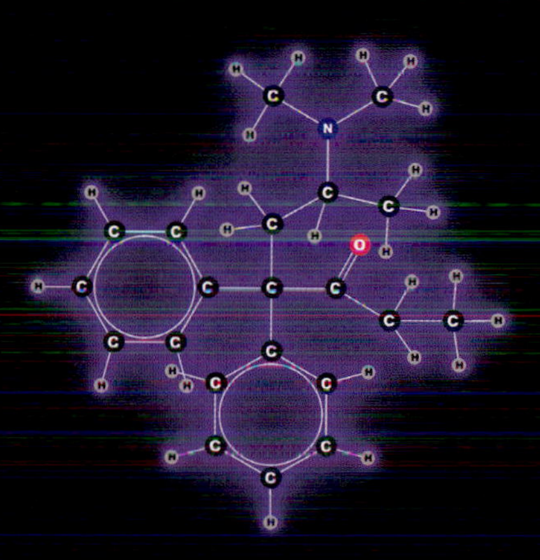

Methadone is not itself chemically similar to any of the opium class of painkillers. It has a very different set of chemical bonds that happen to take a similar overall shape, allowing methadone to fit into and block the same chemical receptors in the nervous system as do morphine, heroin, and other true opioids.

Methadone is used to treat the symptoms of heroin withdrawal. It has a long-lasting effect in the body. By clinging to opiate receptors in the body, a sufficient methadone dose will also block any expected effects from heroine.

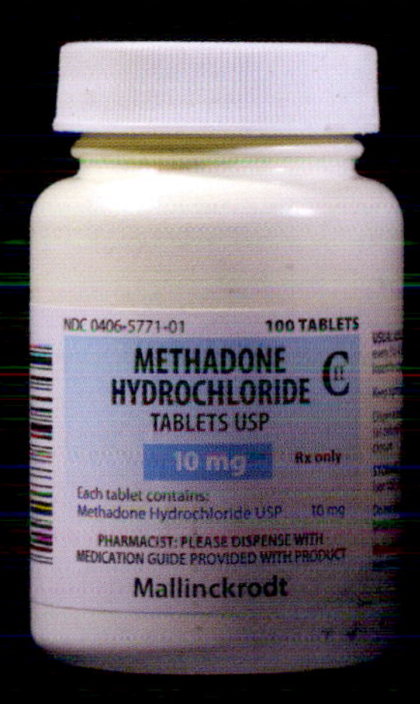

The Power of Peppers

AS POWERFUL AS the drugs from opium poppies are, for the most potent of all painkillers we have to look to entirely different plants: the ones that make peppers and peppercorns.

The sharp taste of black peppercorns comes from the molecule piperine, which contains an unusual and difficult-to-make component: a six-membered ring made of five carbon atoms and one nitrogen. This unit, called piperidine in isolation, forms the basis for some of the most powerful poisons, painkillers, and irritants.

Painkillers and poisons usually go hand in hand. Any painkiller that acts on the central nervous system can be fatal if taken in excess, because to do its job it has to shut down parts of this vital communication system. Along with dulling pain, it will slow heart rate and breathing, potentially to the point of no return.

Painkillers are also close cousins of compounds that cause intense pain or itching. Because both kinds of substances affect the nerves, sometimes all it takes is a tiny change in a molecule for it to go from inhibiting to stimulating nerve action. Sometimes the exact same molecule can cause or relieve pain, depending on where it is present and in what amount.

◀ Piperidine is a simple six-membered ring, with five of its six atoms being carbon and the other being nitrogen. Rings are, relatively speaking, hard to create in synthetic chemistry, so in order to make one of the many compounds that contain this particular ring structure, it's often easiest to start with something that's already got it—hence the popularity of piperidine as a "precursor" or starting point for synthesis.

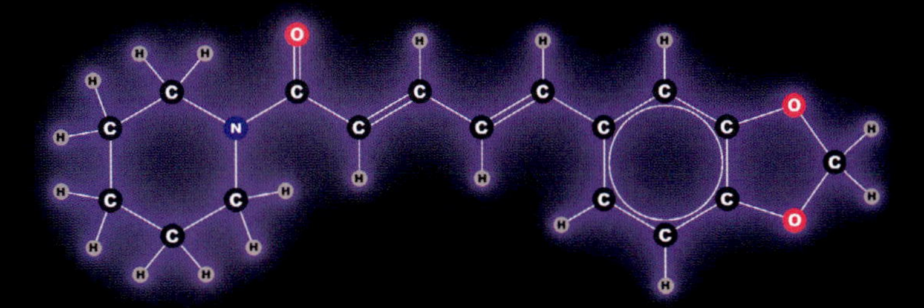

▲ The compound piperine is responsible for the sharp taste of black peppercorns. Despite being related to many painkillers, piperine itself has no such effects: it just has a really, really strong taste.

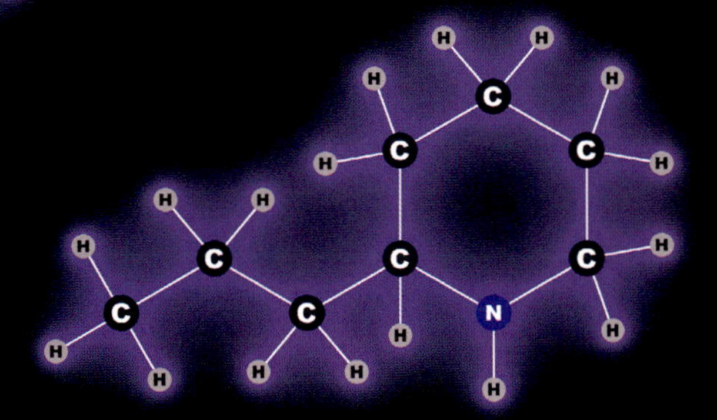

▶ The toxin in poison hemlock is coniine, a very simple variation on piperidine in which a chain of three carbon atoms is attached to its ring. This poison was famously used to carry out the death sentence against Socrates 2,400 years ago. He was convicted of failing to properly acknowledge the official gods, and as is so often the case, the science of the herbalist was called on to carry out the dirty work of the priests and politicians, whose native power is limited to empty words.

▼ Black peppercorns, ground and purified, are the source of pure piperine. As in so many, many cases, when a pure chemical is extracted from its natural source, the end result is a white powder. *So many white powders!* There's actually a good reason for this: color is an unusual property for chemicals and comes about only when a molecule has a particular kind of bonding structure (see Chapter 12).

▼ 1-(1-phenylcyclohexyl)piperidine, often shortened to phencyclidine or PCP. Piperidine's use as a precursor in making PCP accounts for the fact that it is a controlled substance.

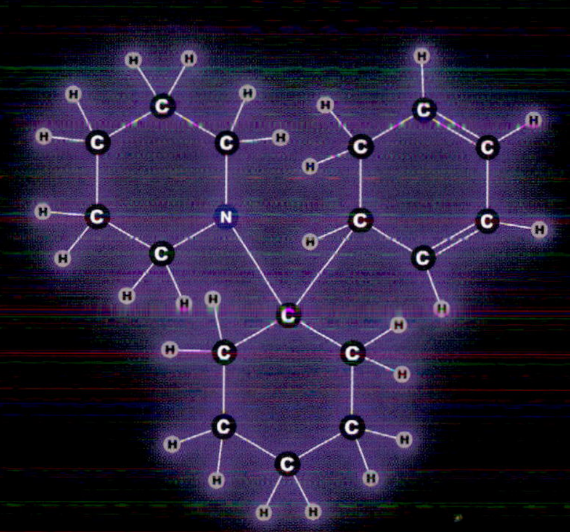

▲ Phencyclidine

▶ Fire ants are so unpleasant (due to their solenopsin production as well as their tendency to eat buildings, electrical wires, and other useful things) that transporting them is illegal, lest they be introduced into areas they don't already plague. So we have photographed this nice metal model instead of the real thing.

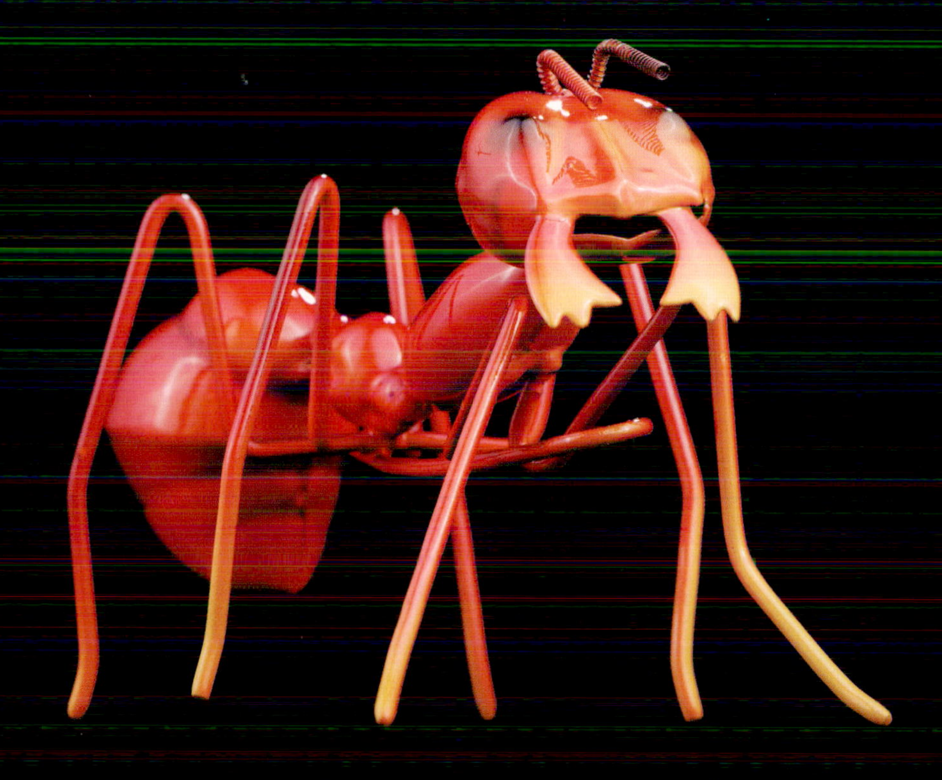

▼ Solenopsin, another piperidine derivative, is what makes the bite of a fire ant extremely painful.

▲ Solenopsin

▶ Hemlock leaves, famous for containing the poison coniine.

▶ Despite the name, the active irritant in pepper spray used for self-defense is not a piperidine variation. It is capsaicin, a quite different molecule that comes from chili peppers rather than the black peppercorns that are the source of piperidine.

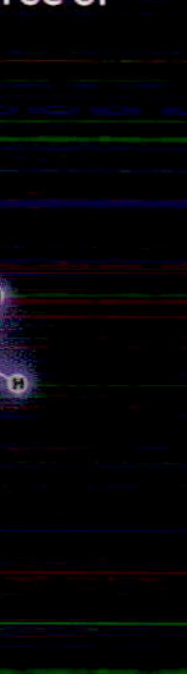

▲ Capsaicin

▶ One of the more surprising twists in the pepper world is that capsaicin, the very same compound used in pepper spray for delivering incapacitating pain to people you wish to disable, also works as a painkiller when rubbed on the skin. These sorts of ointments feel "hot" when first applied, as the capsaicin stimulates nerves. But the effect goes away and is replaced by pain relief as the nerves are fatigued from their intense capsaicin stimulation.

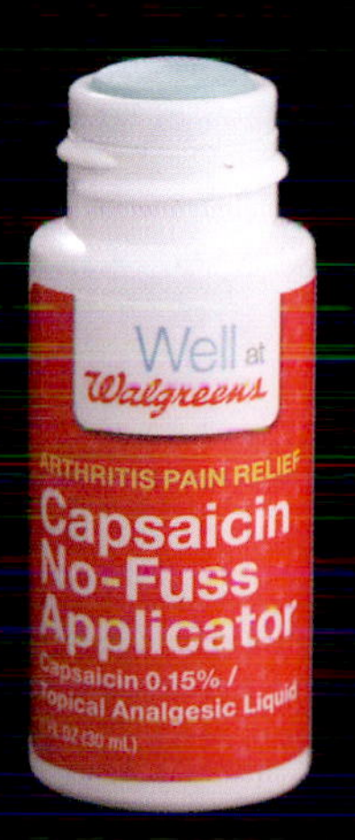

Well at Walgreens
ARTHRITIS PAIN RELIEF
Capsaicin
No-Fuss
Applicator
Capsaicin 0.15% /
Topical Analgesic Liquid

The Power of Peppers

▶ There is a whole zoo of piperidine-related painkillers, each one in use either for human or animal medicine, and each one with its own unique combination of good and bad points. The most powerful ones are typically used only on large animals.

▶ Pethidine

▶ Sufentanil

◀ Anileridine

▲ Alphaprodine

▶ Remifentanil

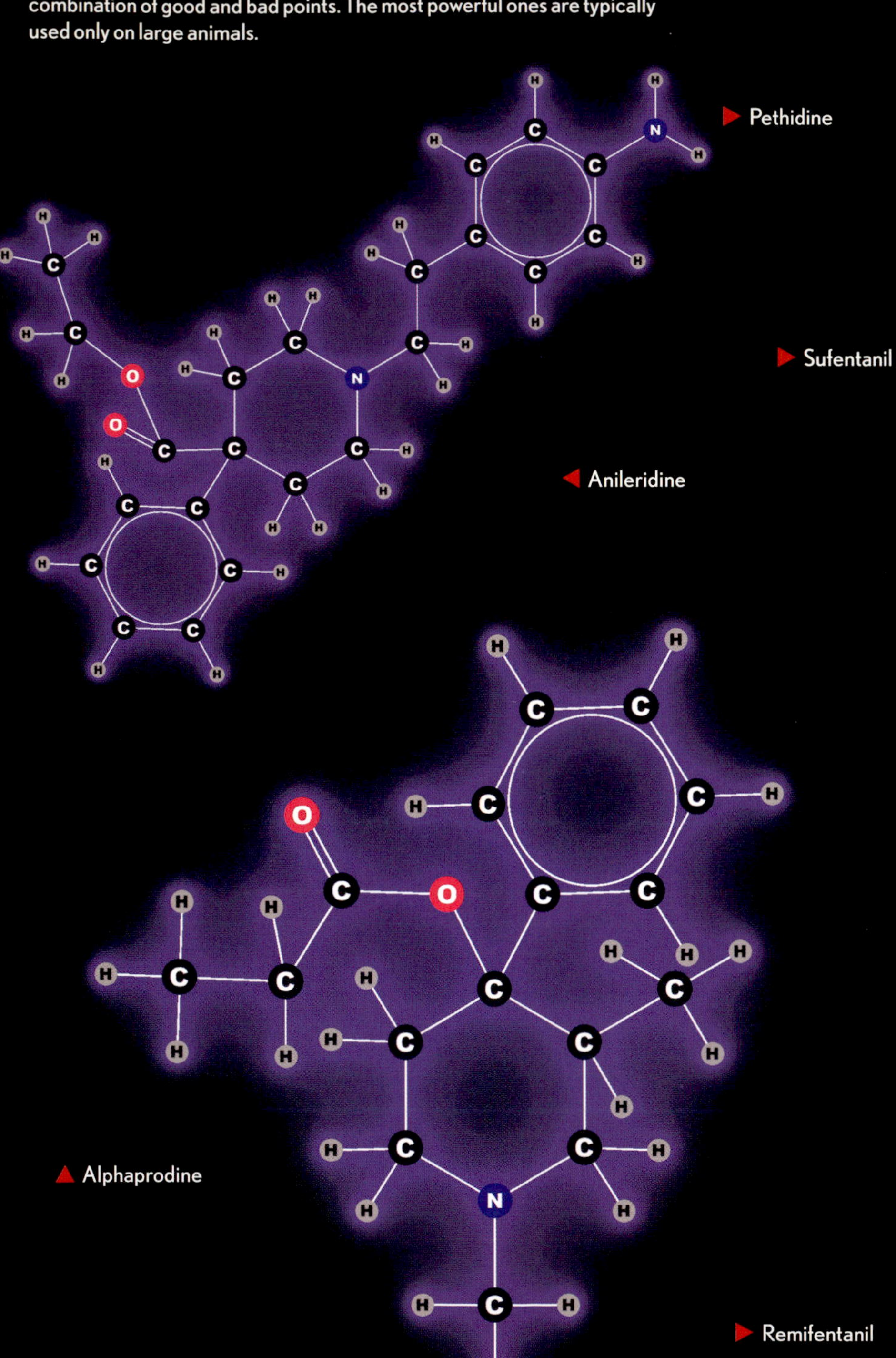

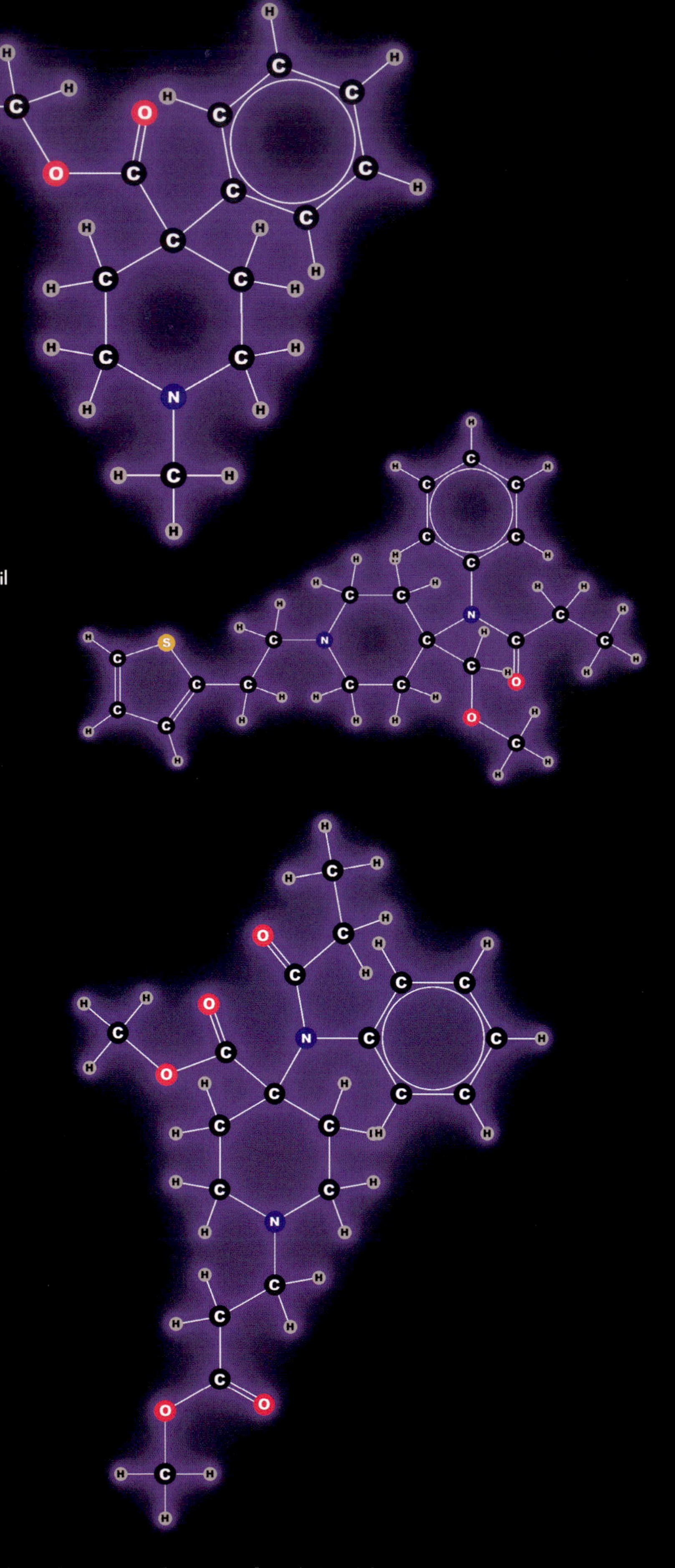

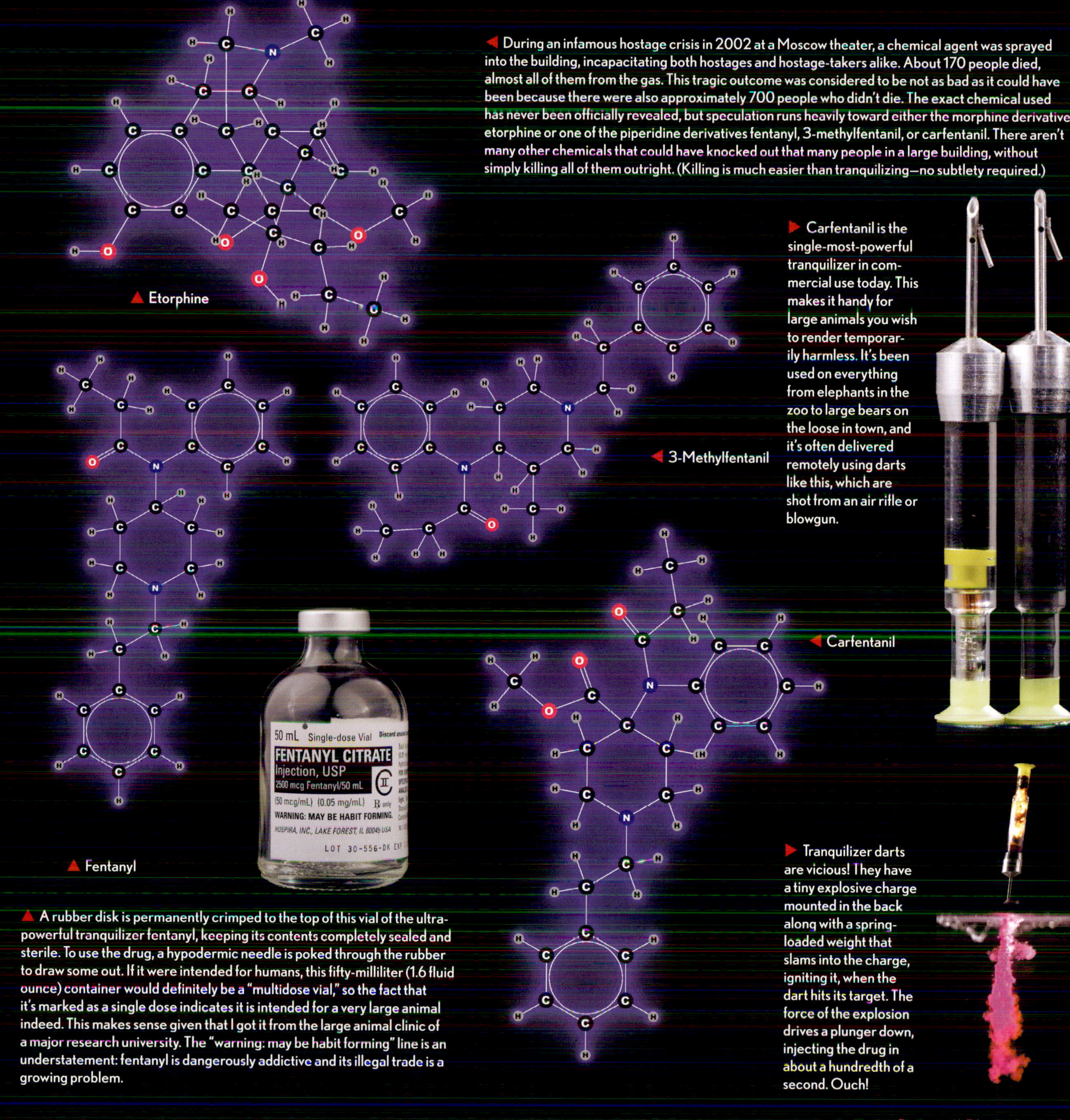

During an infamous hostage crisis in 2002 at a Moscow theater, a chemical agent was sprayed into the building, incapacitating both hostages and hostage-takers alike. About 170 people died, almost all of them from the gas. This tragic outcome was considered to be not as bad as it could have been because there were also approximately 700 people who didn't die. The exact chemical used has never been officially revealed, but speculation runs heavily toward either the morphine derivative etorphine or one of the piperidine derivatives fentanyl, 3-methylfentanil, or carfentanil. There aren't many other chemicals that could have knocked out that many people in a large building, without simply killing all of them outright. (Killing is much easier than tranquilizing—no subtlety required.)

Carfentanil is the single-most-powerful tranquilizer in commercial use today. This makes it handy for large animals you wish to render temporarily harmless. It's been used on everything from elephants in the zoo to large bears on the loose in town, and it's often delivered remotely using darts like this, which are shot from an air rifle or blowgun.

A rubber disk is permanently crimped to the top of this vial of the ultra-powerful tranquilizer fentanyl, keeping its contents completely sealed and sterile. To use the drug, a hypodermic needle is poked through the rubber to draw some out. If it were intended for humans, this fifty-milliliter (1.6 fluid ounce) container would definitely be a "multidose vial," so the fact that it's marked as a single dose indicates it is intended for a very large animal indeed. This makes sense given that I got it from the large animal clinic of a major research university. The "warning: may be habit forming" line is an understatement: fentanyl is dangerously addictive and its illegal trade is a growing problem.

Tranquilizer darts are vicious! They have a tiny explosive charge mounted in the back along with a spring-loaded weight that slams into the charge, igniting it, when the dart hits its target. The force of the explosion drives a plunger down, injecting the drug in about a hundredth of a second. Ouch!

The Use and Abuse of Cocaine

COCAINE HAS DESTROYED countless lives and blighted entire neighborhoods, but throughout much of history, it was considered a useful drug. The Inca people chewed coca leaf, the main source of cocaine, to give them energy. Sigmund Freud, the famous psychoanalyst, used cocaine and recommended it to his patients. Some of the early popularity of Coca-Cola as an energy drink was no doubt due to the fact that it contained cocaine, as its name indicated. (Cocaine was removed from Coca-Cola in 1903.)

Even today, cocaine is commonly and widely used as a topical anesthetic (which means it's applied to the surface of the skin, not taken internally). Dentists use it to numb your gums before they stick that long, long needle in (fat lot of good it does). Amusingly, one of the side effects listed for dental cocaine is "unusual feelings of wellbeing," which is not a problem I've ever had after visiting the dentist!

Cocaine is like any other chemical: it does what it does without intent or purpose, and whether good or evil comes from that is up to the people around it.

▶ Teabags containing cocaine in the form of ground coca leaf are readily available in South America.

◀ Dried coca leaf has been chewed in South America for millennia and is used to make tea for visiting tourists. The cocaine content of these leaves makes them illegal in most parts of the world.

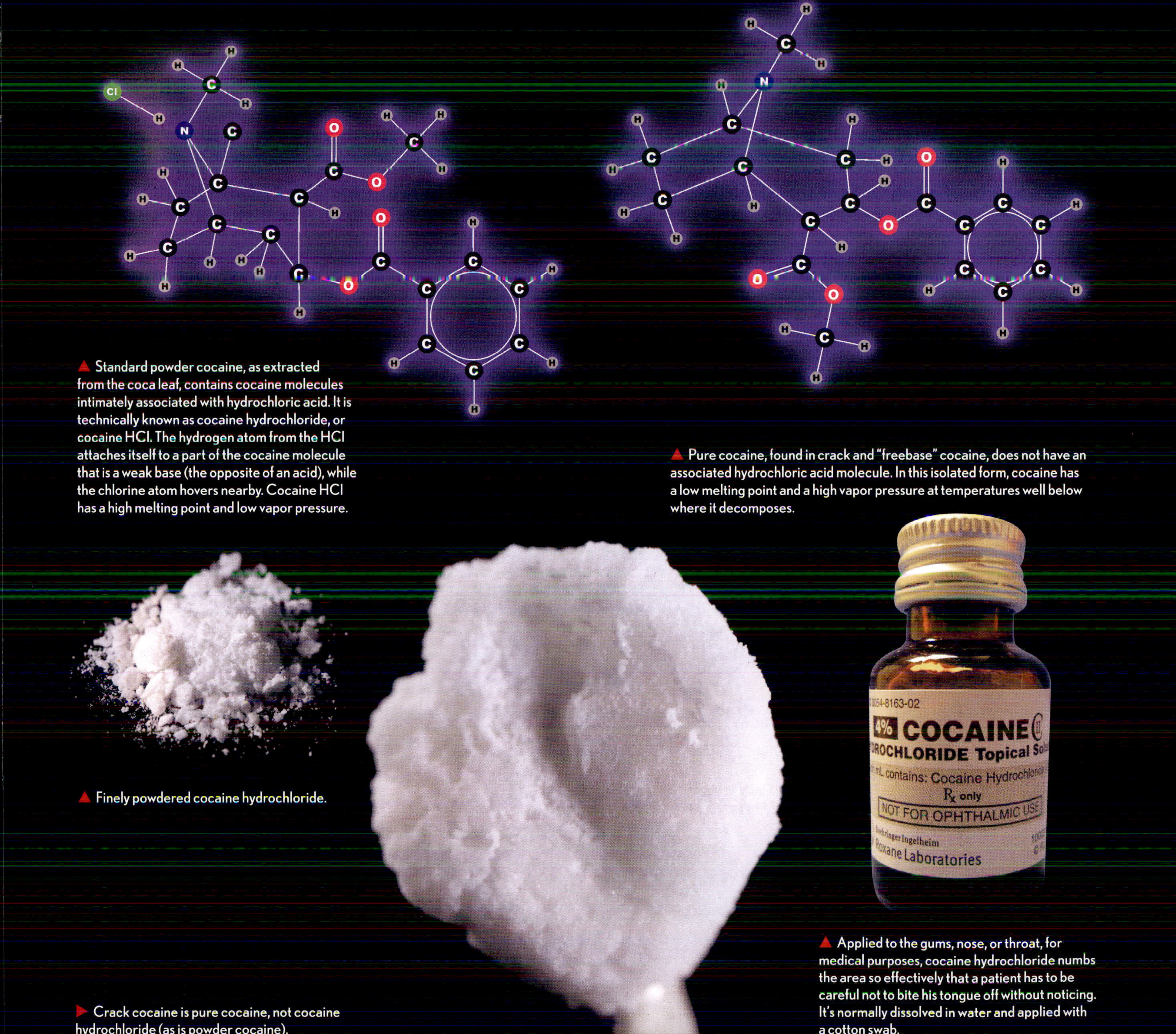

▲ Standard powder cocaine, as extracted from the coca leaf, contains cocaine molecules intimately associated with hydrochloric acid. It is technically known as cocaine hydrochloride, or cocaine HCl. The hydrogen atom from the HCl attaches itself to a part of the cocaine molecule that is a weak base (the opposite of an acid), while the chlorine atom hovers nearby. Cocaine HCl has a high melting point and low vapor pressure.

▲ Pure cocaine, found in crack and "freebase" cocaine, does not have an associated hydrochloric acid molecule. In this isolated form, cocaine has a low melting point and a high vapor pressure at temperatures well below where it decomposes.

▲ Finely powdered cocaine hydrochloride.

▲ Applied to the gums, nose, or throat, for medical purposes, cocaine hydrochloride numbs the area so effectively that a patient has to be careful not to bite his tongue off without noticing. It's normally dissolved in water and applied with a cotton swab.

▶ Crack cocaine is pure cocaine, not cocaine hydrochloride (as is powder cocaine).

The Use and Abuse of Cocaine

Three extremely common anesthetics, lidocaine, novocaine, and benzocaine, sound a lot like they must be cocaine derivatives. In fact, their chemical structures are completely different, with none of cocaine's peculiar multi-ring structure. Like cocaine, they act as topical anesthetics and are used, also like cocaine, in dentistry and minor surgery. Unlike cocaine, they have very little potential for abuse and are thus available over the counter without restriction.

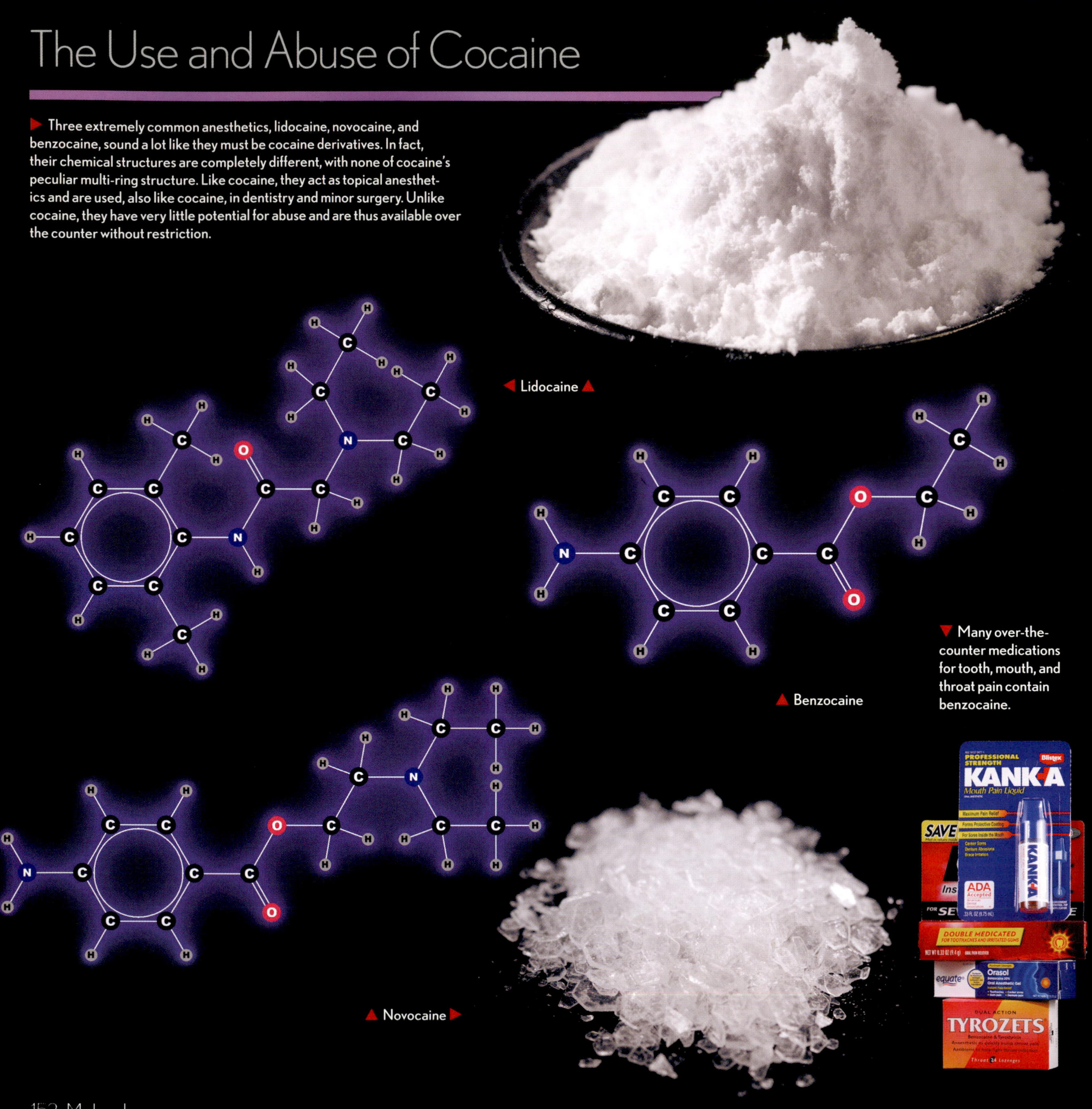

Lidocaine

Benzocaine

Novocaine

Many over-the-counter medications for tooth, mouth, and throat pain contain benzocaine.

Oddball Painkillers

PAIN IS SUCH A STRANGE, subjective thing that it's perhaps not surprising that there are some oddball compounds that work as painkillers. Another reason is that there are many points in the body where pain can be attacked, so quite a wide range of compounds are potential candidates for the job.

Most painkillers (indeed most drugs) are fairly simple, made up of reasonably small molecules, containing perhaps a few dozen to a hundred atoms at most, and built out of robust substructures like benzene rings. In large part, this isn't because that kind of molecule is better in its biological function but rather because that is the kind of molecule that can survive in the stomach long enough to make it into the bloodstream. That is, of course, a prerequisite for any drug that is meant to be swallowed.

Certain proteins, for example, are potentially excellent painkillers. However, "protein" is another name for "food" as far as the stomach is concerned, so they are promptly digested. As a result, drugs derived from proteins can usually be given only by injection or inhalation. Despite this, some of the most promising new painkillers are proteins from unlikely sources.

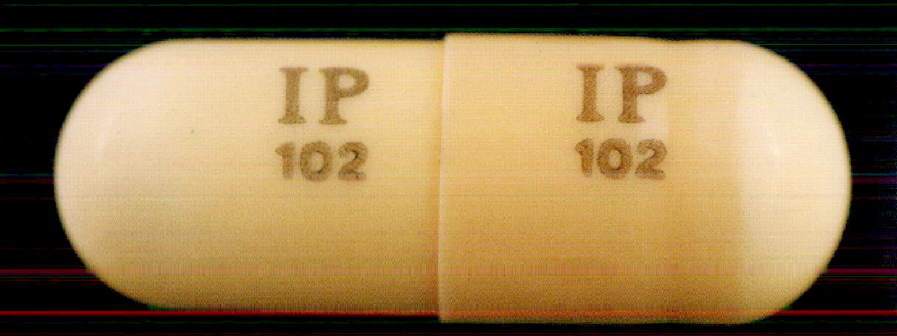

▲ Gabapentin (brand name Neurontin) is not a member of any of the common classes of painkillers we've discussed so far. It's designed to mimic the action of one of the neurotransmitters in the brain, *gamma*-aminobutyric acid, to which it bears a certain structural resemblance. In the normal world, it's given for nerve pain, but in prisons it's used in many situations where a narcotic such as codeine might be more appropriate. That's because it has zero potential for abuse: rather than make people feel good, it tends to make them feel miserable, just with less actual hurting.

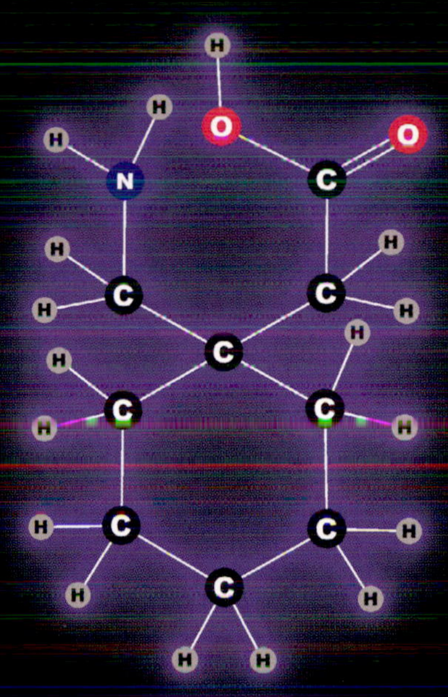

▲ Gabapentin-90-F

▼ Gamma amino Butyric acid

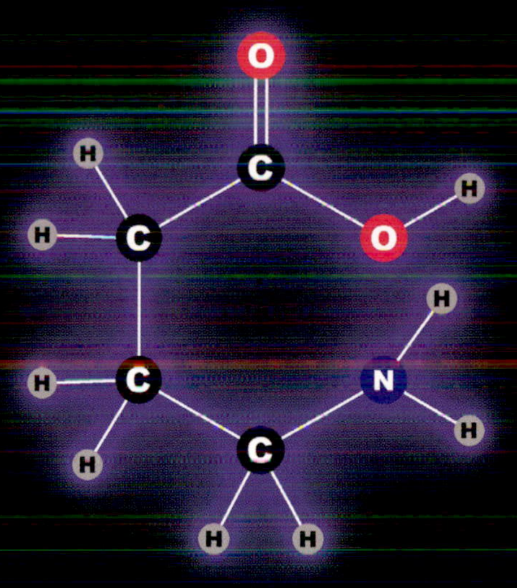

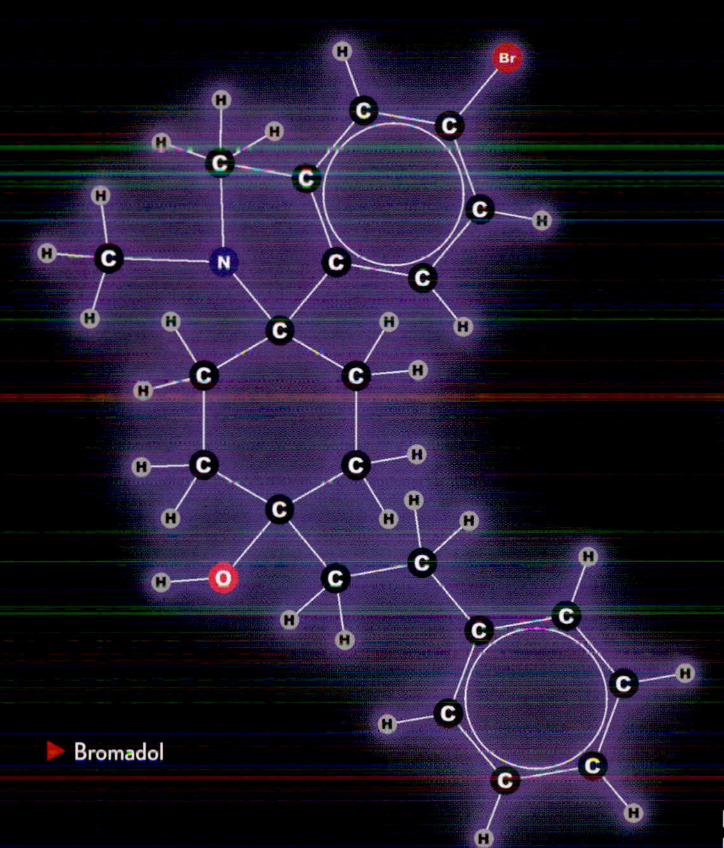

▶ Bromadol

◀ Bromadol doesn't look like any other painkiller, and it's not clear whether it's useful as a human drug. This hasn't stopped criminal gangs from trying to sell it as a narcotic.

▶ Ziconotide, sold under the brand name Prialt is a small synthetic copy of one of the cone snail toxins. It is given by injection directly into the spinal fluid, and used only to treat the most severe and persistent pain.

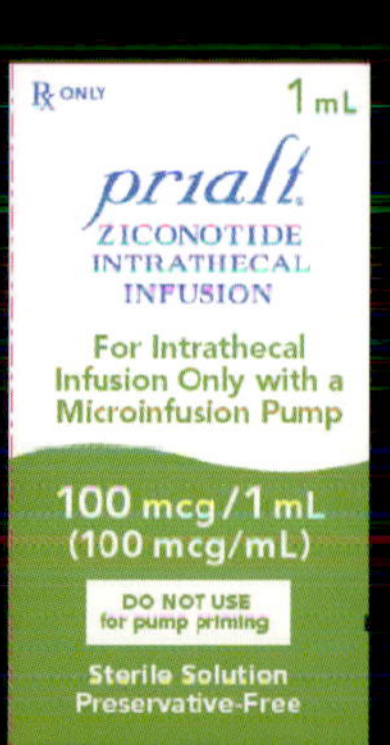

Oddball Painkillers

▲ The shell of the sea snail *Conus gloriamaris* (Glory of the Sea) was once the most valuable in the world, thought to be the rarest and most beautiful of all sea shells. Then scuba diving was invented, and it was discovered that they are dirt common; they just like to live deeper down than people could easily reach before. They are still beautiful and interesting because they, along with many other members of the cone snail family, produce an astonishing array of incredibly toxic proteins. Many species can kill a human with one sting, which makes them tremendously interesting as potential drug sources.

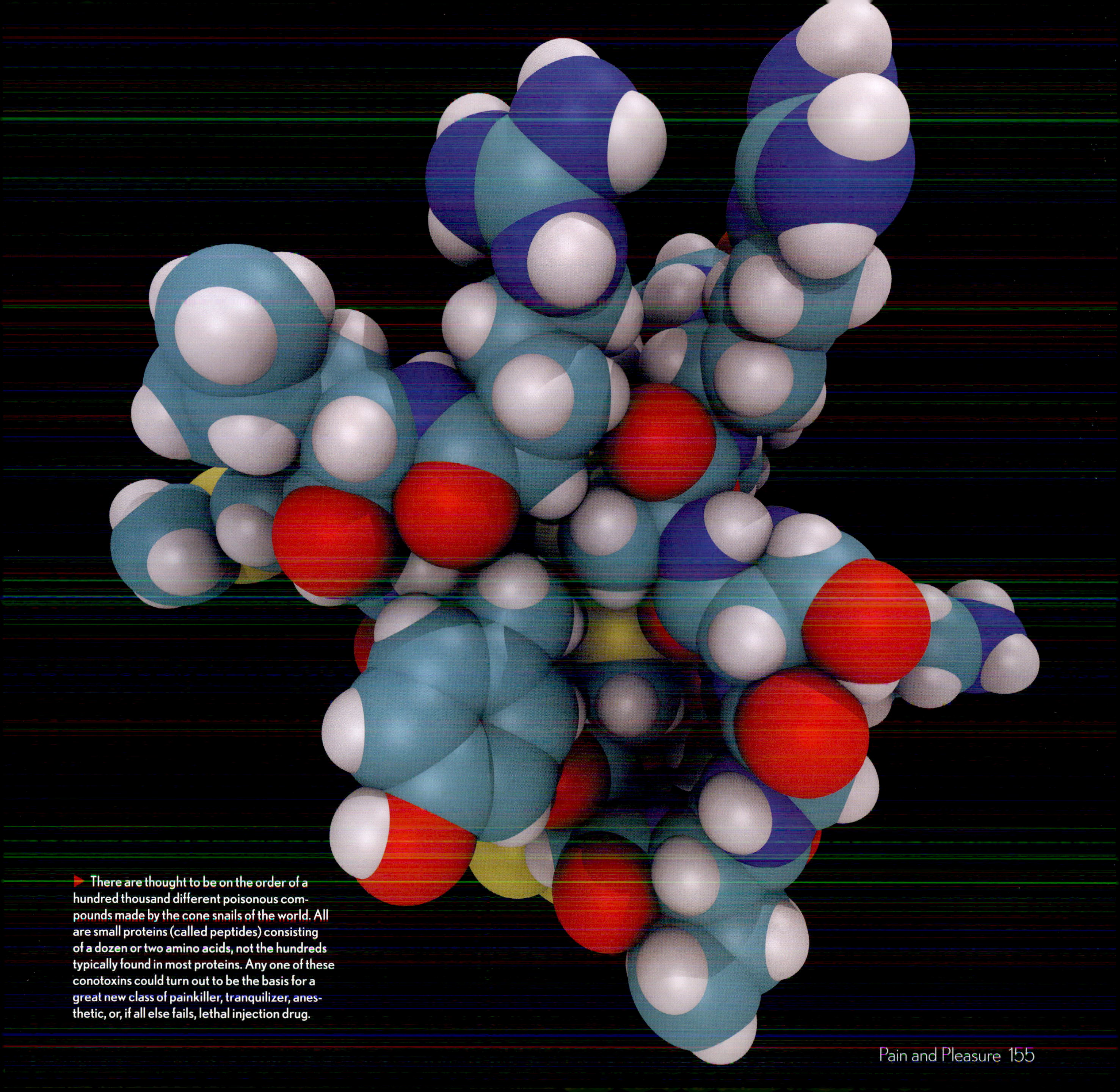

▶ There are thought to be on the order of a hundred thousand different poisonous compounds made by the cone snails of the world. All are small proteins (called peptides) consisting of a dozen or two amino acids, not the hundreds typically found in most proteins. Any one of these conotoxins could turn out to be the basis for a great new class of painkiller, tranquilizer, anesthetic, or, if all else fails, lethal injection drug.

Chapter 9

Sweet and Double Sweet

MANY MOLECULES make things taste sweet. Unfortunately, we tend to abuse the ones we like best. Common table sugar (sucrose) and its close relatives glucose and fructose are all toxic in the amounts we eat them. They contribute to diabetes, heart disease, tooth decay, macular degeneration, peripheral neuropathy, kidney disease, high blood pressure, and stroke. Were they artificial, all of them would have been banned long ago.

Healthier alternatives are available in the form of high-intensity natural and synthetic sweeteners, many of which are so fantastically sweet that only minute quantities are needed. The two biggest problems with them are that many people don't like their specific taste, and the fear of brain cancer (though even the most suspect of artificial sweeteners doesn't even begin to touch any natural sugar in terms of the harm it does to human health). Lower-calorie compounds known as sugar alcohols offer a taste closer to that of table sugar, with only occasional gastric distress.

That hundreds of millions of tons (literally) of these substances are willingly, even enthusiastically, consumed every year is proof of just how *much* we crave the sweet, sweet taste of these marvelous, if slightly evil, molecules.

◀ **Honey, like high-fructose corn syrup, is a roughly fifty–fifty mixture of fructose and glucose.**

The three natural, simple sugars, called monosaccharides, are fructose, glucose, and the less familiar galactose. As you can see from their structures, they are very similar. In fact, glucose and galactose are so similar you can't even see a difference when they are drawn in two dimensions. Their bonding structure is identical, and they differ only in the direction some of their bonds are pointed (their stereochemistry, as it is called). However, galactose is only about half as sweet-tasting as glucose. Taste is a very, very sensitive detector of the shape and chemical features of molecules.

Pears have a relatively high ratio of fructose to glucose: more than three to one compared to the roughly equal ratio found in most fruits. Fructose tastes about 1.7 times sweeter than glucose (which means it takes that much less fructose dissolved in water to reach a level that 50 percent of people can just detect).

Glucose does not occur as the sole component of any common natural source; it's usually combined with fructose. It was first isolated in pure form from raisins, which, like most fruits, contain a mixture of glucose and fructose (about half and half in the case of raisins). Glucose is sometimes known as grape sugar. (Raisins, of course, are dried grapes.)

Galactose is probably the least well known of the simple sugars, so it's fitting that the natural food highest in galactose is also not the first thing you'd think of as an example of a sugary food; it's celery.

Double Sugar

COMMON TABLE SUGAR is not one of the simple sugars we've just looked at. It is instead made of one molecule of glucose and one molecule of fructose (two monosaccharides) bonded to each other and is called a disaccharide. Lactose, the sugar found in milk, is a combination of glucose and galactose. Maltose, the sugar found in malted grains, is a combination of two glucose units. Many, many other variations occur in nature and in factories, based on which simple sugars are connected, how many are connected, and where on the molecules the connections are made. Each combination has a unique chemical structure, taste, and health profile.

A lot of the work done by the sweetener industry boils down to transforming one of these sugars into another. For example, to make high fructose corn syrup, you split maltose into its two individual glucose units and then convert some of the glucose into fructose, resulting in a mixture of glucose and fructose. (Almost exactly the same mix can be found in many fruits and in honey, but it's cheaper to make it out of corn.)

When you look at the ingredient list for a food product, remember that it doesn't much matter where the sugar comes from. Each ingredient, be it cane sugar, agave syrup, honey, high-fructose corn syrup, maltodextrin, or even some nonsugars such as starch, can be decoded into a particular mixture of specific simple sugars. The monosaccharide ratio, and the total quantity of sugar, determines the health profile of the food. The particular sweetener used may contribute delightful flavors and colors, but it doesn't make any difference to nutrition or health other than through its total quantity and its ratio of simple sugars.

▼ Common table sugar, sucrose, comes in a delightful array of forms and colors. Each is nutritionally identical, though differences in particle size and various impurities may dramatically affect how it tastes and feels in the mouth. All sugar starts out brown: the white table sugar we are used to is brown sugar from which molasses has been removed.

▲ We like the taste of sugar because it provides a lot of energy, and from an evolutionary point of view, it's good for creatures to consume foods that give us energy. It's only in our temporary state of overabundance that this is a problem.

Sucrose

▶ This kind of palm sugar ball is sold as a source of sugar, not to eat straight.

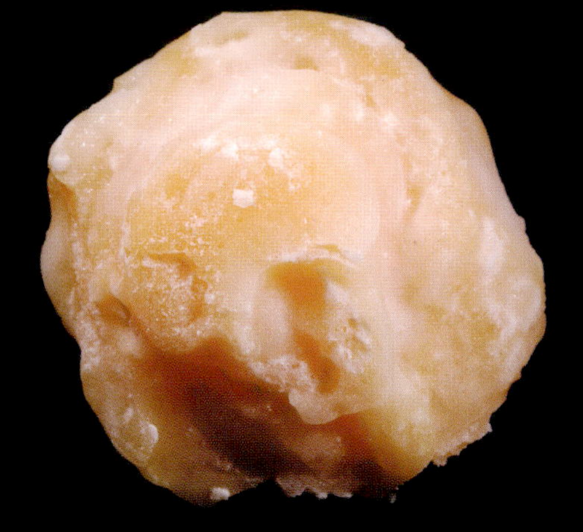

▶ Some candy doesn't even try to pretend it isn't a sugar delivery mechanism. Maple sugar is mostly sucrose with a few percent of glucose and fructose. This candy contains essentially nothing else.

▼ Jaggery is highly unrefined sugar from India. The material may be derived from several sources, including sugar cane, palms, or dates. It includes not only sugar but also things that are normally separated out of purified sugars, including proteins and plant fibers.

▼ The majority of table sugar comes from sugar cane, which is a type of grass (very, very big grass). You can buy fresh-cut sections of sugar cane to chew on as a snack.

▶ Good, old-fashioned sucrose table sugar, packaged for convenient delivery straight to the bloodstream.

▶ The second largest source of table sugar after sugar cane is sugar beets. The type of beet used for sugar production is much bigger than the kind you find in grocery stores, and it is white on the outside.

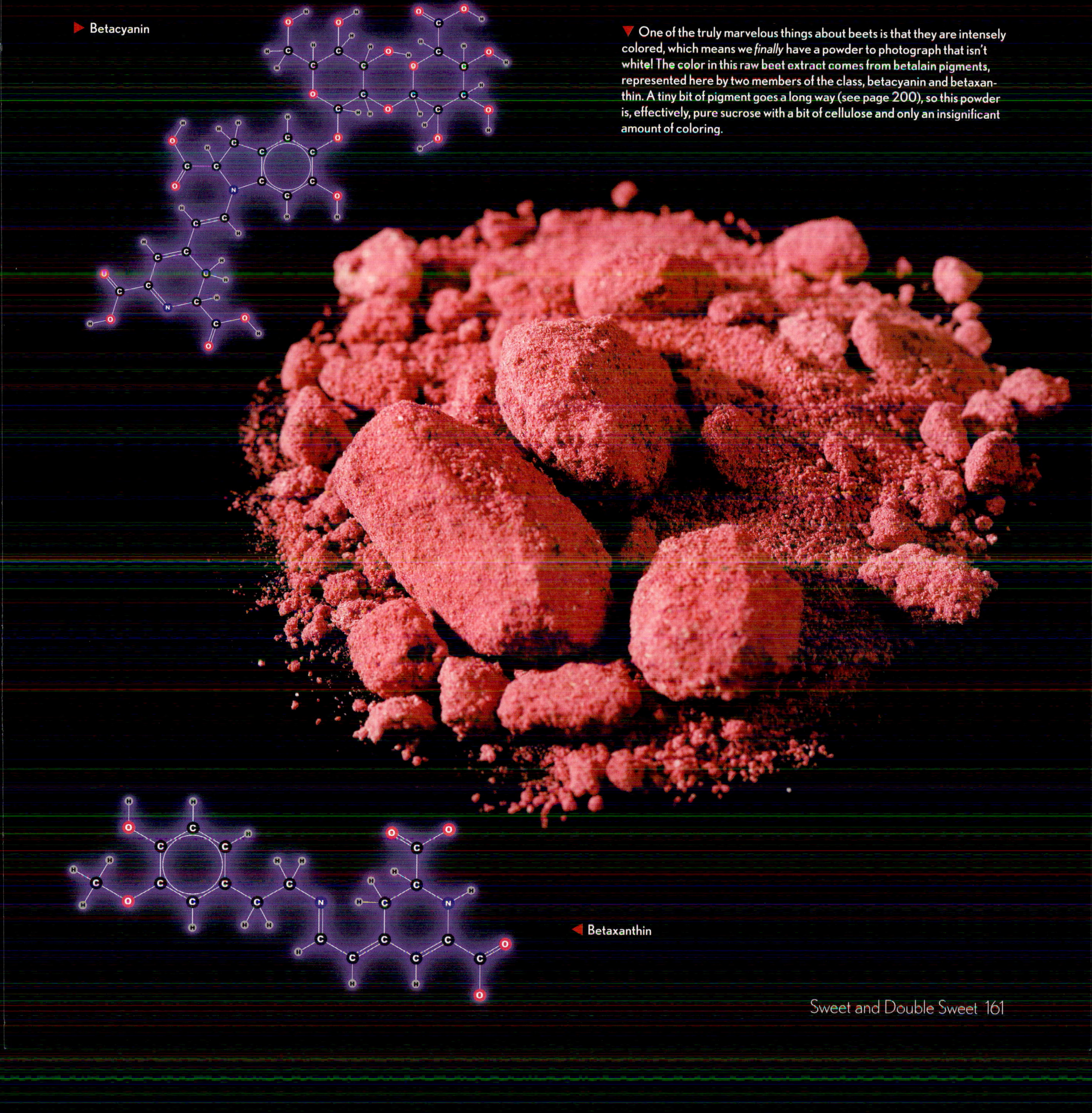

One of the truly marvelous things about beets is that they are intensely colored, which means we *finally* have a powder to photograph that isn't white! The color in this raw beet extract comes from betalain pigments, represented here by two members of the class, betacyanin and betaxanthin. A tiny bit of pigment goes a long way (see page 200), so this powder is, effectively, pure sucrose with a bit of cellulose and only an insignificant amount of coloring.

Lactose

▼ Lactose, the sugar of milk, is a combination of glucose and galactose.

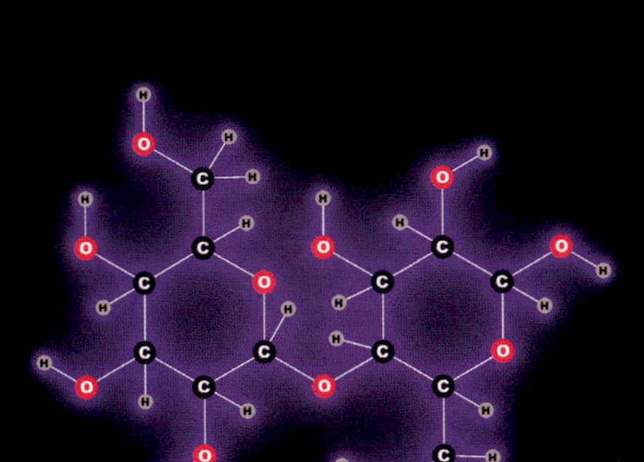

▲ Pure lactose isn't commonly used by individuals, but it is available and looks a lot like table sugar.

▲ The ability to digest lactose in adulthood is the result of a genetic mutation that appeared in humans about ten thousand years ago and has so far spread to only about a third of the population worldwide (though this percentage is much higher in some parts of the world). Nonmutant humans have trouble drinking milk beyond childhood, so pills are available that contain the enzyme necessary to break down lactose, allowing even genetically normal people to enjoy milk and ice cream.

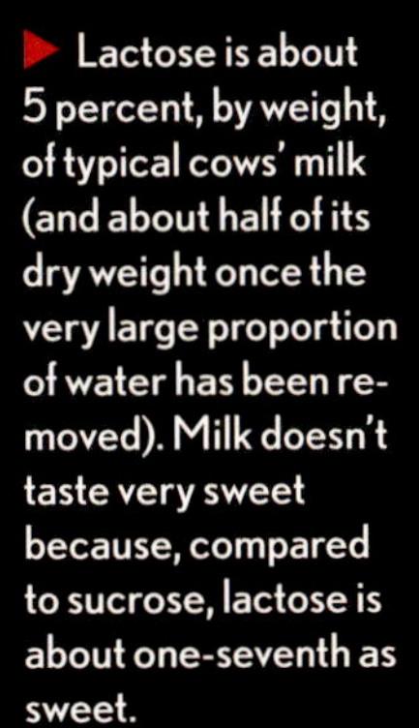

▶ Lactose is about 5 percent, by weight, of typical cows' milk (and about half of its dry weight once the very large proportion of water has been removed). Milk doesn't taste very sweet because, compared to sucrose, lactose is about one-seventh as sweet.

Maltose

▼ Maltose, the sugar of malted grains, is a combination of two glucose units.

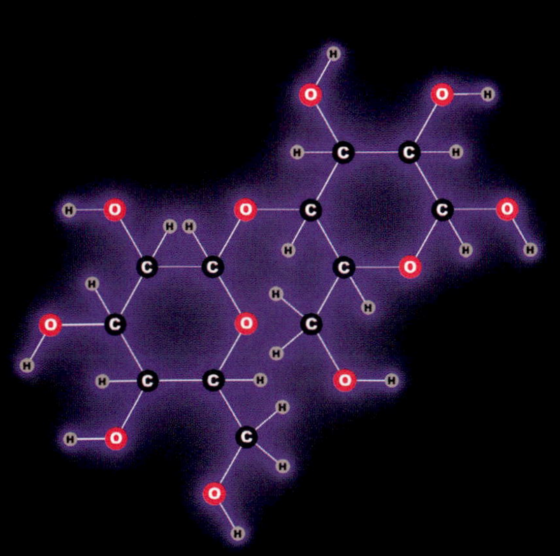

▶ Maltose in pure form is another white powder like sucrose and lactose, but it's more commonly sold as a very, very, very thick syrup. (In a cold room, it's more like a rock than syrup.) This material is about 70 percent maltose, and the fact that it's not white tells you that it's not pure.

▶ Malted grain (corn in this case) is seeds of grain that have been allowed to germinate but not grow beyond the germination stage. Malted grain contains a high percentage of the sugar maltose and is rather tasty. Powdered malted corn extract is used for its sugar content and for brewing purposes (where bacteria convert the sugar into alcohol for making beer and other drinks, or for use as a fuel).

◀ Maltose comes from malted corn, and corn syrup is made almost entirely of maltose, unless it has been processed into high-fructose corn syrup, a very different substance described a bit later on.

Other Mixtures of Sugars

▼ Maltotriose is the simplest example of the class of food ingredients called maltodextrins (commonly found on all kinds of food labels). Maltose is two glucose units, maltotriose is three glucose units, and other maltodextrins are simply more and more glucose units connected in the same way, up to about twenty units. (Beyond that, the substance is called starch.) Below is a pile of commercial maltodextrin powder.

▶ High-fructose corn syrup (HFCS) is very, very widely used, especially in the United States, where, for tax and agricultural policy reasons, it is cheaper than table sugar. Most of it is about half fructose and half glucose. It's difficult to buy pure HFCS in less than trainload quantities, but that's okay because many commercial pancake syrups are made of little else. They are labeled as "light" because, on an equal-sweetness basis, they contain fewer calories than a syrup sweetened with table sugar.

◀ Honey, like high-fructose corn syrup, is also a roughly fifty–fifty mixture of fructose and glucose. Both are made by using enzymes to transform other sugars into the desired fructose-glucose mix. (Honey bees do it in their stomachs; people do it in vats.) Honey also contains a few other sugars and tiny amounts of high-intensity organic compounds that give each variety its unique color and flavor. From an aesthetic and taste point of view, honey is very different from high-fructose corn syrup, but from a nutritional and health point of view, it's hard to justify claims that they are any different. In fact, commercial honey is sometimes adulterated by the addition of high-fructose corn syrup, which is cheaper. Whether this has occurred to a particular bottle of honey is *impossible to determine through any chemical analysis* because the sugars in the two products are not distinguishable, neither by a laboratory nor by your body. (Interestingly, it *is* possible to tell by careful analysis of the carbon-13 isotopic composition of the sugars—but just barely. And the C-13 isotope ratio doesn't affect biological function.)

◀ Invert sugar is table sugar (sucrose) that has been split into its glucose and fructose components (either entirely, giving a fifty–fifty mix of glucose and fructose, or partially, giving a mixture of glucose, fructose, and sucrose). It is thus chemically very similar to high-fructose corn syrup and to honey. The main difference is that it is made from cane or beet sugar rather than from malted corn (in the case of high-fructose corn syrup) or the sugar that bees drink in the nectar from flowers (in the case of honey). In other words, it's more of an economic difference than a chemical or nutritional one. I was surprised to find how similar it tastes to honey; I would have expected more of the taste of honey to come from the minor components, but it seems like a lot of it comes purely from the fructose–glucose mix. That mix is also present in pancake syrups made with high-fructose corn syrup, but there it's masked by artificial flavors not present in this commercial invert sugar paste.

▼ Starch is simply multiple glucose sugar units connected end-to-end in very long chains, like maltodextrin only longer. The bonds between multiple sugar units in both starch and the common double sugars are easily broken by enzymes in the stomach, so when you eat any sugary or starchy food, what you are exposed to nutritionally is a mixture of the simple sugars. For diabetics worried specifically about glucose, starch is actually *worse* than table sugar because table sugar is glucose + fructose, while starch is pure glucose. And it doesn't even taste sweet.

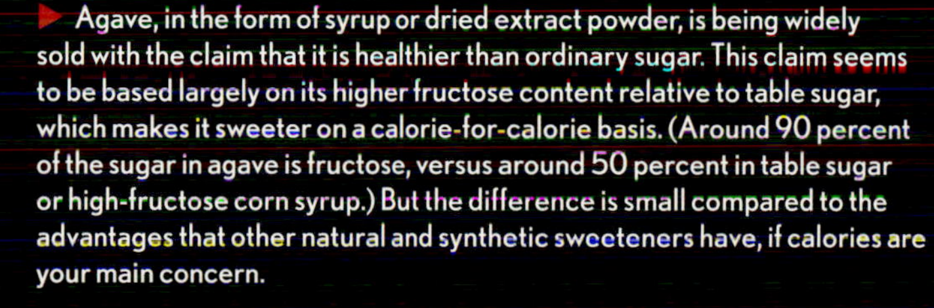

▶ Agave, in the form of syrup or dried extract powder, is being widely sold with the claim that it is healthier than ordinary sugar. This claim seems to be based largely on its higher fructose content relative to table sugar, which makes it sweeter on a calorie-for-calorie basis. (Around 90 percent of the sugar in agave is fructose, versus around 50 percent in table sugar or high-fructose corn syrup.) But the difference is small compared to the advantages that other natural and synthetic sweeteners have, if calories are your main concern.

Sugar Alcohols

◀ An alcohol has an -OH (oxygen plus hydrogen) group attached to a carbon atom, which doesn't have another oxygen atom connected to it. If a molecule has this kind of group, it's an alcohol; if not, it isn't.

AS WE'VE LEARNED, the word *alcohol* refers to any organic compound that contains an alcohol group (see page 38), which consists of an oxygen and a hydrogen atom bonded in a certain way (see the example of ethanol that follows). Alcohols such as methanol, ethanol, and isopropanol have just a single alcohol group, but there's no reason a molecule can't have more.

Look at any of the sugar molecules on the previous pages, and you will see that they are *stuffed* with alcohol groups! Table sugar has fully eight of them. But in addition to being alcohols, sugars are also rings joined up with an ester linkage.

It turns out that molecules very similar to sugars but without the ester ring complication also taste sweet. The simple "sugar alcohols" erythritol and xylitol are widely used as artificial sweeteners in "sugar-free" products. They are not sugar, they don't promote tooth decay (in fact, xylitol appears to prevent, rather than cause, cavities), and they don't raise blood sugar levels.

They vary in sweetness, but overall their level of sweetness is about the same as sugar. They do contribute calories to a food, so using them instead of sugar largely benefits diabetics, not dieters. For losing weight, there are better alternatives.

▼ Several common sugar alcohol sweeteners are simple, fully-occupied sugar alcohols. For example, erythritol has four carbon atoms, and each one of them has its own alcohol group attached. Xylitol has five carbons, each with an alcohol group. There are two variations of the six-carbon example, sorbitol and mannitol. Both have six carbons, and both have one alcohol on each carbon. They differ only in the orientation of one of the bonds, a property that can be seen only in a three-dimensional view of the molecule.

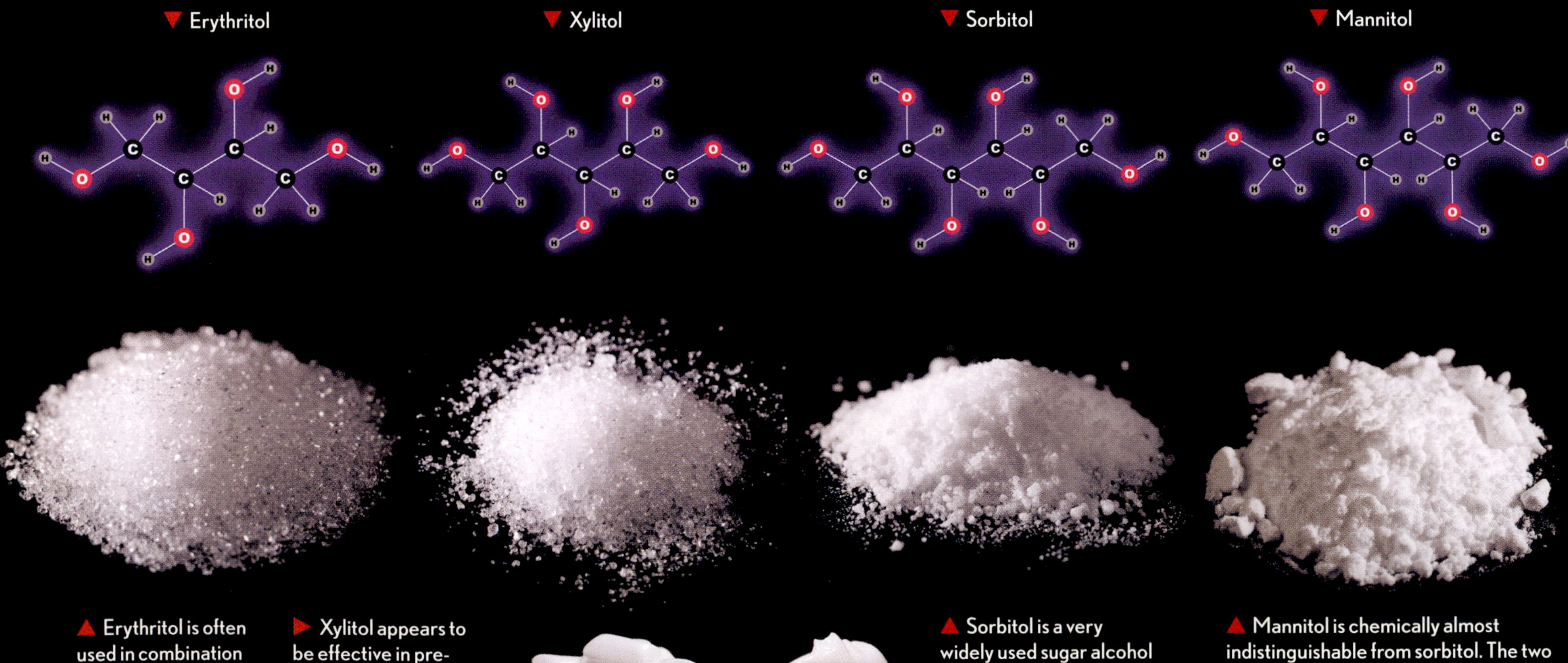

▲ Erythritol is often used in combination with more intense artificial sweeteners to balance their flavor. Unlike other sugar alcohols, it does not cause "gastric distress."

▶ Xylitol appears to be effective in preventing cavities, which makes it absolutely ideal as a sweetener for sugar-free chewing gum and toothpaste.

▲ Sorbitol is a very widely used sugar alcohol sweetener. Although sweet, it tends to cause what is referred to as "gastric distress." For a hint of what this might mean, consider that another common use for sorbitol is as a laxative.

▲ Mannitol is chemically almost indistinguishable from sorbitol. The two powders look different because the appearance of a powder is a function of how fine it is, how much moisture it contains, how it's been arranged, and how desperate we were to make this white powder look just a little different from every other white powder.

Maltitol and Isomalt

Maltitol and isomalt are both combinations of glucose and a sugar alcohol, but connected at a different place on the sugar alcohol. Maltitol is a combination of glucose and sorbitol, while isomalt is a combination of glucose and mannitol.

These gummy bears acquired a certain degree of Internet fame from creative writing exercises that reviewers on an online commerce website engaged in, describing the intestinal effects of eating handfuls of them. These reports may have been slightly exaggerated for comic effect, but the fact is that the main ingredient, Lycasin, is a mixture of various sugar alcohol and sugar combinations dominated by maltitol, which is an effective laxative in larger doses.

Maltitol is a widely used sugar substitute, and I was rather shocked to learn, literally in the course of writing this very caption, that it, and the similar compound isomalt, are not pure sugar alcohols, but rather a combination of a sugar alcohol and an actual sugar!

Isomalt seems to be less widely used in the United States than maltitol, but they are, in any case, very similar.

Not only does virtually every kind of sugar-free chocolate contain maltitol, but maltitol is also the number one listed ingredient in most of them (meaning it's the ingredient of which there is the most by weight). It turns out that my beloved sugar-free chocolates are not actually as harmless as I'd imagined, given that maltitol has about half as many calories as sugar (i.e., still quite a lot) and does in fact affect blood sugar to an appreciable degree because stomach enzymes break it down into glucose and sorbitol.

Super Sweets

SUGARS AND SUGAR alcohols are bulk sweeteners. You need a substantial quantity of them to make food or drink taste sweet. In very sweet foods such as candy or breakfast cereal, they are sometimes the first ingredient listed, meaning that by weight there is more sweetener than anything else in the food.

But some compounds are in a whole other league of sweetness. These substances are hundreds or thousands of times sweeter than sugar, so a small fraction of a gram is all it takes. On the one hand, this is great because it means no matter what the substance is, it can't contribute any significant number of calories; there's just so little of it present. On the other hand, it's a problem because sugar and bulk sugar substitutes contribute important properties to food other than just sweetness (such as sugar's famous stickiness that binds other ingredients, its attractive browning at high temperatures, its mouth feel, and its preservative properties, not to mention its sheer bulk).

When using super-powerful sweeteners, other substances must be found to take the place of sugar without introducing undesirable tastes or textures and without putting right back in all the calories you were trying to avoid.

Most of these sweeteners are synthetic compounds, but two of the most powerful ones in commercial use, stevia and mogroside (from monk fruit), are natural plant extracts. Of course, saying a given molecule is of natural or synthetic origin tells you nothing about whether it's going to taste good or be safe to eat, but it does make a *big* difference with respect to labeling. If a food item is made with a plant extract, it can be labeled as "All Natural!" or words to that effect.

Saccharin

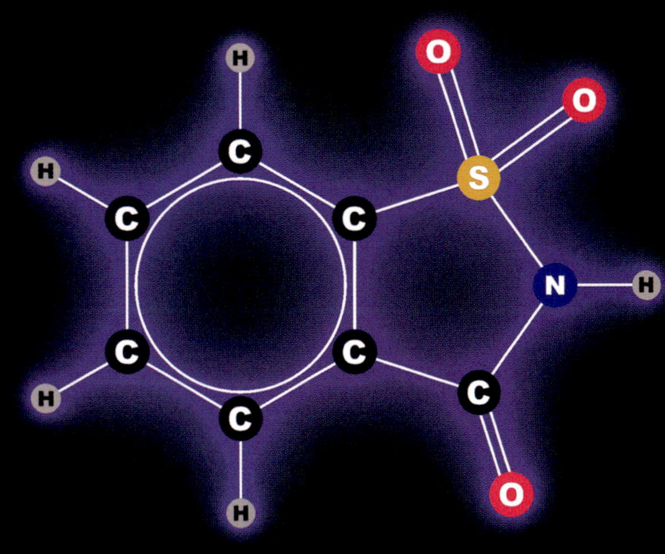

▲ The first nonpoisonous, commercially successful artificial sugar substitute was saccharin. It went through some hard times. First, it was branded as a fraudulent, cheap substitute for sugar that had no nutritional value (back when getting calories was considered a *good* thing). Then, from the 1970s to 1990s, it was suspected of causing bladder cancer, and warning labels were added. By the year 2000, it had become clear that the substance doesn't actually cause problems in humans, and the warning labels were removed.

▼ Saccharin is about three hundred times sweeter than table sugar (sucrose)—and old enough that some nice saccharin-related antiques now exist. I've placed this saccharin bowl on top of a similar sugar bowl as a sign of how much more potent saccharin is. A sugar bowl that size would just be silly since it could hold only enough sugar to sweeten one or two coffees. Meanwhile, a saccharin bowl the size of the sugar bowl could hold a lifetime supply!

▲ Saccharin is used in a huge number of products, but it's also found in restaurants and for home use in little packets like this. I grew up always seeing warning labels on anything with saccharin, so it's a bit of a surprise to now discover that it's in loads of things and requires no special labels. The warnings were mistaken, but they have probably left an entire generation with a queasy feeling about saccharin. Isn't it remarkable that the generic comes in exactly the same shade of pink as the brand-name product? This is the case for most sugar substitutes and their generics—and is the kind of thing that drives merchandisers to the point of insanity trying to trademark particular colors.

▼ One of the problems with super sweeteners is that, in their pure form, it's very difficult to measure how much you should use for, say, a single cup of coffee. Since most people don't carry around milligram scales to measure out a few grains of powder, sweeteners like saccharin are almost always mixed with a larger amount of some kind of filler so their potency is closer to that of sugar. An alternative is to press them into tiny little pills of a consistent size. Even the pills are diluted with filler, but not as much as the powder needs to be. This pretty little saccharin pill box comes with a tiny pair of tongs for picking up the pellets, each of which is equivalent to a teaspoon (five grams) of sugar.

▶ This handsome antique tin of saccharin was probably sold for commercial use: it's difficult to deal with a sweetener this intense in pure powdered form unless you're making a very large batch of something.

◀ For some reason, I ordered half a kilo (about one pound) of saccharin to photograph. What am I supposed to do with it now? That's equivalent to 330 pounds (150 kilograms) of sugar!

▼ An unexpected use for a very intense sweetener: this is a kit for testing whether your dust filter mask fits properly. You breathe a mist of saccharin solution while wearing the mask, and if you can taste the sweetness of the saccharin, the mask needs to be fixed or adjusted. One could use any number of intensely flavored compounds—say, capsaicin powder (i.e., pepper spray)—but I suppose saccharin is a fairly pleasant way to find out your mask isn't working.

Cyclamate

Both saccharin and this molecule, cyclamate, show off their artificial origin in the form of a sulfur atom bonded to two oxygen atoms. That's not something you see very often in natural molecules, and it's not present in any natural sweeteners. But for some reason, we seem to like the taste of this configuration; the same group occurs in several other artificial sweeteners.

Na

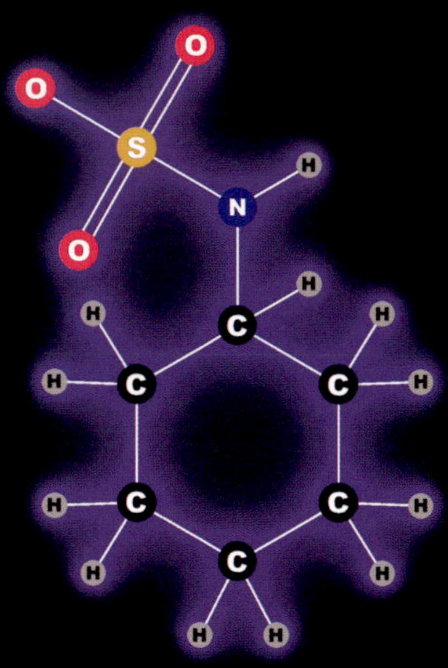

Can't we all just get along? In the United States, cyclamate is banned, and saccharin is legal, so here Sweet'N Low is made with saccharin. In Canada, cyclamate is legal, and saccharin is banned, so those notorious contrarians up north make their Sweet'N Low with cyclamate.

In countries where both cyclamate and saccharin are permitted, including most of Europe, a mixture of the two is preferred because each compound can partly counteract the negative aspects of the other's taste. Cyclamate is about one-tenth as sweet as saccharin, so a ten-to-one ratio of cyclamate to saccharin gives roughly equal prominence to each.

This is a blend of saccharin and cyclamate in a liquid form that is about ten times sweeter than sugar.

Acesulfame Potassium

Acesulfame potassium features the same sulfur-oxygen group as saccharin and cyclamate—as well as their metallic aftertaste, which some people don't much care for. It is used for baked goods because it's quite stable at higher temperatures, unlike some other sweeteners.

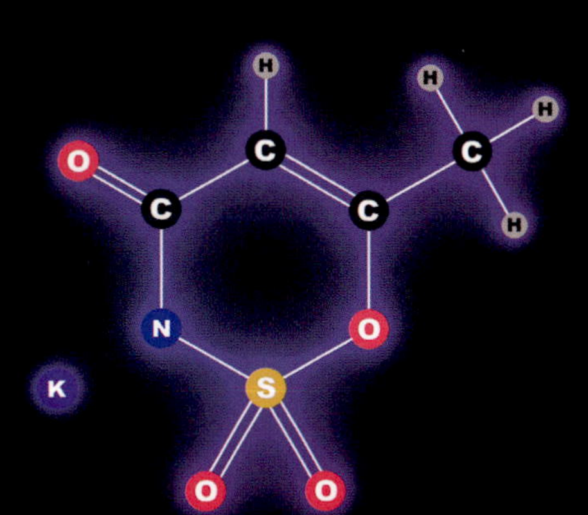

Acesulfame potassium is about two hundred times sweeter than sugar.

Aspartame and Neotame

▼ Aspartame is a combination of two amino acids (the building blocks of all proteins), aspartic acid and phenylalanine. They are joined in a different way than they would be in a protein, but this bond is broken almost immediately in the stomach, so the only exposure from eating aspartame is to two amino acids, both of which are essential nutrients required for a healthy life. It's really quite difficult to see how aspartame could be harmful, and indeed, despite decades of controversy, all indications are that the substance is utterly and completely safe. (The only issue, and the reason you see caution labels on products containing aspartame, is that somewhere around one in ten thousand people have a genetic disease that means they need to limit the amount of phenylalanine they consume in their diet. So, along with maintaining a strict diet to avoid it in all the foods where it occurs naturally, they also need to avoid foods sweetened with aspartame.)

◀ Neotame is a promising new derivative of aspartame in which a dimethylbutyl group (the six carbon and thirteen hydrogen atoms on the top part of this diagram) has been attached to the aspartic acid part of aspartame. This addition makes the substance about fifty times sweeter than aspartame, or ten thousand times sweeter than table sugar! Its breakdown products in the body are pretty harmless even to people with the disorder that makes them sensitive to the phenylalanine that is present in both aspartame and neotame.

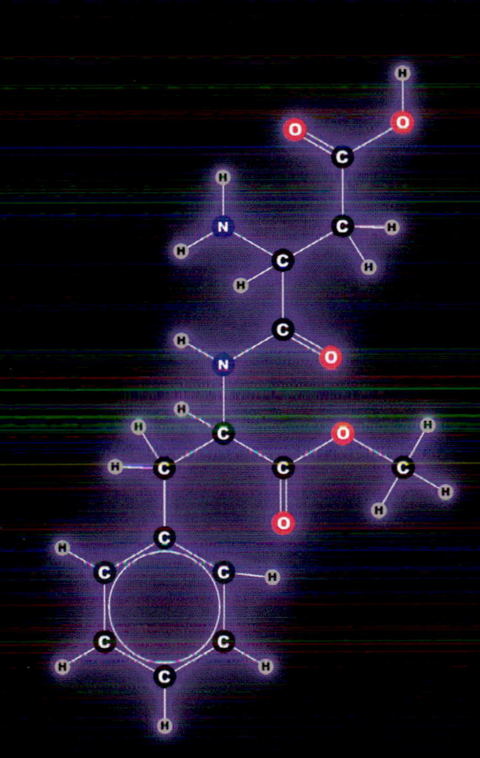

▶ Neotame is the most potent known sugar substitute of any kind, natural or synthetic. The 4.5 grams of neotame powder shown here (i.e., the tiny pile you can barely see at the top) is roughly equivalent in sweetening power to the one hundred pounds of sugar it's sitting on. Those 4.5 grams represent zero calories against the sugar's 171,000 calories! To equal a teaspoon of sugar, you need only 0.4 milligrams of neotame. This extreme potency is one reason to think neotame should be quite safe. To put it in perspective, 0.4 milligrams is such a small amount that even if neotame were as toxic as VX nerve gas, the most toxic synthetic compound known, you'd still have a decent chance of surviving a cup of coffee sweetened with it.

▼ Neotame is astonishingly, startlingly, sweet. Within seconds of opening the bag and starting to scoop it out to make this pile, I could taste sweet in the back of my mouth. It's a fine powder, and even though I was moving slowly and trying not to stir it up, invisible quantities must have floated into my nose. For hours afterward, I could taste it on my mustache. Absolutely invisible amounts result in a burst of not-unpleasant sweetness! Sorry to sound like a commercial for neotame, but this stuff really is *amazing*.

Sucralose

In chemical structure, sucralose is identical to sucrose (table sugar) except three alcohol (-OH) groups are replaced with chlorine atoms. This change renders it both six hundred times sweeter and indigestible, meaning that the tiny amounts needed to sweeten things contribute no calories.

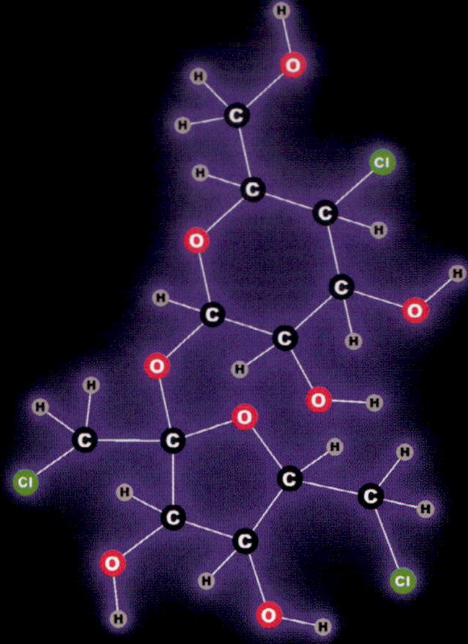

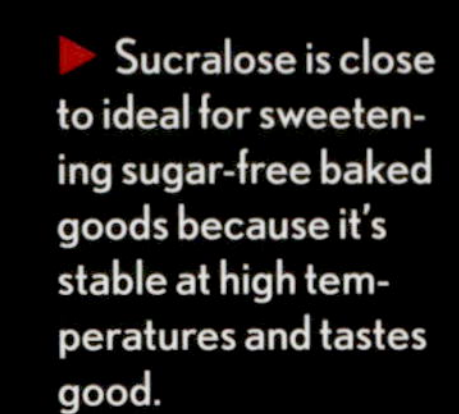

▶ Sucralose is close to ideal for sweetening sugar-free baked goods because it's stable at high temperatures and tastes good.

▼ Splenda and generic equivalents in similarly colored packets are sweetened principally with sucralose. The majority of the powder in the packet is glucose (dextrose) and maltodextrin. It probably has about four calories, but the FDA permits this to be rounded to zero.

Stevia

Stevia leaf extract consists of several compounds known collectively as steviol glycosides, some of which are as powerful as many synthetic sweeteners. (They are about three hundred times as sweet as sugar.) Two of the most important contributors are shown here, rebaudioside-A and stevioside.

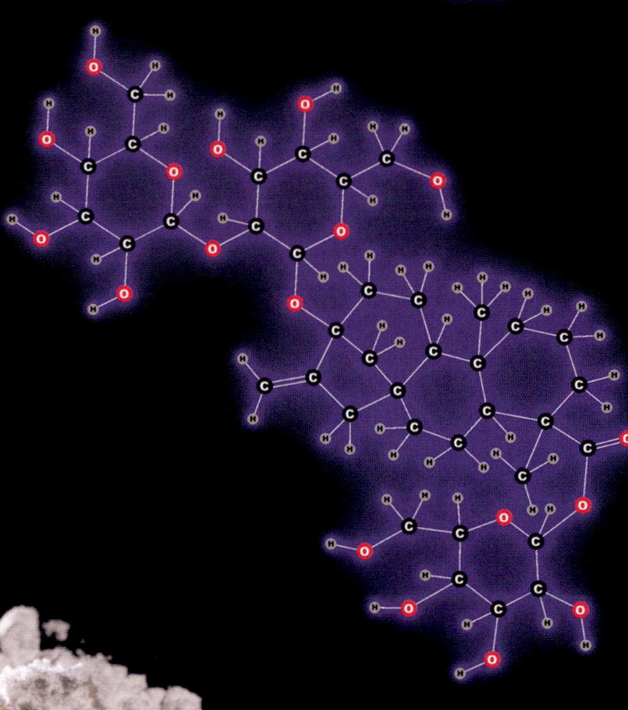

▼ These are raw stevia leaves from which stevia sweetener is extracted. Many people consider stevia to be a perfect sweetener because it's a zero-calorie sweetener that is all natural. Other people hate it because of the taste, which is a good bit different from the familiar taste of sugar.

▲ The pure extract of stevia leaf is a mixture of several related molecules. They are chemical compounds like any other, neither inherently safe nor unsafe because their origin happens to be natural. So far, they seem to be safe, though they have not been studied nearly as well as the more established synthetic sweeteners.

▶ Stevia, like super-intense synthetic sweeteners, is often sold in liquid form because liquid drops are an easy way to dispense a very small amount of material. This type of packaging allows the liquid to be very concentrated and a large number of servings to fit in a small space. This is harder to do with powder.

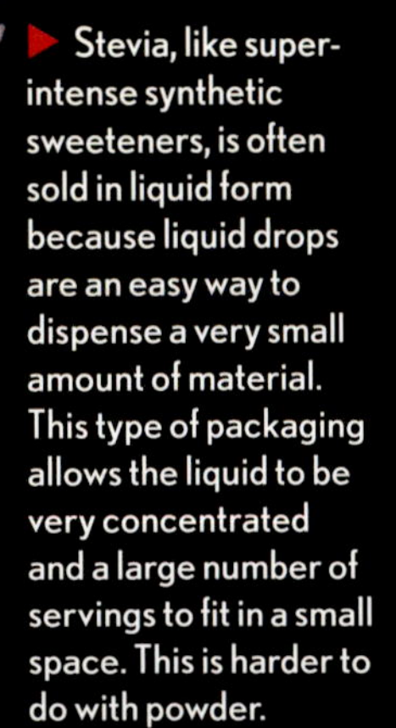

▶ Because stevia is a plant extract, it can be labeled as "all natural" and even as a "dietary supplement," implying that perhaps it's good for you. But this one-gram packet is 96 percent glucose sugar (labeled as its exact synonym, dextrose). The remaining 4 percent is stevia extract, which contributes the majority of the flavor. Being nearly pure sugar is remarkabl common for sugar substitute packets. They can be labeled as zero calories only because the FDA permits anything under five calories to be listed as zero. The one gram of glucose they all contain has four calories. Shame. O the other hand, the packet is equivalent to about two teaspoons of sugar, which would have about thirty-two calories. Therefore, if you use just one packet, you consume about an eighth as many calories as you would if you' used sugar. But not zero calories.

These stevia-based sweeteners use erythritol instead of glucose as their fillers, which is nice because erythritol has far fewer calories than glucose and doesn't contribute to blood sugar, both of which seem like reasonable expectations of a sugar substitute being consumed, in many cases, by people who are overweight, diabetic, or both. But these can't be labeled as "all natural." Stevia is a direct plant extract, but erythritol is created by fermentation of corn under human control and isn't considered "natural" enough by some people.

Monk Fruit

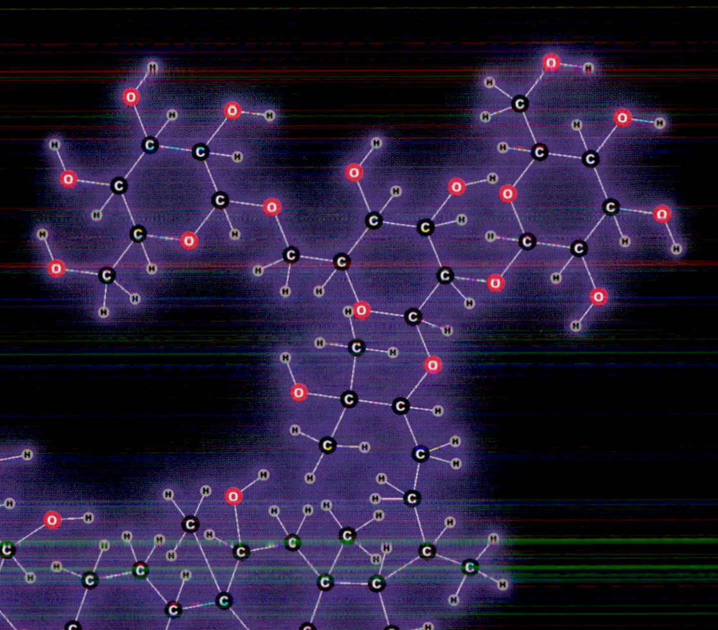

Another one of those complicated, multi-cyclic plant compounds, mogroside comes in several varieties (number 5 is pictured here).

The strong association of this compound with Traditional Chinese Medicine leads to labels like this one for a mogroside-based sweetener.

The fruit from which mogrosides are extracted is called the monk fruit, or luo han guo after its Chinese name.

Pure mixed mogrosides are about three hundred times sweeter than sugar, comparable to most of the high-intensity synthetic sweeteners. This raw extract powder is claimed to be about 7 percent mogrosides, which still makes it many times sweeter than sugar. What is the other 93 percent? That's hard to say without a detailed chemical analysis.

The Mixture Is Better Than the Parts

ESSENTIALLY ALL SWEETENERS other than sugar are judged by most people to be inferior in taste. Whether this is biological or psychological is a bit hard to say. (I, for example, have been trying to train myself for over a decade to prefer the taste of diet soda to regular, with only partial success. My training regimen consists of absolutely never drinking the real thing so I won't be reminded of what I'm missing.)

One way manufacturers have found to improve the taste of artificial sweeteners is to mix several of them together. Often, the off taste, aftertaste, or slow-acting taste of one sweetener can be offset by the flavor profile of another.

▶ This packet is sweetened with aspartame and acesulfame potassium. As usual, the bulk of the material is glucose, labeled as zero calorie only because it's under five calories and regulations allow rounding four down to zero.

▶ The variety of sugar substitutes available in the grocery store, containing mixtures of just about any natural and synthetic sweeteners you can name, is astonishing. For example, this one combines table sugar (sucrose), erythritol, and stevia.

▼ The whole category of tiny, super-concentrated flavor bottles is possible only because of the intense power of super-sweet molecules. (These bottles are less than four inches [ten centimeters] tall.) If sweetened with sugar, bottles this small would only contain enough to make one or two drinks, but in fact, these tiny little bottles contain enough flavor and sweetness to make gallons of flavored water. Each has a particular mix of natural and/or artificial sweeteners.

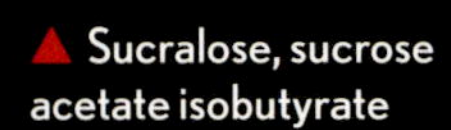

▲ Sucralose, sucrose acetate isobutyrate

▲ Cane sugar, stevia extract

▲ Erythritol, stevia extract

▲ Monk fruit extract

▲ Caffeine, sucralose, acefulfame potassium, sucrose acetate isobutyrate

▶ Sucralose, acesulfame potassium

▶ Baked goods present a special challenge for sugar substitutes. Only substances able to withstand high temperatures for prolonged periods can be used, which limits the selection considerably.

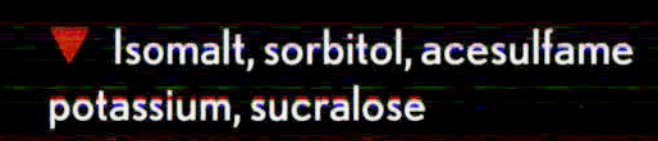

▼ Isomalt, sorbitol, acesulfame potassium, sucralose

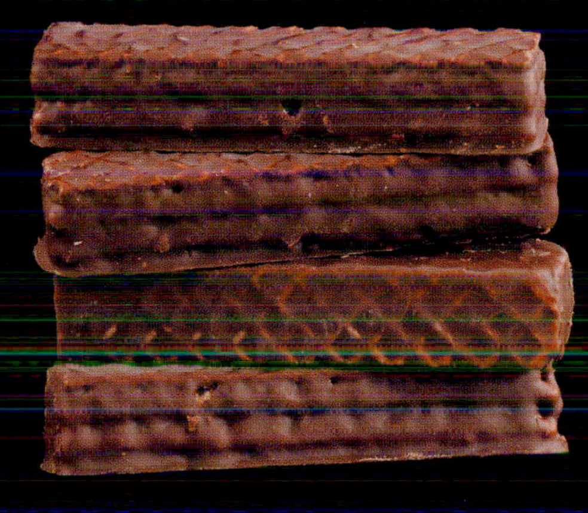

▶ Maltitol, lactitol, sorbitol, acesulfame potassium, sucralose

Chapter 10

Natural and Artificial

IN THE PREVIOUS CHAPTER, we learned about natural and artificial sweeteners. Compounds like saccharin and aspartame are a touchy subject. A lot of people don't trust them, and the most popular ones have all been put through scientific, regulatory, and public opinion wringers. But natural plant extracts such as stevia often get a free ride. People tend to assume they are OK until proven otherwise, and they get considerably less scrutiny from regulators.

You might think, given my generally positive attitude toward chemistry, that I would be eager to try out new synthetic sweeteners. But this is not the case. Despite the best efforts (and sometimes the sloppy or corrupt efforts) of government and industry to test compounds for safety, subtle problems with new molecules may show up only after they have been used by millions of people over many years. I tend to feel that it's best to give these things a few decades to shake out.

But I also don't eat random mushrooms that I find in the forest or use herbal supplements from dodgy natural food suppliers who rest on their "organic" laurels. What makes newly discovered synthetic compounds slightly scary is exactly the same thing that makes strange mushrooms or unregulated supplements scary: *uncertainty*. There isn't any sort of inherently greater danger from synthetic compounds over natural ones.

Sure, there are unhealthy compounds created in labs, but good grief, if you want to find toxic substances, look to the natural world! Plants, in particular, spend an inordinate amount of time synthesizing compounds designed to kill or seriously inconvenience the animals that keep trying to eat them. (Not being able to move, plants have few options beyond chemical weapons.)

Molecules don't know where they came from. They don't know if they are natural or artificial, good or evil, wholesome or poisonous. They just are. Whether they were created in a lab, in the venom gland of a sea snail, in a factory, or in the leaf of an herb simply has no bearing on the question.

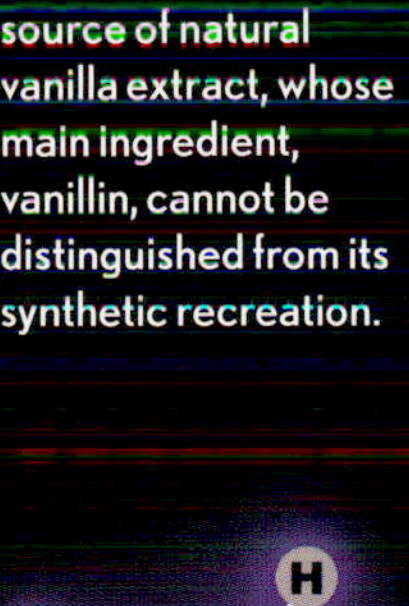

◀ Fermented vanilla beans are the source of natural vanilla extract, whose main ingredient, vanillin, cannot be distinguished from its synthetic recreation.

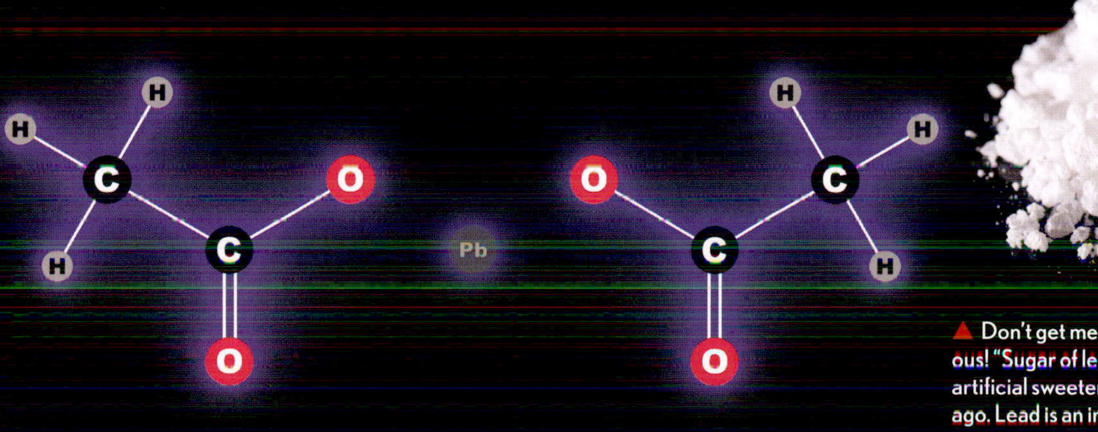

▲ Lead acetate is the lead salt of acetic acid (vinegar).

▲ Don't get me wrong; some artificial sweeteners definitely are poisonous! "Sugar of lead," the alchemical name for lead acetate, was used as an artificial sweeter as early as the Roman empire some two thousand years ago. Lead is an insidious thing—a chronic cumulative neurotoxin (i.e., eating even small amounts over a period of time makes you stupid). No one noticed this for centuries because it was easier to blame witches or demons for insanity than to figure out the real causes.

▶ Despite the problem with lead acetate (i.e., it's a heavy metal poison), it is legally and commonly used, even today, in progressive hair-dye formulas designed for men hoping to hide their gray. The lead is the pigment: it becomes permanently embedded in the hair fiber. This seems to me to be a *very bad idea*, given that a threshold for lead exposure below which there is no harm has not yet been found. I will not use this stuff, solely because of its lead content. (I mean, of course, I wouldn't, if I needed to.)

▲ An investigation in 2013 found that 68 percent of herbal supplements sold in the United States contained plant material from types of plants not listed on the label (i.e., they added some weeds to the fancy stuff on the label). A shocking 32 percent of the products contained *none* of the listed ingredients *at all*. Makers of artificial food ingredients are not inherently more honest, but at least artificial food ingredients are, in theory, regulated and inspected, while natural foods and herbal supplements are completely unregulated. No one is checking any of this stuff. (For example, to get this photo, I gathered random dried leaves from my yard and put them in a capsule. In other words, I did exactly the same thing that the makers of fully a third of herbal supplements sold in the United States do.)

▲ Lead acetate is a slow poison, but other synthetic compounds are much faster. VX nerve agent is fatal in nearly invisible amounts and is the most toxic synthetic compound known. Nonetheless, it comes in a distant fourth in the list of the most toxic substances known. The winners of gold, silver, and bronze in the race to be most poisonous are shown on the following pages. They are Botulinum toxin, maitotoxin, and Batrachotoxin; all are natural compounds with no artificial colors, flavors, or additives.

▶ Some poisons, such as lead acetate, are insidious. They sneak into a population and can kill large numbers of people before being noticed. Not so with poison gases, which are anything but subtle. They have killed millions—but mainly on purpose, not just because no one noticed.

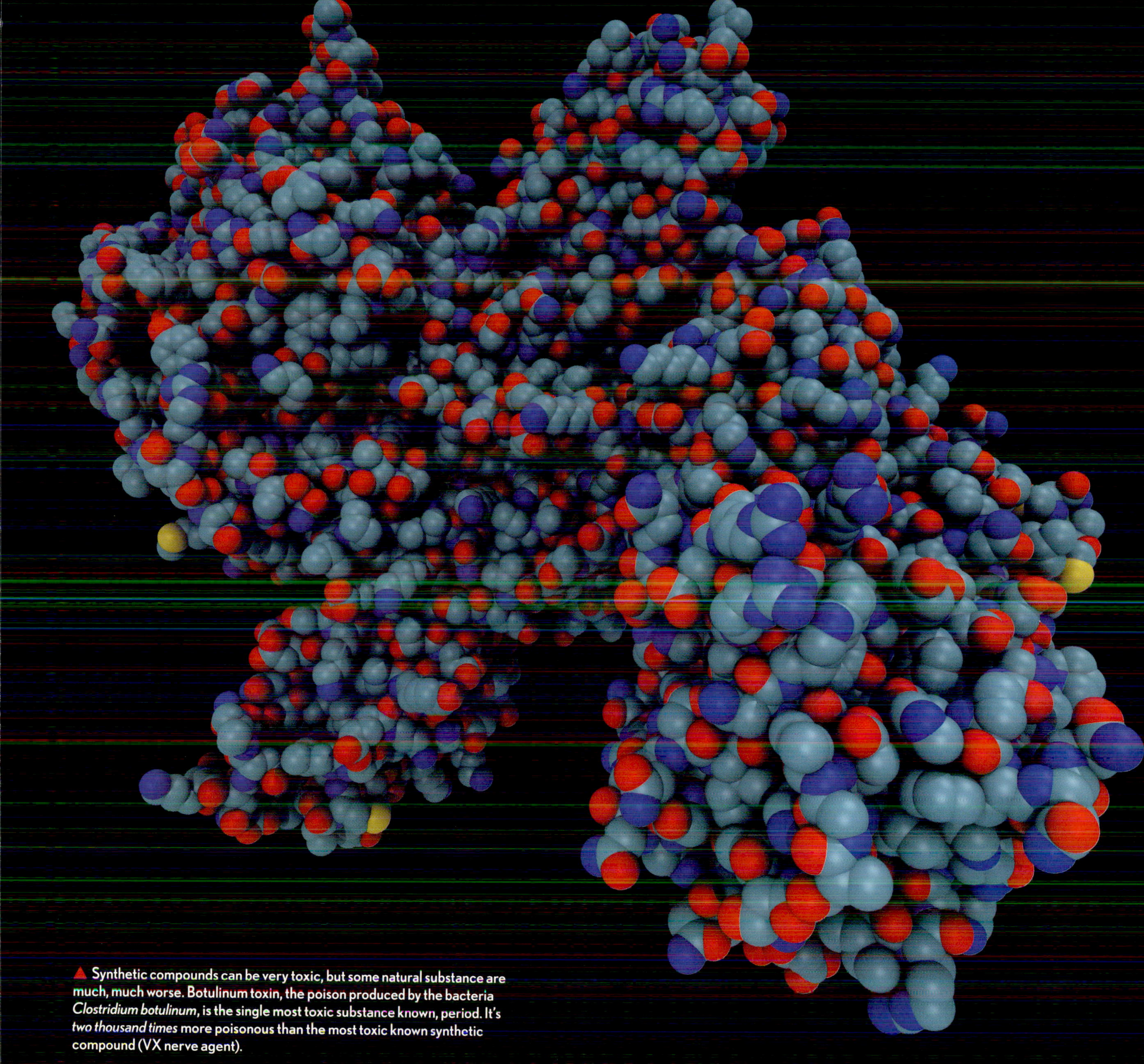

▲ Synthetic compounds can be very toxic, but some natural substance are much, much worse. Botulinum toxin, the poison produced by the bacteria *Clostridium botulinum*, is the single most toxic substance known, period. It's *two thousand times* more poisonous than the most toxic known synthetic compound (VX nerve agent).

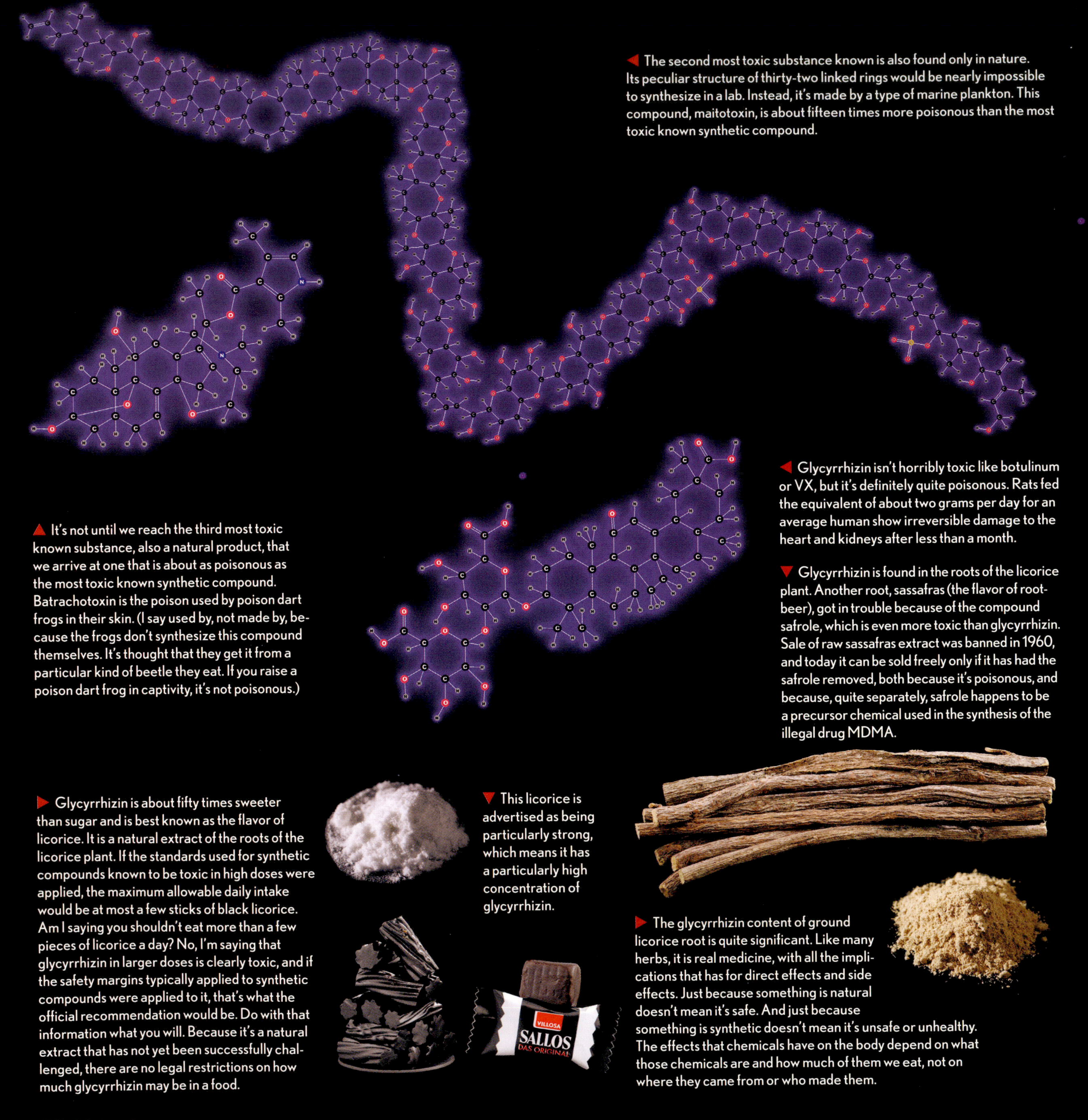

◀ The second most toxic substance known is also found only in nature. Its peculiar structure of thirty-two linked rings would be nearly impossible to synthesize in a lab. Instead, it's made by a type of marine plankton. This compound, maitotoxin, is about fifteen times more poisonous than the most toxic known synthetic compound.

▲ It's not until we reach the third most toxic known substance, also a natural product, that we arrive at one that is about as poisonous as the most toxic known synthetic compound. Batrachotoxin is the poison used by poison dart frogs in their skin. (I say used by, not made by, because the frogs don't synthesize this compound themselves. It's thought that they get it from a particular kind of beetle they eat. If you raise a poison dart frog in captivity, it's not poisonous.)

◀ Glycyrrhizin isn't horribly toxic like botulinum or VX, but it's definitely quite poisonous. Rats fed the equivalent of about two grams per day for an average human show irreversible damage to the heart and kidneys after less than a month.

▼ Glycyrrhizin is found in the roots of the licorice plant. Another root, sassafras (the flavor of root-beer), got in trouble because of the compound safrole, which is even more toxic than glycyrrhizin. Sale of raw sassafras extract was banned in 1960, and today it can be sold freely only if it has had the safrole removed, both because it's poisonous, and because, quite separately, safrole happens to be a precursor chemical used in the synthesis of the illegal drug MDMA.

▶ Glycyrrhizin is about fifty times sweeter than sugar and is best known as the flavor of licorice. It is a natural extract of the roots of the licorice plant. If the standards used for synthetic compounds known to be toxic in high doses were applied, the maximum allowable daily intake would be at most a few sticks of black licorice. Am I saying you shouldn't eat more than a few pieces of licorice a day? No, I'm saying that glycyrrhizin in larger doses is clearly toxic, and if the safety margins typically applied to synthetic compounds were applied to it, that's what the official recommendation would be. Do with that information what you will. Because it's a natural extract that has not yet been successfully challenged, there are no legal restrictions on how much glycyrrhizin may be in a food.

▼ This licorice is advertised as being particularly strong, which means it has a particularly high concentration of glycyrrhizin.

▶ The glycyrrhizin content of ground licorice root is quite significant. Like many herbs, it is real medicine, with all the implications that has for direct effects and side effects. Just because something is natural doesn't mean it's safe. And just because something is synthetic doesn't mean it's unsafe or unhealthy. The effects that chemicals have on the body depend on what those chemicals are and how much of them we eat, not on where they came from or who made them.

▼ "Red licorice" is just a marketing term; it's not really licorice at all. The artificial strawberry flavors in this candy are unrelated to the glycyrrhizin in true black licorice, so you can eat as much as you like.

A Tale of Two Vanillas

NATURAL AND SYNTHETIC substances differ interestingly in the kinds of minor constituents that are most likely to accompany a particular main compound. (You can call these minor components "impurities," "aromas," "contaminants," or "complex flavor components," depending on what they are and what your attitude about them is.)

Because synthetic compounds are often made from mineral or petroleum precursors, you have to watch out for things such as lead or carcinogenic petroleum distillates working their way into the mix as undesirable contaminants. It's also common for the reactions that create the desired compound to also create a number of similar but undesirable compounds at the same time.

Conversely, in the case of plant sources, you have to watch out for the many toxic compounds that plants make for self-defense. Contamination from sloppy handling or toxins in the soil is a constant problem. And the manipulations done to natural products, fermentation and cooking for example, are chemical reactions that may neutralize some of the naturally occurring toxic compounds while creating new and possibly undesirable compounds.

Vanilla is an interesting case study of these differences.

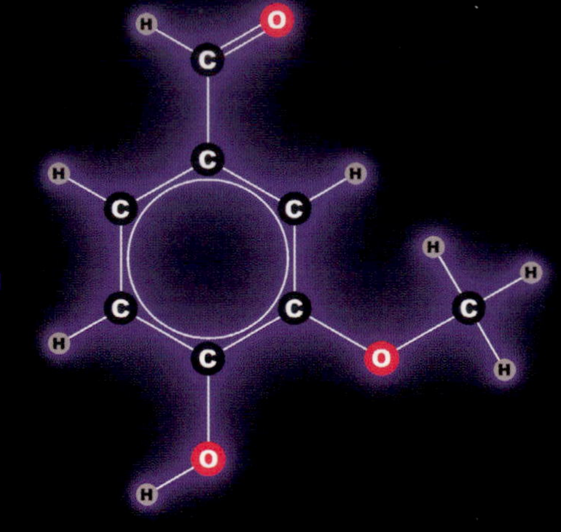

▲ The molecule called vanillin, with systematic name 4-hydroxy-3-methoxybenzaldehyde, is by far the most important flavoring ingredient in the world. Exactly this same molecule provides the bulk of the flavor in both natural and artificial vanilla flavorings. The only difference between the two is in the minor chemicals that are also present in the mix. Cooks insist that various natural vanillas have very different flavors, and they are right. But it's not because there are different kinds of vanillin molecules, only that there are different minor chemicals present based on where the vanilla beans were grown and how they were processed.

Natural Vanilla Extract

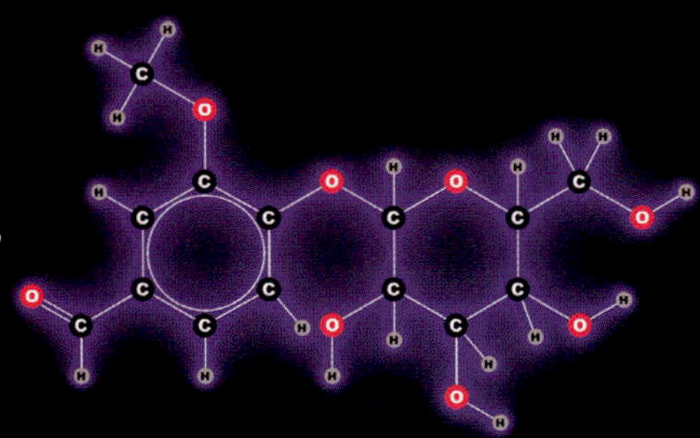

▶ Glucovanillin, which is a vanillin molecule bound to a glucose molecule, is the form in which the vanillin is present in green, unfermented vanilla beans. Fermentation of the beans (i.e., a series of chemical reactions initiated by human intervention but considered natural anyway) allows enzymes to separate these two components, liberating the vanillin.

◀ Vanilla beans, grown in Madagascar and other places I'd like to visit, are the source of natural vanilla flavor. The pods are initially green, but after several weeks of alternating between sun and water (which is what I'd want as well, if I were in Madagascar), they turn dark (as I also would, if I were in Madagascar alternating between sun and water for a few weeks).

▲ Ground up, fermented vanilla beans are about 2 percent vanillin. Vanillin is extracted from such powder using a mixture of alcohol and water, together with at least a hundred other minor components.

▼ Commercially available "pure vanilla extract" is mostly a mixture of alcohol and water. The commercial standard requires it to be at least 35 percent alcohol and contain the extracted components of one hundred grams of dried, fermented vanilla beans per liter of liquid. That means the concentration of vanillin in the liquid you buy is only about a tenth or less of the concentration in the original bean pods, and far lower than in the dried extract powder. About 0.2 percent of the mixture is the main flavor component, vanillin.

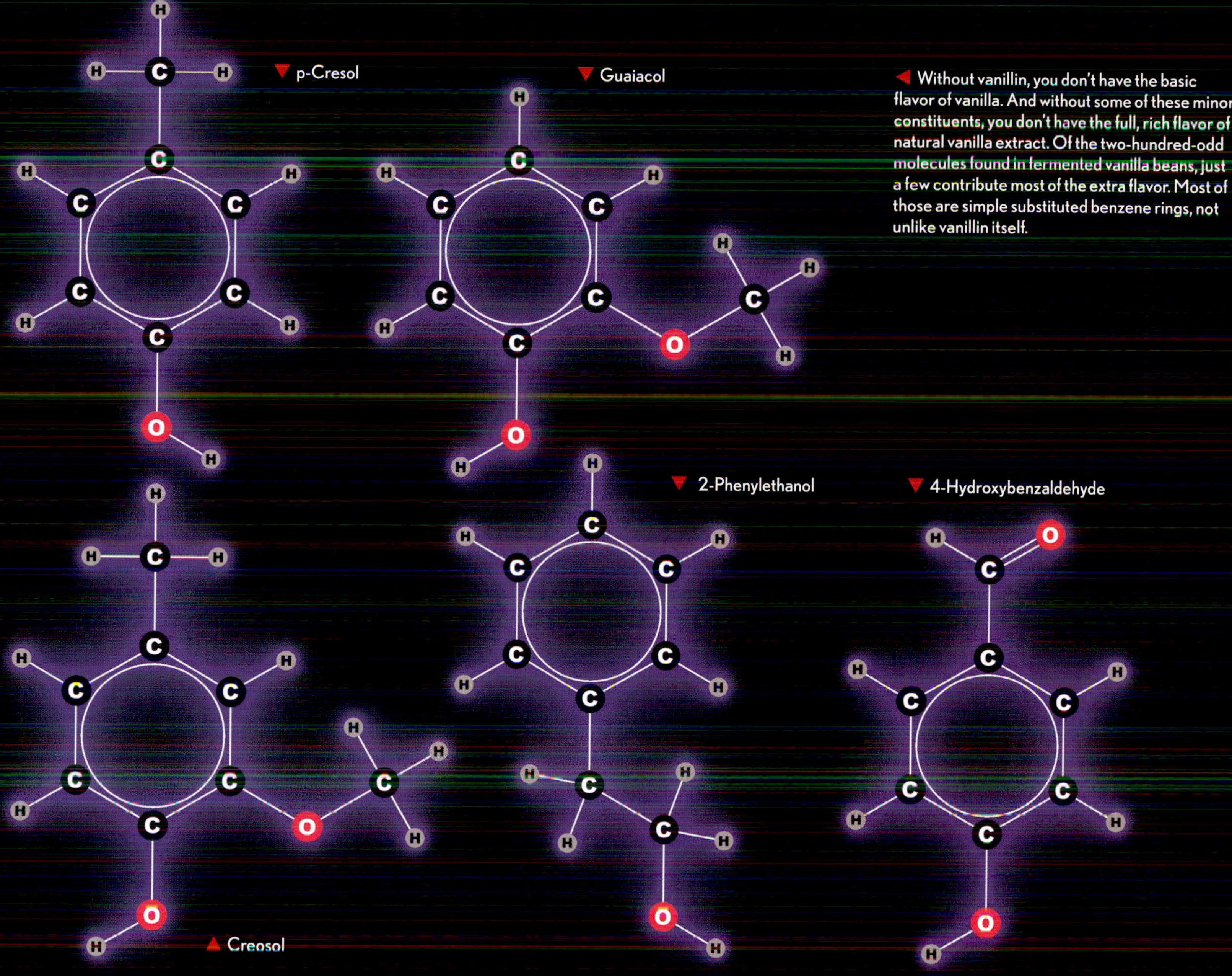

◀ Without vanillin, you don't have the basic flavor of vanilla. And without some of these minor constituents, you don't have the full, rich flavor of natural vanilla extract. Of the two-hundred-odd molecules found in fermented vanilla beans, just a few contribute most of the extra flavor. Most of those are simple substituted benzene rings, not unlike vanillin itself.

Synthetic Vanilla

▶ In the 1930s, a practical process was developed to synthesize vanillin from the lignin left over from processing wood pulp for paper production. This resulted in a sudden and massive drop in the price of vanilla worldwide.

▶ Today, most synthetic vanillin is made from compounds extracted from oil or coal. This fact is responsible for a very amusing way to tell whether synthetic vanilla is being passed off as natural (which is very tempting for companies to do because the natural product sells for a far higher price and cannot be distinguished by any chemical test). Natural vanilla extract, you see, is radioactive, while synthetic vanilla is not. It might sound surprising, but it must be so: *all* substances derived from living plant sources have approximately the same ratio of radioactive carbon-14 as do living plants, about one part per trillion. The carbon-14 comes from the CO_2 the plants absorb from the atmosphere. But over time, carbon-14 decays and becomes nonradioactive (this is the basis for carbon-14 dating). Crude oil and coal, being very, very old, have no carbon-14 radioactivity in them at all, and neither do any compounds derived from them.

Synthetic Vanilla

Synthetic vanilla flavoring, also called imitation vanilla, is chemically identical to natural vanilla as far as the main ingredient, vanillin, is concerned. In some countries, regulations permit a designation of a food additive as "identical to nature" rather than "imitation," which makes a lot of sense, since it allows people who understand chemistry to know that they are in fact getting the real thing, just made in a factory. In the United States, you have to read the ingredient label to see that the word "imitation" doesn't really mean imitation; it means a synthetic recreation of the exact chemical you were looking for.

Natural vanilla is expensive because it has to be pollinated and harvested by hand. But synthetic vanilla is cheap, just a few dollars a pound (about $10 per kilogram). One dollar's worth of synthetic vanillin is enough to make about thirteen gallons (fifty liters) of the kind of extract you buy in the store! Synthetic vanillin contains fewer random components than natural vanilla extract does and is thus inherently more predictable, though with a less complex flavor. Whether that's a good or bad thing depends on your goal. If you're cooking at home, adding natural vanilla extract is a quick and easy way to add a large number of different chemicals to your food at the same time, and if you like the taste of those chemicals (which many people do), that's great. For example, I use natural vanilla when making liquid nitrogen ice cream. But if you're making a commercial food product with a carefully controlled taste profile, you may prefer to use synthetic vanillin, not only because it's cheaper but also because if you wanted a particular secondary flavor, you would add it separately in precisely the desired quantity rather than rely on a fairly random natural mix, which may change from batch to batch. The very variety of natural vanilla flavors that a top chef enjoys is nothing but trouble for a commercial chef.

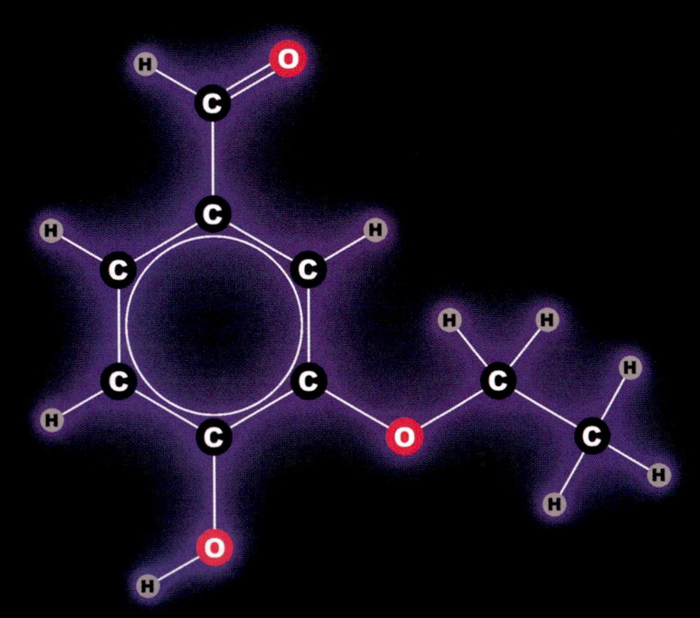

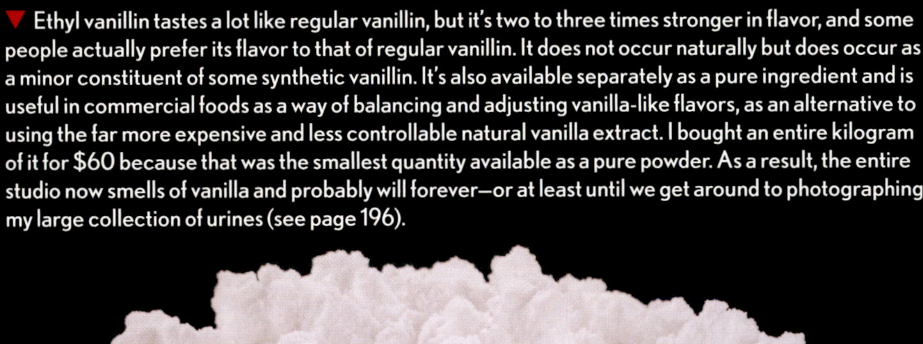

Ethyl vanillin tastes a lot like regular vanillin, but it's two to three times stronger in flavor, and some people actually prefer its flavor to that of regular vanillin. It does not occur naturally but does occur as a minor constituent of some synthetic vanillin. It's also available separately as a pure ingredient and is useful in commercial foods as a way of balancing and adjusting vanilla-like flavors, as an alternative to using the far more expensive and less controllable natural vanilla extract. I bought an entire kilogram of it for $60 because that was the smallest quantity available as a pure powder. As a result, the entire studio now smells of vanilla and probably will forever—or at least until we get around to photographing my large collection of urines (see page 196).

Ethyl vanillin is the same as vanillin, except the single-carbon (methyl) group on the far right is replaced by a two-carbon (ethyl) group. It occurs as a sort of accident in the artificial synthesis of vanillin. In that sense, it's a contaminant of synthetic vanillin that is not present in natural vanilla. But, like some of the "contaminants" in natural vanilla, it actually tastes pretty good.

Intentional Food

THE INGREDIENT LISTS on some packaged foods seem shockingly long. Why do they need to put so *many* different chemicals into our food? But the real question isn't why these lists are so long, it's why they are so short.

The lists of ingredients in unprocessed natural foods are, on the whole, *much* longer, you just don't see them because there's no requirement to list the chemicals in natural foods. Apple pie just lists "apple" as the ingredient, not the two hundred or so chemicals that make up the apple.

Human-made foods with very long lists of ingredients are trying to do exactly the same thing nature does in putting together something like an apple. Nearly every chemical in an apple helps that apple fulfill its function in some way. The sugars are a lure to entice animals to eat, and thus transport, the seeds. The cellulose holds the apple together. The acids and toxic chemicals fend off insects and mold. Dyes and scents advertise the delicious contents to animals and birds who might be available to carry the seeds.

When food designers assemble a man-made food, they add chemicals for similar reasons: sugars for taste and nutrition; starch, cellulose, or protein to hold it together and give it structure, lightness, or a good feeling in the mouth; toxic chemicals to fend off mold; and dyes and scents to attract customers.

You might think processed foods should be *healthier* on the whole since we're actually *trying* to make food intended for human consumption. With the one single exception of mother's milk, everything nature makes that we eat has not been made with us in mind, and only a small fraction of plants are even edible at all! (Though it's true that fruits want to be eaten, they have an agenda: getting you to transport seeds for the plant. Your long-term health is not their concern.)

Unfortunately, most of the time the power to engineer foods is pointed toward making them sell better and taste better, not be better for you. But there are exceptions, and over time, as people have come to realize how unhealthy the modern Western diet has become, things have actually gotten a bit better in the unnatural foods area.

◀ Water, Cellulose, Sugar, Thiophene, Thiazole, Vanillin, Asparagusic acid, Quercetin, Rutin, Hyperoside, Diosgenin, Quercetin-3-glucuronide, Asparagine, Arginine, Tyrosine, Kaempferol, Sarsasapogenin, Shatavarin I-IV, Asparagosides A-I, Sucrose 1-fructosyltransferase, Spirostanol glycoside, 1-methoxy-4{5-(4-methoxyphenoxy)-3-pentene-1-ynyl}-benzene, 4{5-(4-methoxyphenoxy)-3-pentene-1-ynyl}-phenol, Capsanthin, Capsorubin, Capsanthin 5, 6-epoxide, 3-O-[α-L-rhamnopyranosyl-(1→2)-α-L-rhamnopyranosyl-(1 → 4)-β-D-glucopyranosyl]-(25S)-spirost- 5-ene-3β-ol, 2-hydroxyasparenyn 4′-trans-2-hydroxy-1-methoxy-4-5(4-methoxyphenoxy)-3-penten-1-ynyl-benzene, Adscendin A, Adscendin B, Asparanin A-C, Curillin G, Epipinoresinol, 1,3-O-diferuloylglycerol, 1,2-O-diferuloylglycerol, Linoleic acid, Blumenol C, Asparagusic acid oxide methyl esters, 2-Hydroxyasparenyn, Asparenyn, Asparenyol, Monopalmitin, Ferulic acid, 1,3-O-Di-p-coumaroylglycerol, 1-O-Feruloyl-3-O-p-coumaroylglycerol, Inulin, Officinalisins I and II, Beta-sitosterol, Dihydroasparagusic acid, S-acetyldihydroasparagusic acid, Alpha-amino-dimethyl-gamma-butyrothetin, Succinic acid, Sugars, Daidzein, p-Hydroxybenzoic acid, p-Coumaric acid, Gentisic acid, Asparagusate dehydrogenase I and II, Lipoyl dehydrogenase

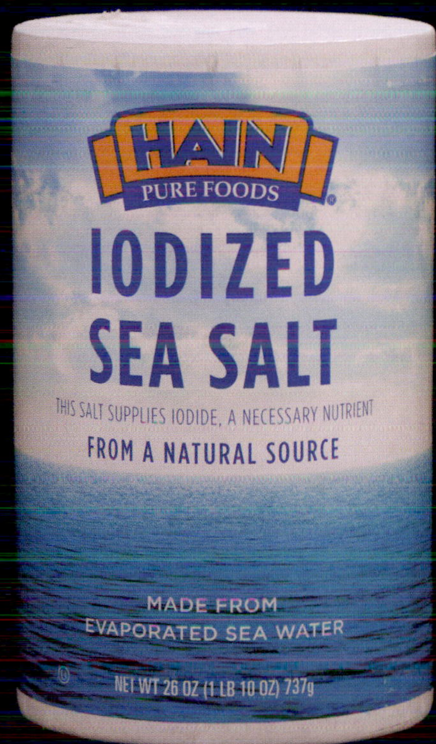

◀ Iodized salt is an early and nearly universal success story in redesigning a natural food to be better for you. To be healthy, humans require a certain amount of iodine in their diet, and in many parts of the world, they get it through the iodine content of ordinary foods. But in areas where the soil is naturally very low in iodine, a normal diet may not be enough. It was thus decided that it would be a good idea to add a small amount of iodine to salt, a natural potential carrier for this nutrient. This widely implemented good idea has virtually eliminated the diseases that used to be caused by iodine deficiency.

▶ Milk is almost universally fortified with vitamin D for the same reason salt has artificially added iodine: it's good public health policy. Vitamin D deficiency used to be shockingly common among children but is now essentially unheard of, largely because of fortified milk.

Intentional Food

One area where people seem to approve of long ingredient lists is added vitamins. No matter how much you dislike chemicals, you can't live without these particular ones. In this array, you can see about two grams of each pure vitamin (except for vitamin B12, which is so expensive in pure form that I got only a single gram of it). Next to each image is a sometimes quite astonishing number: how long two grams of this vitamin would last if you ate the official recommended daily allowance every day. It ranges from 22 days for vitamin C up to 2,280 *years* for vitamin B12. The daily dose of B12 is a mere 2.4 micrograms, an amount that would make about one respectable speck of dust. Such tiny amounts are needed because vitamins often serve in a catalytic role: they work with enzymes that transform one chemical into another in the body, without being consumed themselves. So a supply of the vitamin in your body may last a long time and be sufficient even if replenished only infrequently.

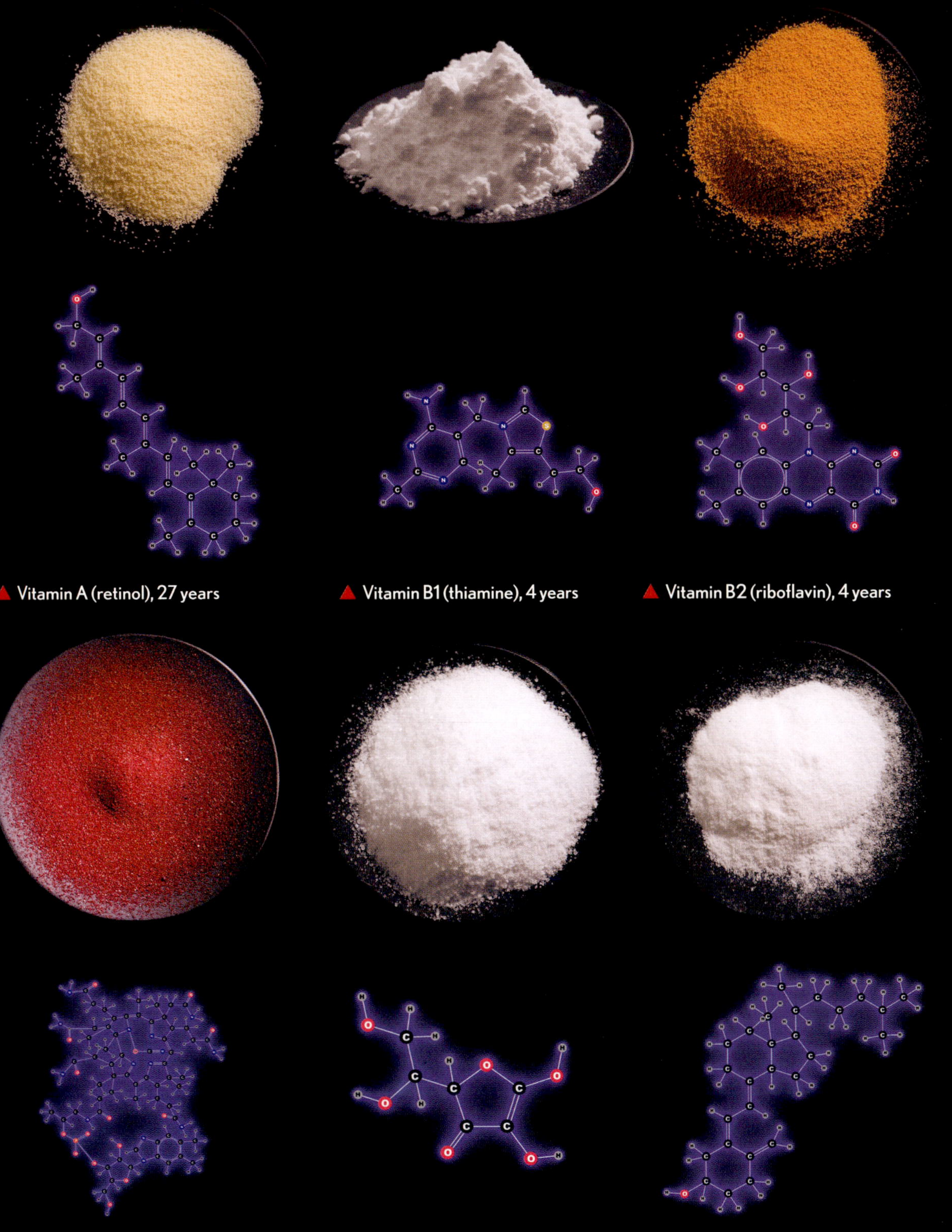

▲ Vitamin A (retinol), 27 years

▲ Vitamin B1 (thiamine), 4 years

▲ Vitamin B2 (riboflavin), 4 years

▲ Vitamin B9, (folic acid), 14 years

▲ Vitamin B12, (cyanocobalamin), 2,280 years (if this were 2g; it's actually 1g)

▲ Vitamin C, (ascorbic acid), 22 days

▲ Vitamin D3, (cholecalciferol), 548 years

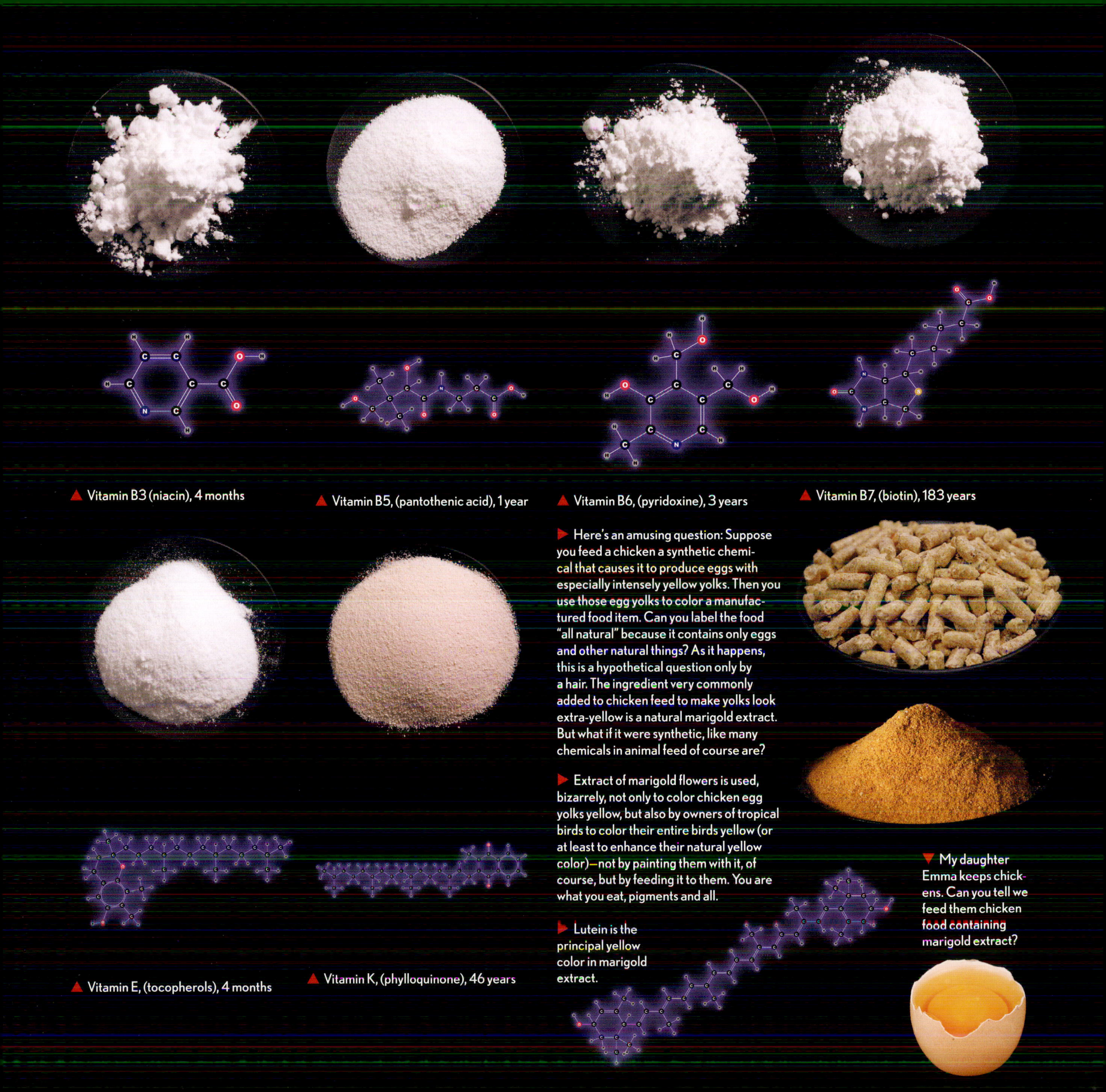

▲ Vitamin B3 (niacin), 4 months

▲ Vitamin B5, (pantothenic acid), 1 year

▲ Vitamin B6, (pyridoxine), 3 years

▲ Vitamin B7, (biotin), 183 years

▲ Vitamin E, (tocopherols), 4 months

▲ Vitamin K, (phylloquinone), 46 years

▶ Here's an amusing question: Suppose you feed a chicken a synthetic chemical that causes it to produce eggs with especially intensely yellow yolks. Then you use those egg yolks to color a manufactured food item. Can you label the food "all natural" because it contains only eggs and other natural things? As it happens, this is a hypothetical question only by a hair. The ingredient very commonly added to chicken feed to make yolks look extra-yellow is a natural marigold extract. But what if it were synthetic, like many chemicals in animal feed of course are?

▶ Extract of marigold flowers is used, bizarrely, not only to color chicken egg yolks yellow, but also by owners of tropical birds to color their entire birds yellow (or at least to enhance their natural yellow color)—not by painting them with it, of course, but by feeding it to them. You are what you eat, pigments and all.

▶ Lutein is the principal yellow color in marigold extract.

▼ My daughter Emma keeps chickens. Can you tell we feed them chicken food containing marigold extract?

Chapter 11

Rose and Skunk

SCENTS ARE MOLECULES as messengers. They come into the nose, spend a bit of time bound to smell receptors, and then wash out with the next breath. Although some things smell with no particular purpose, many smells exist in order to communicate a specific piece of information.

There is one universal fact about all smell molecules: they must be fairly small and simple. Why? Because in order to *be* a smell, a molecule has to reach your nose, and in order to reach your nose, it has to evaporate. As a general rule, the larger a molecule is, the higher its boiling point, and the less of it will evaporate at temperatures below the boiling point.

But within that constraint, there is a great deal of room for interesting molecules.

◀ Menthol opens the nose, but despite being a volatile plant extract, at room temperature it forms beautiful large crystals.

▶ The perfume industry dates back thousands of years to when, due to limited personal hygiene, perfume was a lot more necessary. Wine and perfume makers share the same ridiculous language for describing taste and smell—lots of "fruity notes" and the like. But these fancy bottles all come down to a few dozen volatile compounds, which, as it happens, are mostly esters.

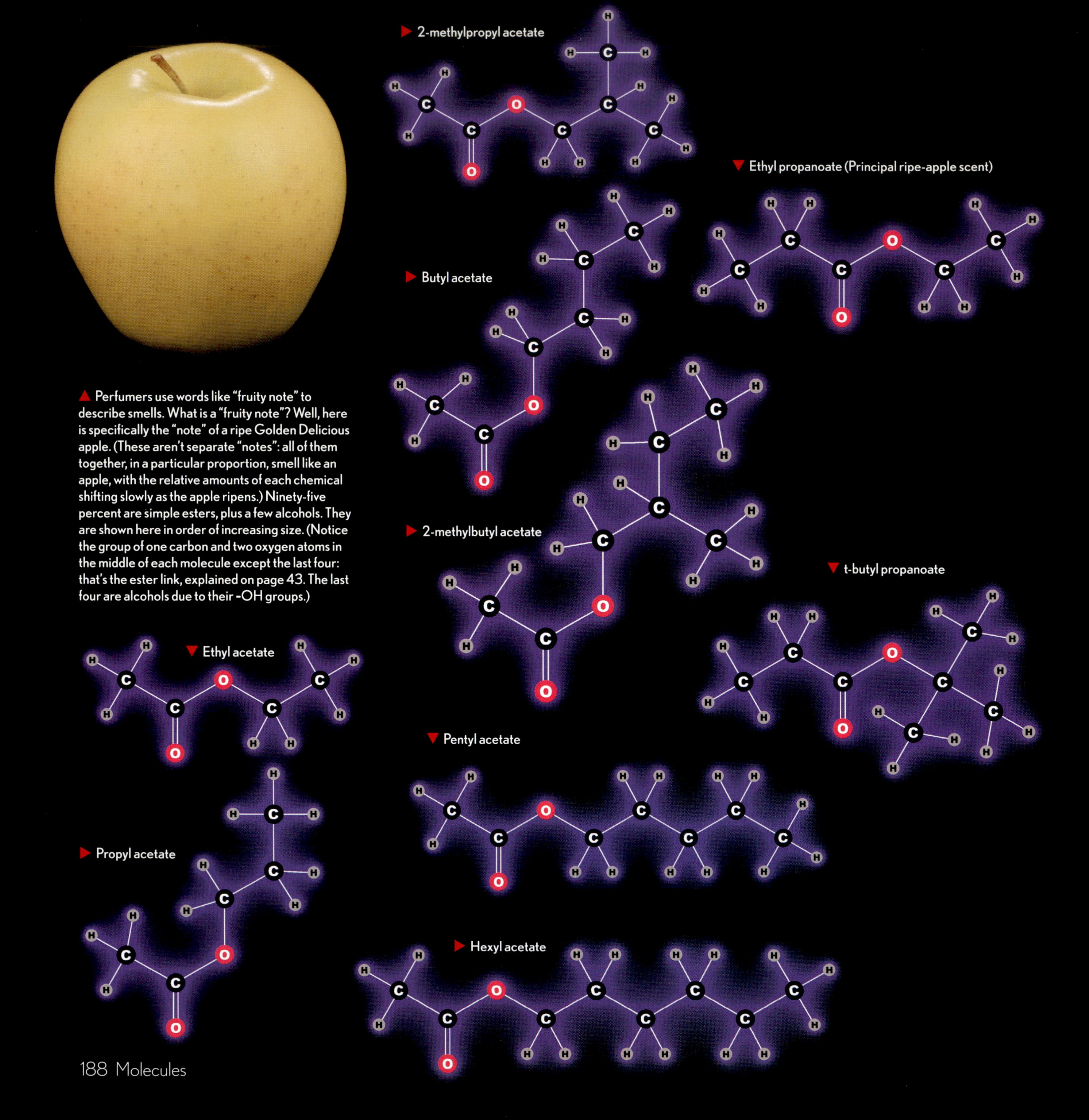

▲ Perfumers use words like "fruity note" to describe smells. What is a "fruity note"? Well, here is specifically the "note" of a ripe Golden Delicious apple. (These aren't separate "notes": all of them together, in a particular proportion, smell like an apple, with the relative amounts of each chemical shifting slowly as the apple ripens.) Ninety-five percent are simple esters, plus a few alcohols. They are shown here in order of increasing size. (Notice the group of one carbon and two oxygen atoms in the middle of each molecule except the last four: that's the ester link, explained on page 43. The last four are alcohols due to their -OH groups.)

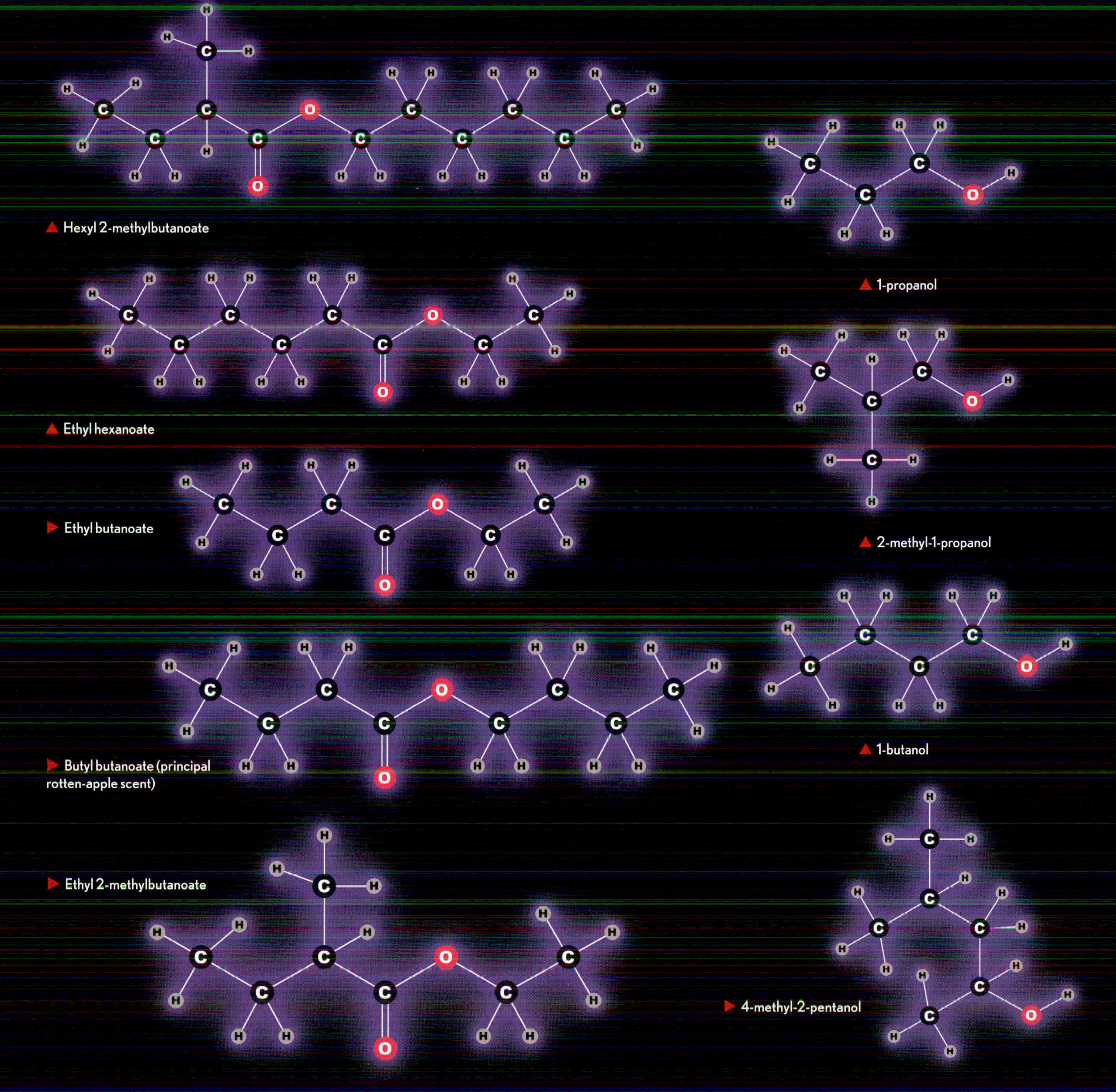
Hexyl 2-methylbutanoate
1-propanol
Ethyl hexanoate
2-methyl-1-propanol
Ethyl butanoate
1-butanol
Butyl butanoate (principal rotten-apple scent)
Ethyl 2-methylbutanoate
4-methyl-2-pentanol

Let's be honest here: ninety percent of perfume is about attraction. The perfume industry has given the field of human pheromones (scents used to attract mates) a bit of a quack feel because their claims are highly exaggerated. Nonetheless, the fact is that humans, like all animals and even plants, do signal each other using smell molecules. While you may have to be as dumb as a potato to respond to someone wearing too much perfume, the beauty of perfume is that sometimes, no matter how smart you are, it'll make you think like a potato.

Questionable claims abound in the pheromone industry. This product says it's high in androstadienone and several related compounds, for which there is *some* indication that they *might* be involved in attraction between humans.

Androstadienone

When it comes to insects, there is absolutely no doubt that chemical pheromones control their lives. This powerful pheromone, which is called bombykol and is used by silk moths, is an alcohol with a long hydrocarbon chain. A vanishingly small amount of it will attract months from hundreds of meters away. (OK, I admit, this isn't a silk moth as it should be to represent bombykol. Instead it's a Giant Atlas moth, which totally lives up to its name, pictured here approximately life-size.)

Bombykol

Humans can, at least sometimes, resist the urge to be lured into dangerous situations by the smell of an attractive potential mate. Insects, being slightly less clever, have essentially no power to resist this urge, making insect pheromones popular as baits in traps—talk about bait and switch!

THE FEMALE GIANT ATLAS MOTH

Ants use a range of straight-chain hydrocarbons with between twenty-three and thirty-one carbons as scent markers. The ants of a given colony lay down a unique pattern of these molecules that allows them to recognize their nest and return to it efficiently after foraging. Ants don't have enough neurons in their tiny little brains to get nostalgic about anything, but if they did, these are the compounds that would represent the comforts of home, just like other, more complex and varied molecules give us a visceral sense of comfort and security when we return to a familiar place of safety. For ants, these molecules are the smell of home.

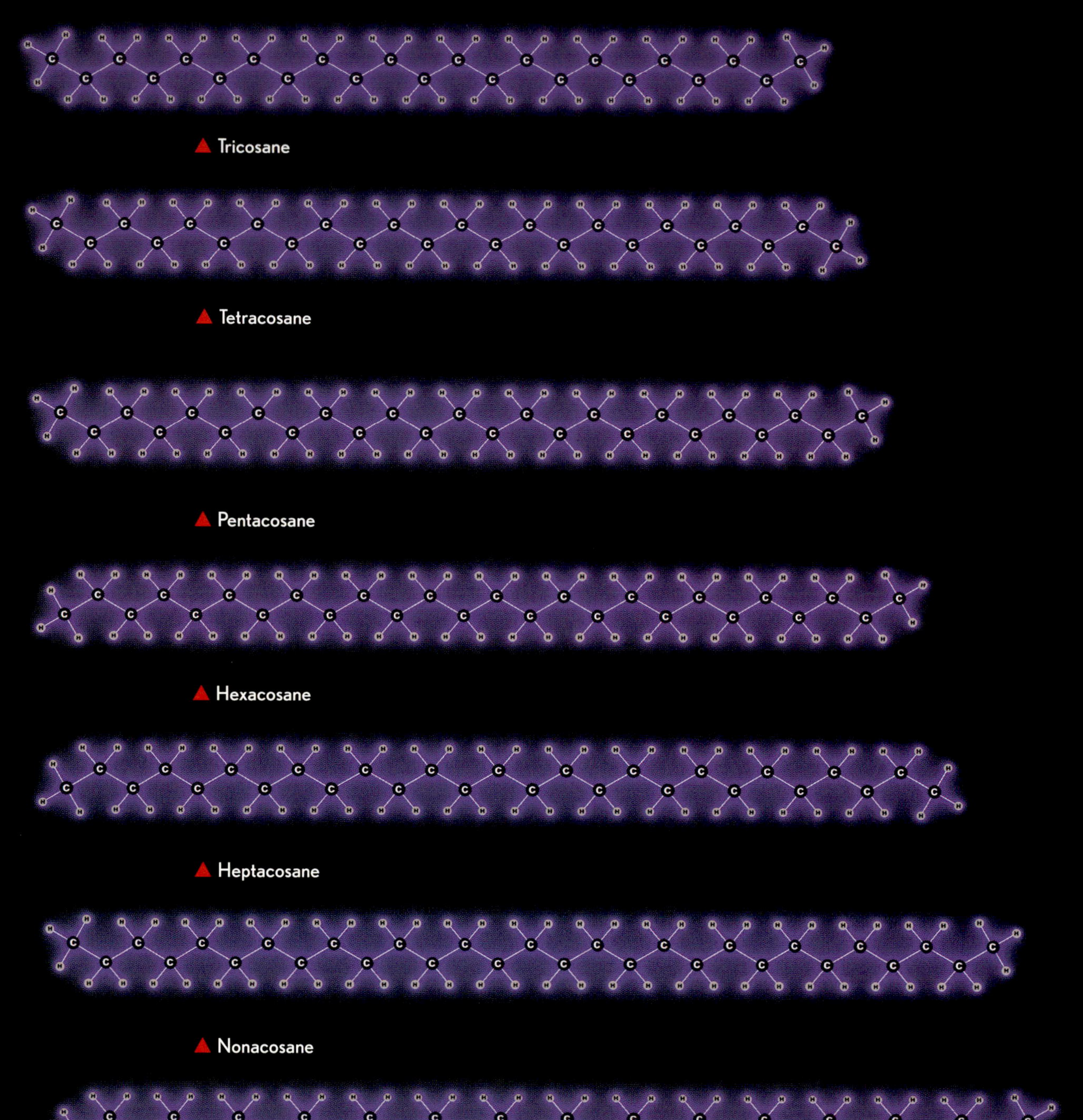

Tricosane

Tetracosane

Pentacosane

Hexacosane

Heptacosane

Nonacosane

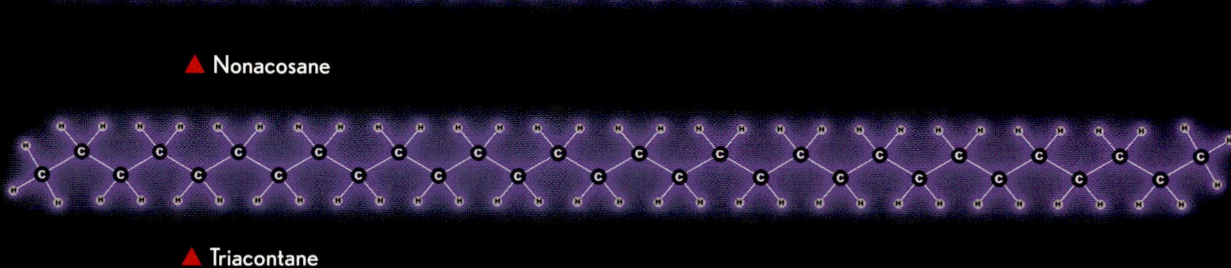

Triacontane

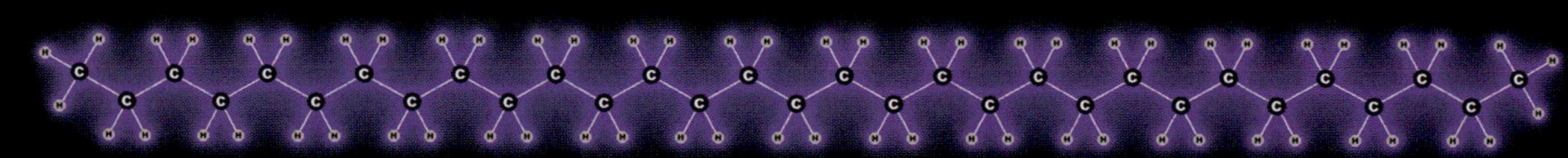

Hentriacontane

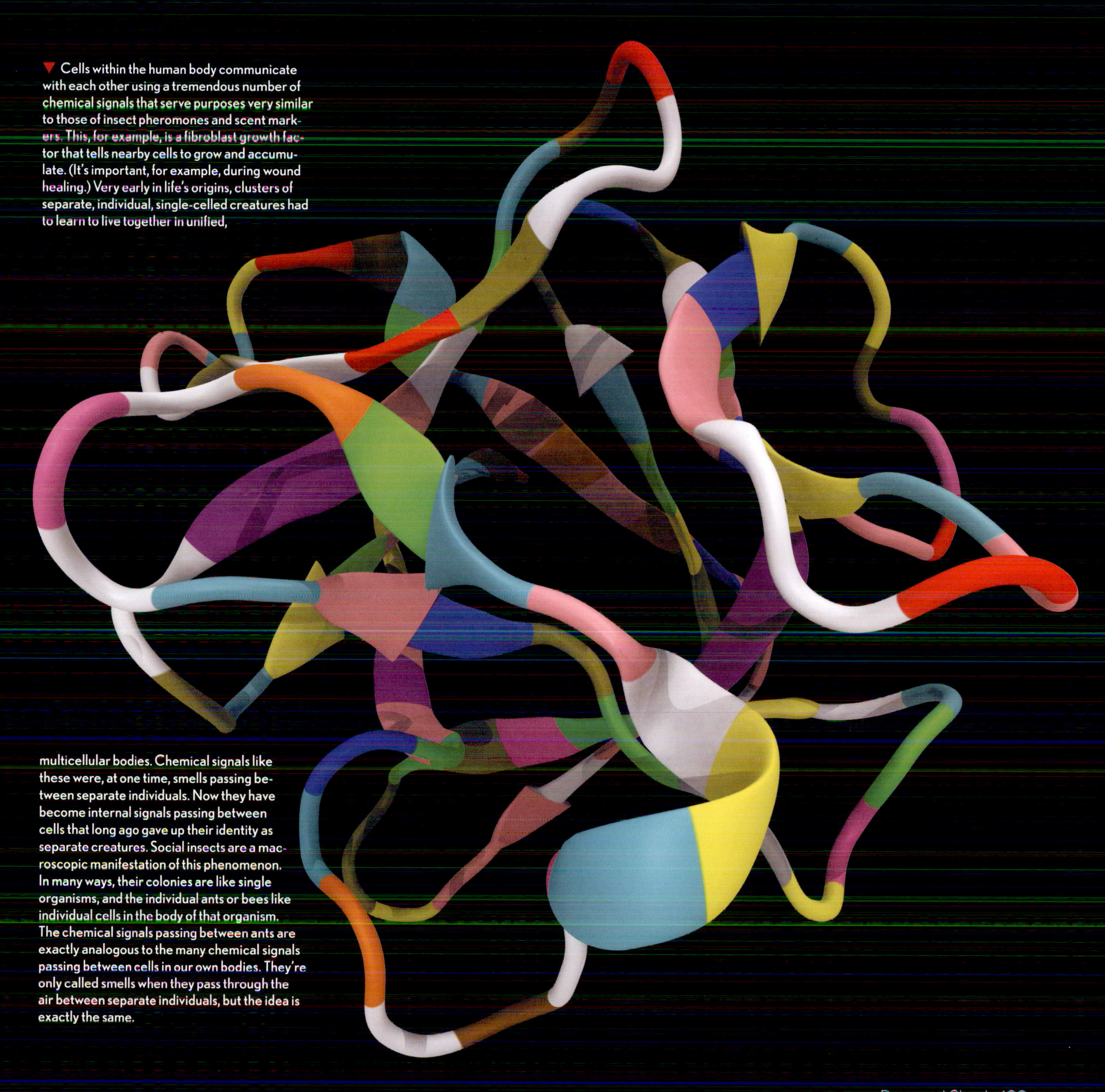

▼ Cells within the human body communicate with each other using a tremendous number of chemical signals that serve purposes very similar to those of insect pheromones and scent markers. This, for example, is a fibroblast growth factor that tells nearby cells to grow and accumulate. (It's important, for example, during wound healing.) Very early in life's origins, clusters of separate, individual, single-celled creatures had to learn to live together in unified, multicellular bodies. Chemical signals like these were, at one time, smells passing between separate individuals. Now they have become internal signals passing between cells that long ago gave up their identity as separate creatures. Social insects are a macroscopic manifestation of this phenomenon. In many ways, their colonies are like single organisms, and the individual ants or bees like individual cells in the body of that organism. The chemical signals passing between ants are exactly analogous to the many chemical signals passing between cells in our own bodies. They're only called smells when they pass through the air between separate individuals, but the idea is exactly the same.

Many perfumes, candles, incense sticks, and other smelly things derive their smells from "essential oils." These are extracts from flowers, seeds, leaves, herbs, and so forth that contain various mixtures of volatile organic compounds. For example, bisabolene is part of the smell of bergamot, ginger, and lemon oil, while eucalyptol makes up part of the smell of lavender, peppermint, and eucalyptus. With an extract, the flower, or other fragrant source, is soaked in some mixture of solvents—usually including alcohol—to pull out the soluble components. The solvent is then distilled or evaporated, concentrating the molecule of interest. The same compounds keep coming up over and over again in these oils in slightly different proportions.

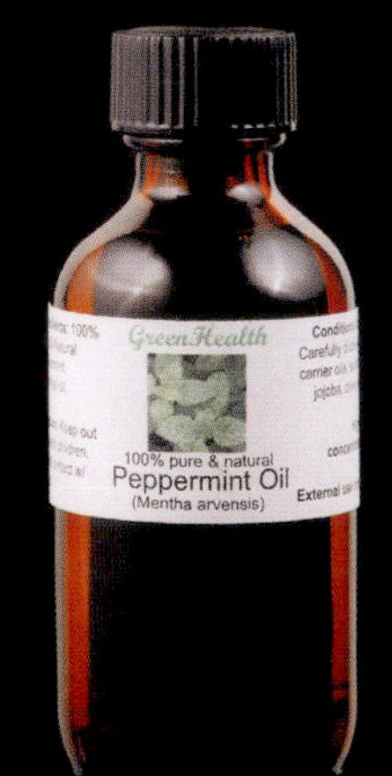

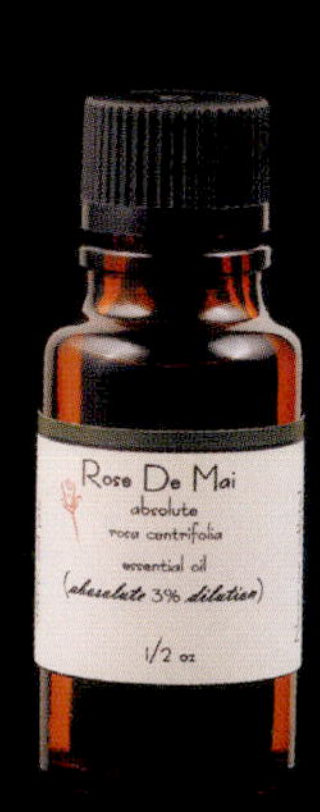

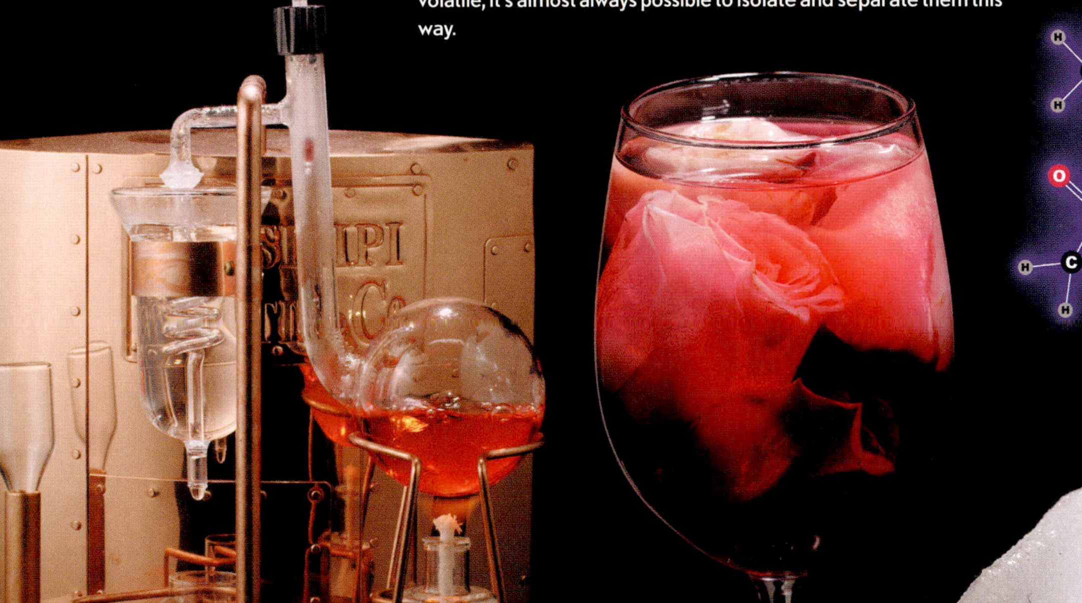

Distillation of plant extracts is a delicate art that can be done on a hobby scale with a distillation apparatus like this. Distillation is a process of controlled evaporation and condensation that allows you to separate out, from a mixture, components that boil at different temperatures. Because scent molecules are necessarily volatile, it's almost always possible to isolate and separate them this way.

Camphor is a component of lavender and rosemary oils.

Eucalyptol, like camphor, is a double-cyclic compound, but small differences in its structure make it a liquid at room temperature. However, it's just as good as camphor at opening the nose.

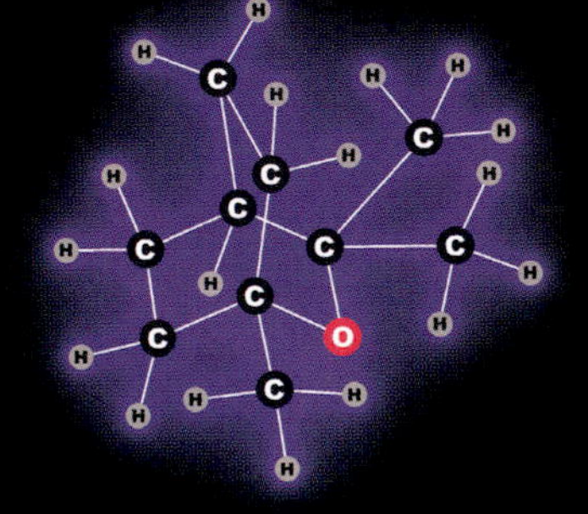

Eucalyptol is structurally similar to camphor and occurs in lavender, peppermint, and, of course, eucalyptus oils.

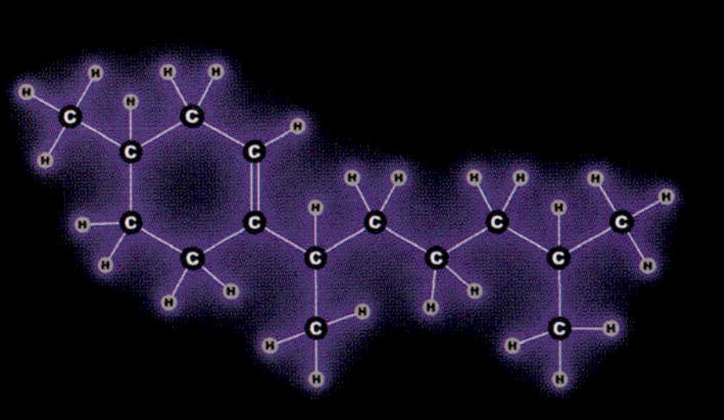

Bisabolene is found in bergamot, ginger, and lemon oils and contributes to the scent of each. Each specific essential oil is composed of a dozen or more compounds of this general nature. Some are characteristic of just one essential oil, while others occur across many.

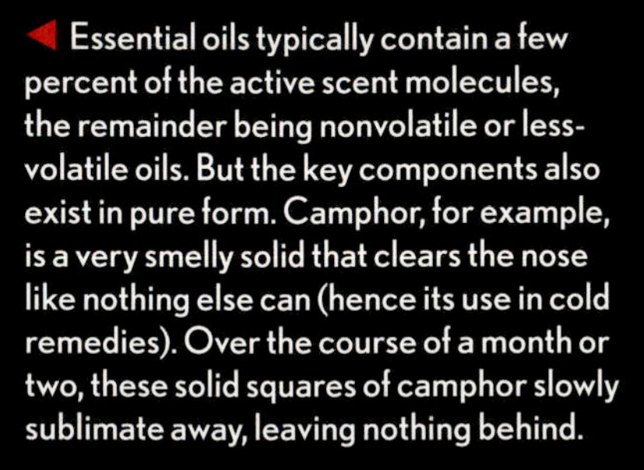

Essential oils typically contain a few percent of the active scent molecules, the remainder being nonvolatile or less-volatile oils. But the key components also exist in pure form. Camphor, for example, is a very smelly solid that clears the nose like nothing else can (hence its use in cold remedies). Over the course of a month or two, these solid squares of camphor slowly sublimate away, leaving nothing behind.

▶ Menthol is a component of peppermint and spearmint oils as well as menthol cigarettes.

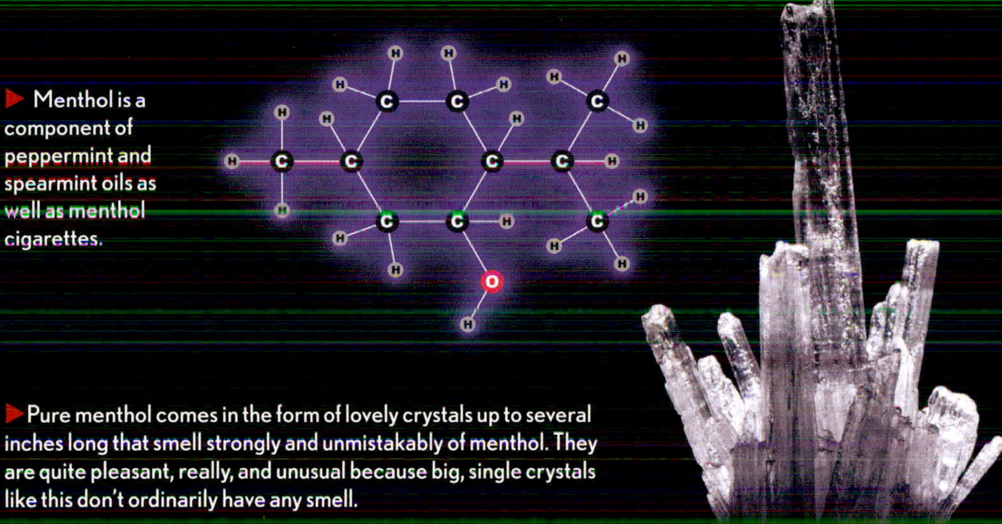

▶ Pure menthol comes in the form of lovely crystals up to several inches long that smell strongly and unmistakably of menthol. They are quite pleasant, really, and unusual because big, single crystals like this don't ordinarily have any smell.

▶ Thymol gives the herb thyme its distinctive aroma.

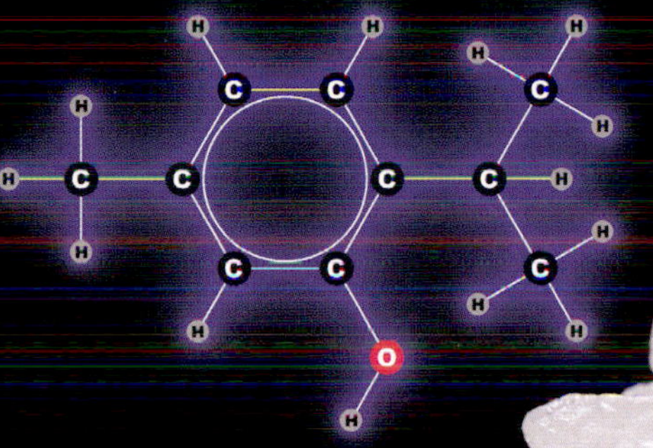

▶ These lumps smell very strongly of something herbal. Not surprising since they are thymol, the essence of the herb thyme—extracted, distilled, and crystallized.

▼ Smell molecules, almost by definition, have to be small so they are volatile enough to reach the nose. This is an example of one that pushes the limit, with a total of forty-two atoms arranged in a very unusual extra-large ring structure. (The vast majority of organic rings have six members, and a fair number have five or seven, but few have more or less than that. This one has seventeen atoms forming its ring.) I have no idea what the thing smells like, and as is typical with descriptions of smells by smell professionals, I still have no idea after reading its description: "It is a superb fixative and highly substantive, and yet exalts the top note of a fragrance in an exceptional manner" (from Givaudan, a smell and taste supplier).

▶ This complex alcohol, ambrein, is the fanciest smell of all according to the perfume industry. It is the principal scent component of the rare and remarkable substance called ambergris.

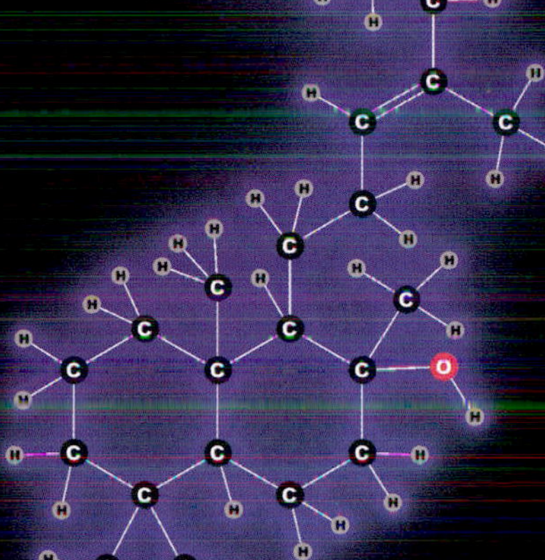

▲ Ambrein is extracted from sperm whale vomit (ambergris). It's expensive stuff that smells kind of "meh" to me, but apparently when combined with other expensive perfume ingredients, it's fabulous. Many of the most famous classic scents include ambrein.

▼ Ambergris is a waxy substance produced in the sperm whale's stomach, perhaps to help expel sharp objects such as squid beaks (see page 120). The best quality material floats in the sea for years after coming out of one or the other end of the whale. It's collected when it occasionally washes up on beaches and is then sold to perfume companies for tens of thousands of dollars a pound. (This single gram cost me $150.) You just can't make this stuff up.

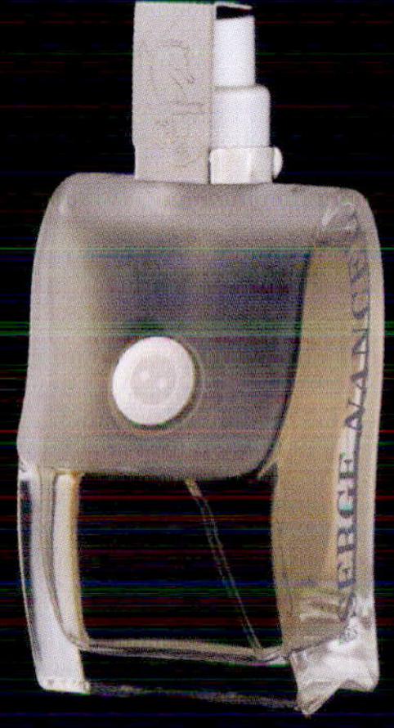

▶ It's not just whale snot that interests the perfume industry: they have their eye on beaver butt as well—specifically on the castoreum excreted by beavers through their anal glands. Castoreum is used by these fine animals to mark their territory.

▼ Many animals use their pee as perfume—by which I mean they use it for the same purpose as humans use perfume: to signal availability or interest and to influence the behavior of other animals. And because humans are also sometimes interested in influencing the behavior of animals, you can buy an amazing array of bottled animal pee. I don't recommend it. (This stuff actually does have a purpose: it's used for attracting animals to hunters, for scaring animals away from gardens, and for initiating mating in domestic animals. For example, if you're an unsportsmanlike hunter wishing to shoot a male deer, you might use the pee of a female dear in heat to attract the stag. If you're a gardener plagued by rabbits, you might waft the pee of their natural predators to scare them away.)

◀ I had to buy this rubber-gasket-sealed metal ammo case from an army surplus store just to house my animal pee collection. The smell is *awful* and comes right through the tightly closed plastic bottles. I can't imagine the smell in the factory where they bottle this stuff.

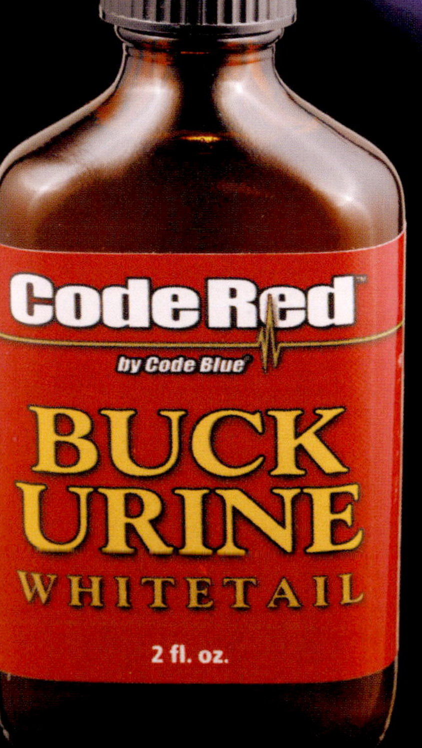

▲ Hydrogen sulfide, like many compounds containing sulfur, smells terrible. Specifically, it's the smell of rotting eggs and volcanos.

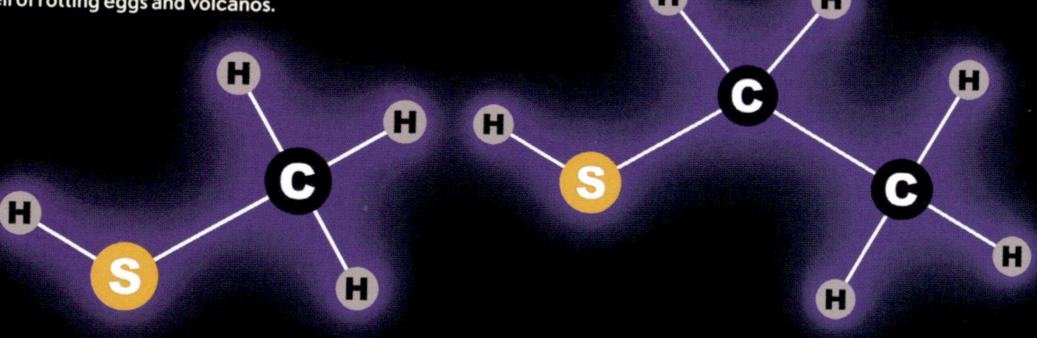

▲ Horrible smells sometimes serve a very important purpose. Methyl mercaptan and ethyl mercaptan, despite their names, don't contain any mercury. Instead, they are organosulfur compounds that do their best to uphold the reputation of this class of molecules as being incredibly foul. Methyl mercaptan is basically the smell of farts. Ethyl mercaptan can be smelled by humans at less than half a part per billion, and they don't like it. This is precisely why it's added to natural gas and propane, which would otherwise be odorless. The entire concept of "reporting a gas leak" exists only because of the smell of ethyl mercaptan. Without it, the first sign of trouble would be a massive explosion. As it is, such explosions do happen with some regularity but usually when no one is in the house, because people either fix the problem or flee when they smell the distinctive odor of ethyl mercaptan.

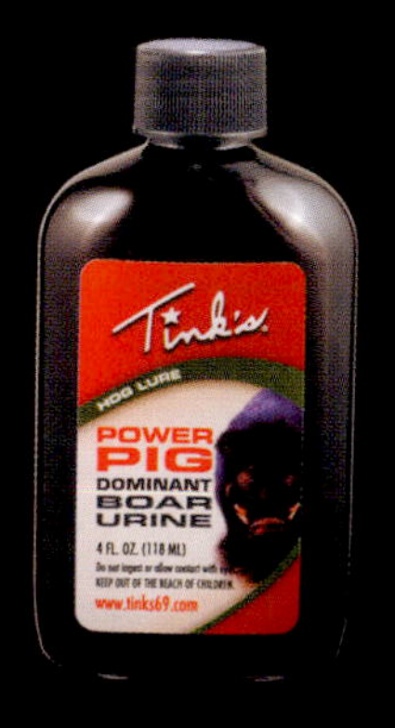

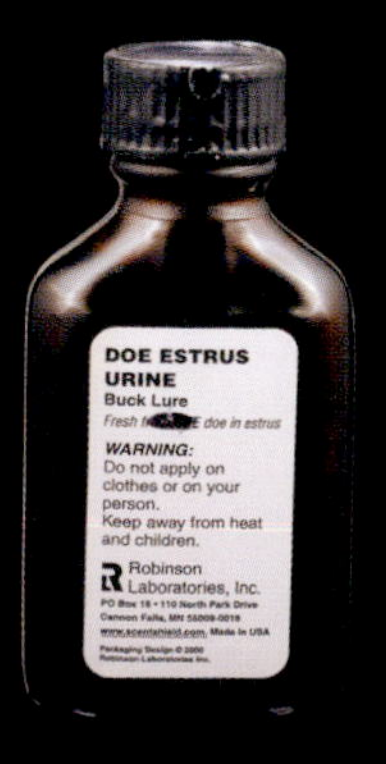

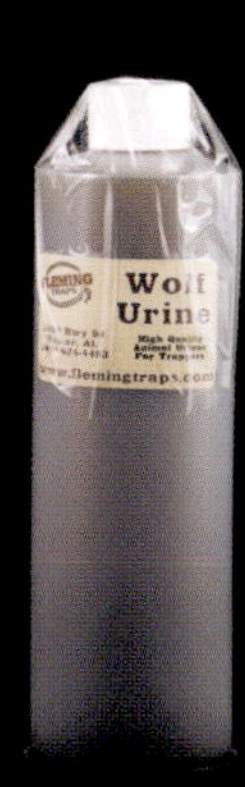

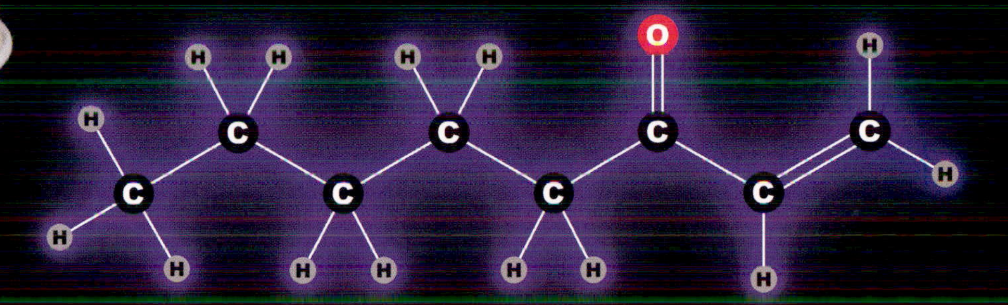

▲ This molecule, amyl vinyl ketone, is the smell of money. It doesn't come from the money, it comes from the skin of all those who have handled it.

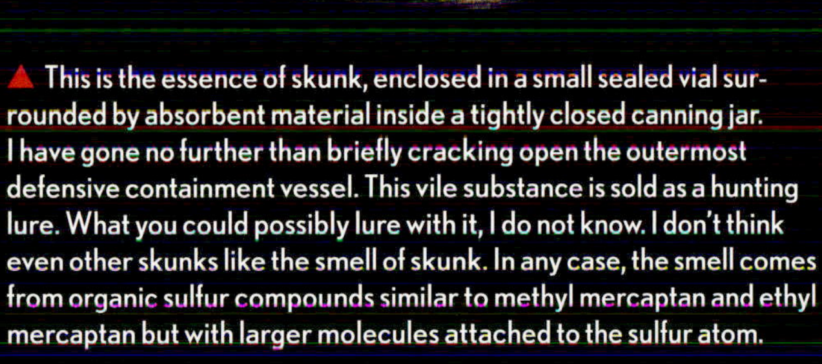

▲ This is the essence of skunk, enclosed in a small sealed vial surrounded by absorbent material inside a tightly closed canning jar. I have gone no further than briefly cracking open the outermost defensive containment vessel. This vile substance is sold as a hunting lure. What you could possibly lure with it, I do not know. I don't think even other skunks like the smell of skunk. In any case, the smell comes from organic sulfur compounds similar to methyl mercaptan and ethyl mercaptan but with larger molecules attached to the sulfur atom.

◀ There's some dispute about exactly why your pee smells funny after eating asparagus. And if you don't think asparagus makes your pee smell funny, it's because not everyone can smell the breakdown products of asparagus metabolism at very low levels. They tested 328 people to determine this fact.

▲ The "smell of money" (coins specifically) cannot possibly be the money itself. Metal is absolutely nonvolatile and so cannot reach your nose. After considerable research, it has been determined that the highly characteristic smell of coins, and other metal surfaces, comes from the catalytic breakdown of skin oils into a few smaller volatile compounds. It's interesting that animals evolved the ability to distinguish a characteristic smell of metal, since metals basically do not exist free in nature. One theory is that a very similar smell is produced by the iron in blood. If that's true, then it could be said that the lust for money is indeed a bloodlust.

Chapter 12

Color Me Chemical

YOU MAY HAVE NOTICED an inordinate presence of white powders throughout this book. And that's despite my very best efforts to find something, anything, to photograph other than yet another white powder. The sad fact is that nearly all pure, organic compounds are white. This isn't too surprising when you think about what's needed to make a substance colorful.

White light is a mixture of all colors, and when we say a compound is colored, we mean that it reflects light of one color (which corresponds to light photons of a particular range of wavelengths) more strongly than light of other colors.

For example, if a molecule absorbs mostly blue light, then it will look yellow because there is more yellow light reflected after the blue has been absorbed.

But visible light spans only a tiny fraction of the full range of possible wavelengths. A molecule might absorb photons from anywhere across the whole electromagnetic spectrum, from radio waves through hard X-rays, but it will be colorful only if there is a difference in how strongly it absorbs one *visible* wavelength over another.

That turns out to be fairly uncommon. Most molecules absorb light only above the visible spectrum, in the ultraviolet range. The world looks colorful to us not because there are many different colored compounds but because a few go a long way. Being colorful is a specialist occupation; a few particular molecular structures come up over and over again in colored compounds, and, of course, our eyes have evolved to see the common ones in the natural world around us.

◀ Realgar is arsenic sulfide, an example of a classic painter's pigment that is also a wee bit poisonous.

▼ The electromagnetic spectrum is very wide, spanning over fifteen orders of magnitude (a factor of more than 1,000,000,000,000,000) from radio waves all the way to high-energy gamma rays. Only by drawing the spectrum on a logarithmic scale can you even see the fraction occupied by visible light. We tend to focus on this particular part of the spectrum, but molecules don't care. They can absorb photons across a much wider range, from microwaves (which is why microwave ovens work) through x-rays (which is why medical x-rays work). Only atomic nuclei are dense enough to absorb the energy in the higher-energy ranges.

▶ Flowers are known for sometimes looking dramatically different under ultraviolet (UV) light. That's because bees see further into the UV spectrum than we do, and the colors and patterns of flowers are for their benefit, not ours. It turns out that many organic compounds absorb some light in the UV range that bees can see, so while nearly all organic compounds are white to us, to bees more of them look colored. What colors are they? We have no words for these hues. Their names can be spoken only in the dance language of the bees as they tell their tales of the paths they have flown and the flowers they have seen.

Radio Waves
Microwaves
Terahertz Waves
Infrared Light
Visible Light
Ultraviolet Light
X-Rays
Gamma Rays

Your German Molecules

THE MOST VIBRANT, rich, and diverse coloring agents come from specialist organic compounds, both natural and synthetic. Many organic dyes are astonishingly intense. I have a small lake that holds about four million gallons of water, and every year I pour into it a solution containing about five pounds of a special blue–green dye mixture designed to control algae. (Otherwise, the lake gets gross.) A concentration of about 150 parts per *billion* is all it takes to turn the whole lake a lovely shade of aqua blue!

Organic molecules absorb light when a photon of light interacts with the electrons holding the molecule together, temporarily bumping one out of place. This takes energy, and because the energy content of a photon depends on its color, different electrons can be knocked out by different colors of light, depending on how tightly those electrons are bound into the molecule. Red photons have the least energy, followed by green, then blue, and finally violet photons, which have the most energy in the visible light range. Ultraviolet photons have even more energy. X-rays are photons with so much energy that we don't even call them light anymore.

Electrons that are bound very tightly can be dislodged only by high-energy ultraviolet light or even X-rays. Most electrons in most compounds are bound *that* tightly, which is why the compounds are white. But molecules can be constructed that have nearly any binding strength you like, including in the right range to selectively absorb some, but not other, colors of light.

There are a few particularly common molecular structures with electrons in the right range of binding strength. These have given rise to families of dyes. By changing which atoms are arranged around the active center, the binding strength—and thus the color—can be tuned across the visible spectrum.

▶ Indigo played a part in cultural traditions around the world and was a major item of global trade for centuries. This kimono shows its use in Japanese traditional dress.

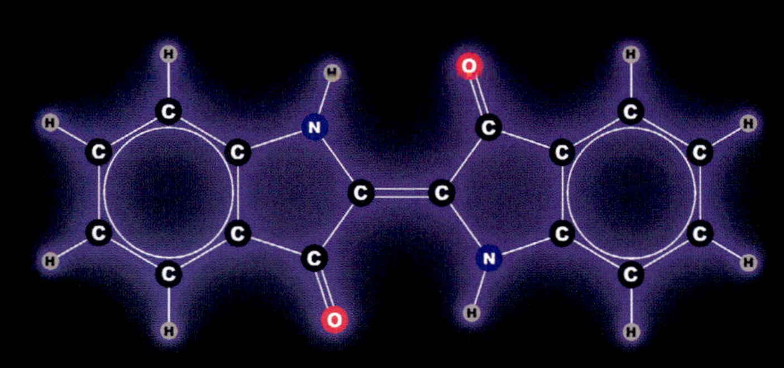

▲ Indigo, the classic natural dye, derives its color from a set of three double-bonds in the center of its lovely, symmetrical structure. The hydrogen and oxygen atoms that point at each other from opposite sides are not strongly bonded to each other (thus no line is drawn). However, they do form what is called a "hydrogen bond," which helps keep the molecule flat and the three double bonds all in the same plane, allowing electrons to migrate between them with only a minimal bump in energy. The result conveniently corresponds to the color of orange light. (Absorb orange light, and you have indigo-blue light left over.)

▲ Indigo comes, historically, from the *Indigofera tinctoria* plant (and some related species) grown in the tropics. It was a driver of trade in the era of sailing ships due to the hunger of European customers for its rare, rich blue color. Even today, natural ground *Indigofera tinctoria* leaves can be ordered directly from India (through eBay, of course, and it arrives by airplane, not tall-masted clipper ship). The raw leaf powder is green and contains not indigo, but a related compound called indican glycoside. When the powder is heated with water, this compound is converted to indoxyl, a colorless compound that is soluble in water. Fabric in contact with the powder will absorb some of the water containing indoxyl. Contact with air then oxidizes the indoxyl into indigo, which is not soluble in water and thus remains fixed in the fabric.

▲ Virtually all indigo used today is synthetic. About as much is being manufactured today as was harvested from plants in 1897, just before that market collapsed. (You might think far more would be made today, given how many more people there are. But today there are also far more options for dyes. Back then, indigo was about the only game in town for blue.) The push to develop economical ways of synthesizing indigo and novel synthetic dyes was *the* major factor in the development of the organic chemical industry in the second half of the 1800s, and it worked: within fifteen years of the first commercially viable synthesis in 1897, the plant-based material was essentially gone as a significant economic force.

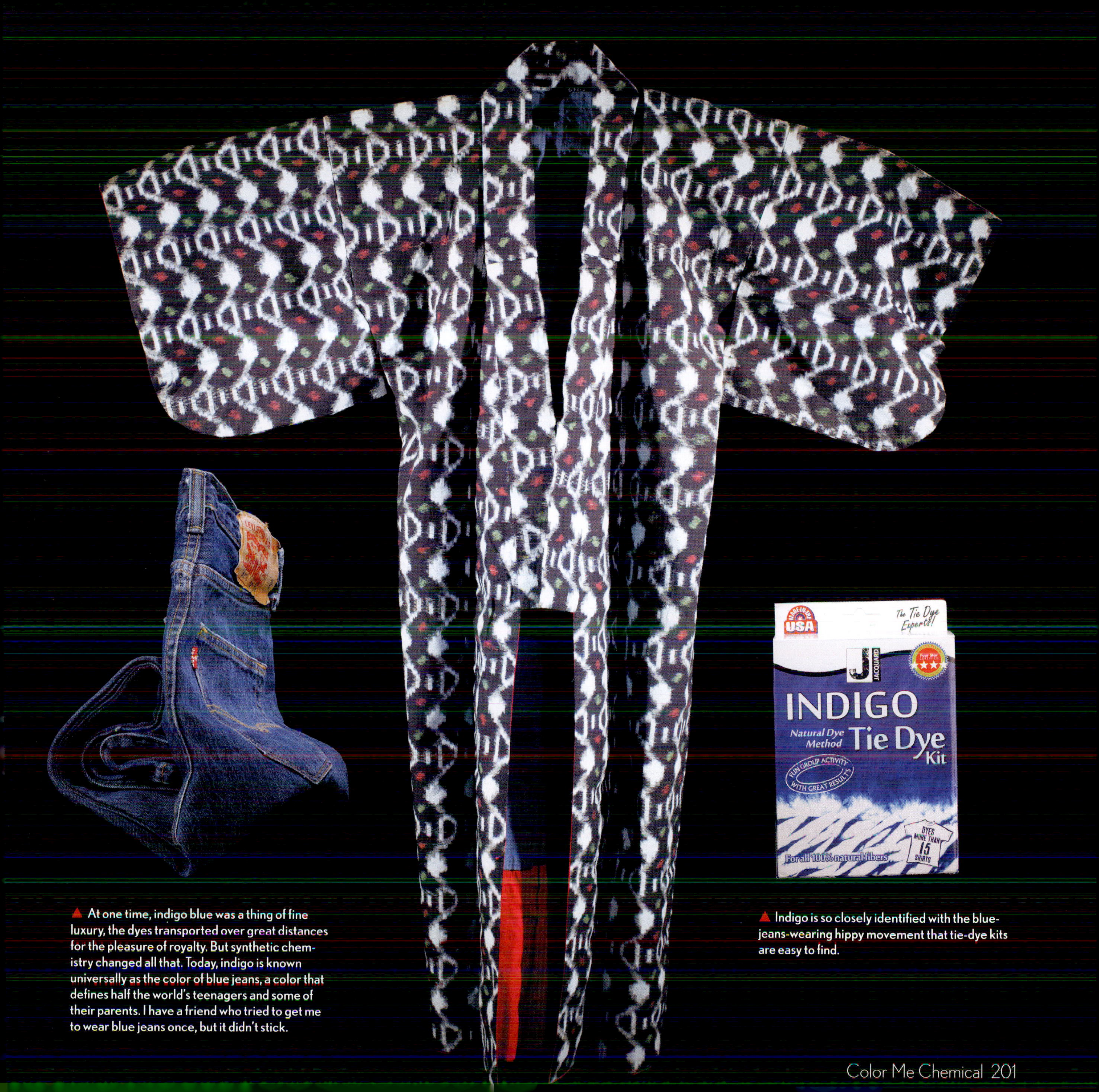

▲ At one time, indigo blue was a thing of fine luxury, the dyes transported over great distances for the pleasure of royalty. But synthetic chemistry changed all that. Today, indigo is known universally as the color of blue jeans, a color that defines half the world's teenagers and some of their parents. I have a friend who tried to get me to wear blue jeans once, but it didn't stick.

▲ Indigo is so closely identified with the blue-jeans-wearing hippy movement that tie-dye kits are easy to find.

Draw Me with One of Your German Molecules

▼ Mauveine (this structure plus three other very similar ones) was the very first synthetic organic dye, called an aniline dye because the starting point for its synthesis was the chemical aniline. Its accidental discovery in 1856 touched off a massive explosion of activity, scientific and industrial, in organic chemistry in Germany, leading to the dominance of Germany as the center of a chemical industry that persists to this day.

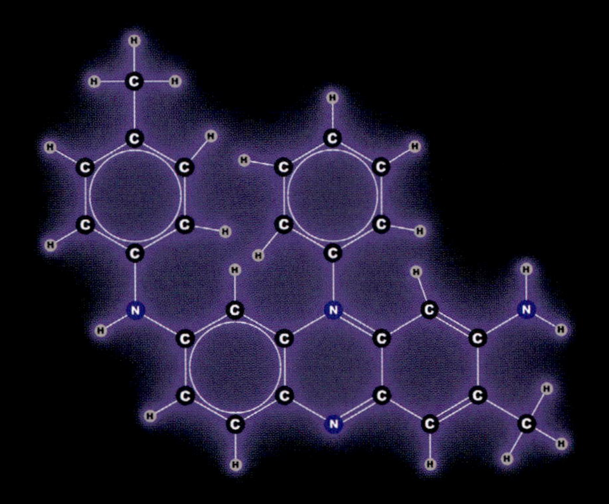

◀ Mauveine caused a sensation in Victorian England when no less a personage than Her Majesty Queen Victoria, for whom the era is named, wore a dress dyed with the fashionable new color.

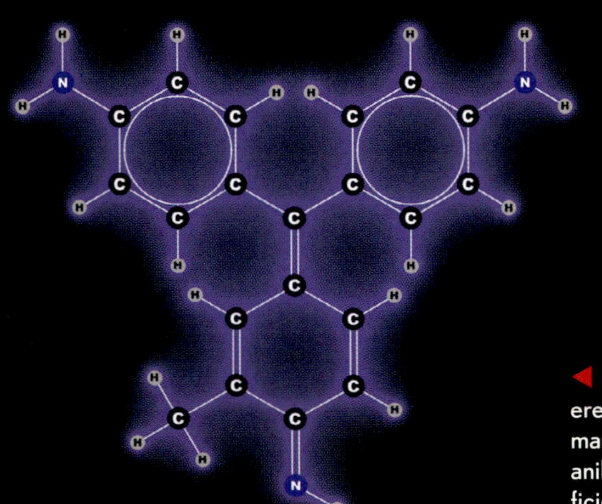

◀ Fuchsine, discovered not long after mauveine, is another aniline dye. The efficient production of aniline from coal tar led to a great many different chemicals becoming available cheaply through various synthetic routes.

◀ Fuchsine was the first synthetic dye made by Friedrich Engelhorn, founder of what became the BASF Corporation, now the largest chemical company in the world. Germany in the 1860s was to organic chemistry what Silicon Valley is to computers today, so of course he carried out this synthesis in his kitchen. (Cars hadn't been invented yet, so he didn't have a garage to work in.) Although it's known as a pink-red dye, fuchsine in dry form is actually green and turns red only when dissolved.

▶ Fushsine is particularly useful for dying silk, which in turn is particularly useful for making ties. Ties are not useful.

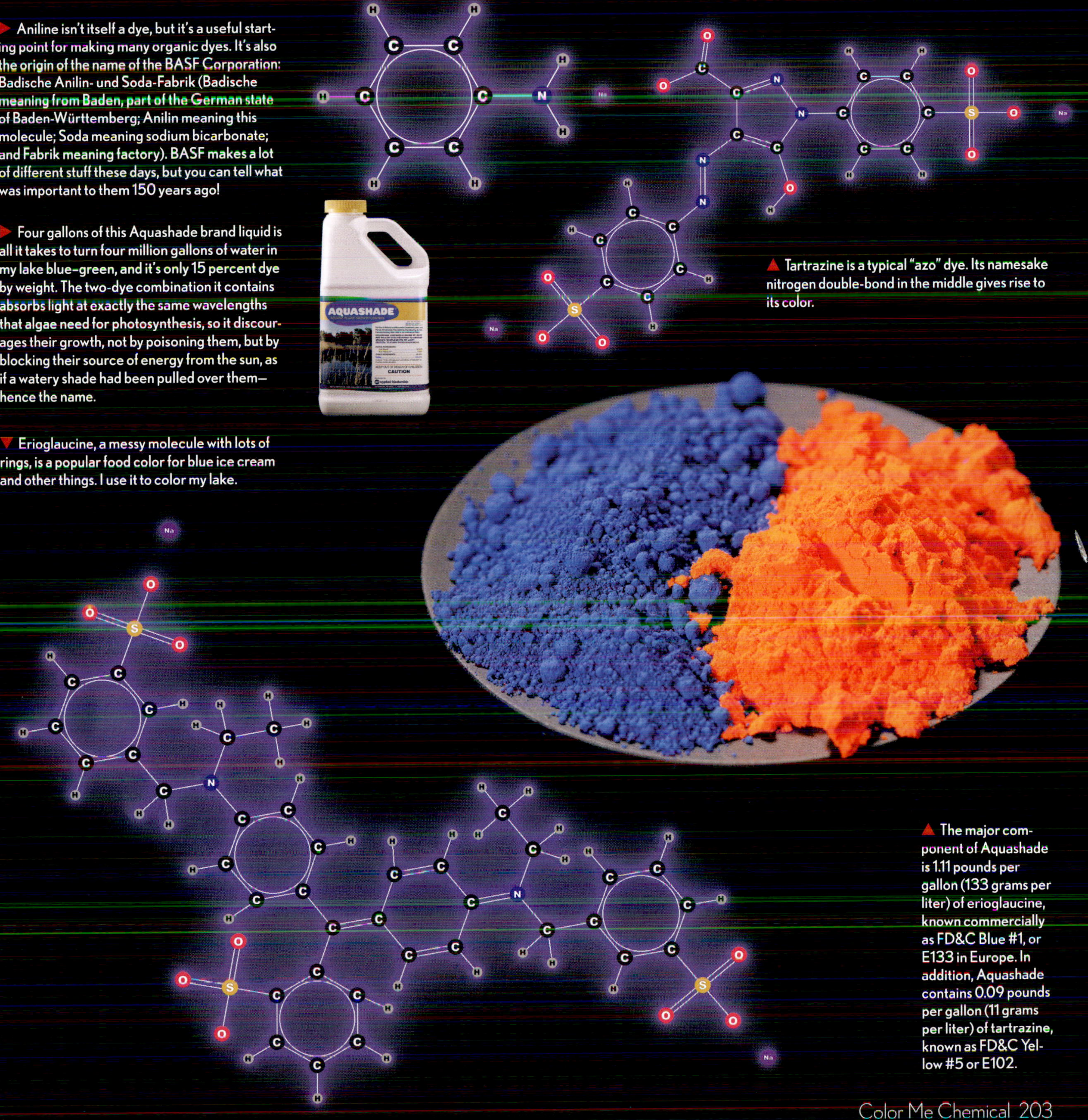

Aniline isn't itself a dye, but it's a useful starting point for making many organic dyes. It's also the origin of the name of the BASF Corporation: Badische Anilin- und Soda-Fabrik (Badische meaning from Baden, part of the German state of Baden-Württemberg; Anilin meaning this molecule; Soda meaning sodium bicarbonate; and Fabrik meaning factory). BASF makes a lot of different stuff these days, but you can tell what was important to them 150 years ago!

Four gallons of this Aquashade brand liquid is all it takes to turn four million gallons of water in my lake blue-green, and it's only 15 percent dye by weight. The two-dye combination it contains absorbs light at exactly the same wavelengths that algae need for photosynthesis, so it discourages their growth, not by poisoning them, but by blocking their source of energy from the sun, as if a watery shade had been pulled over them—hence the name.

Erioglaucine, a messy molecule with lots of rings, is a popular food color for blue ice cream and other things. I use it to color my lake.

Tartrazine is a typical "azo" dye. Its namesake nitrogen double-bond in the middle gives rise to its color.

The major component of Aquashade is 1.11 pounds per gallon (133 grams per liter) of erioglaucine, known commercially as FD&C Blue #1, or E133 in Europe. In addition, Aquashade contains 0.09 pounds per gallon (11 grams per liter) of tartrazine, known as FD&C Yellow #5 or E102.

Draw Me with One of Your German Molecules

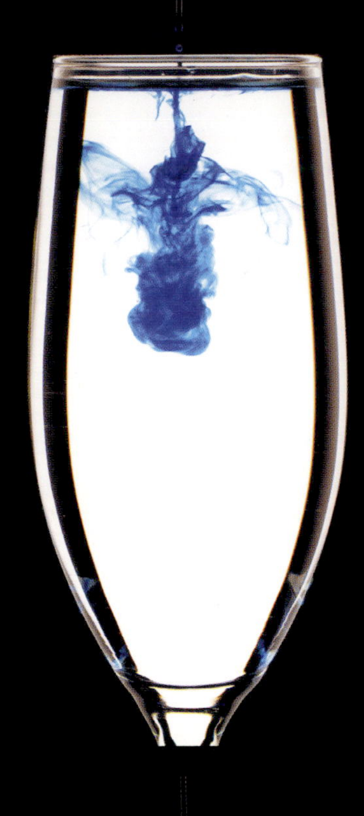

▶ Most of the assorted organic dyes shown in this sequence being squirted into water were already highly diluted from their original concentrations. If not, the water would pretty much turn black immediately.

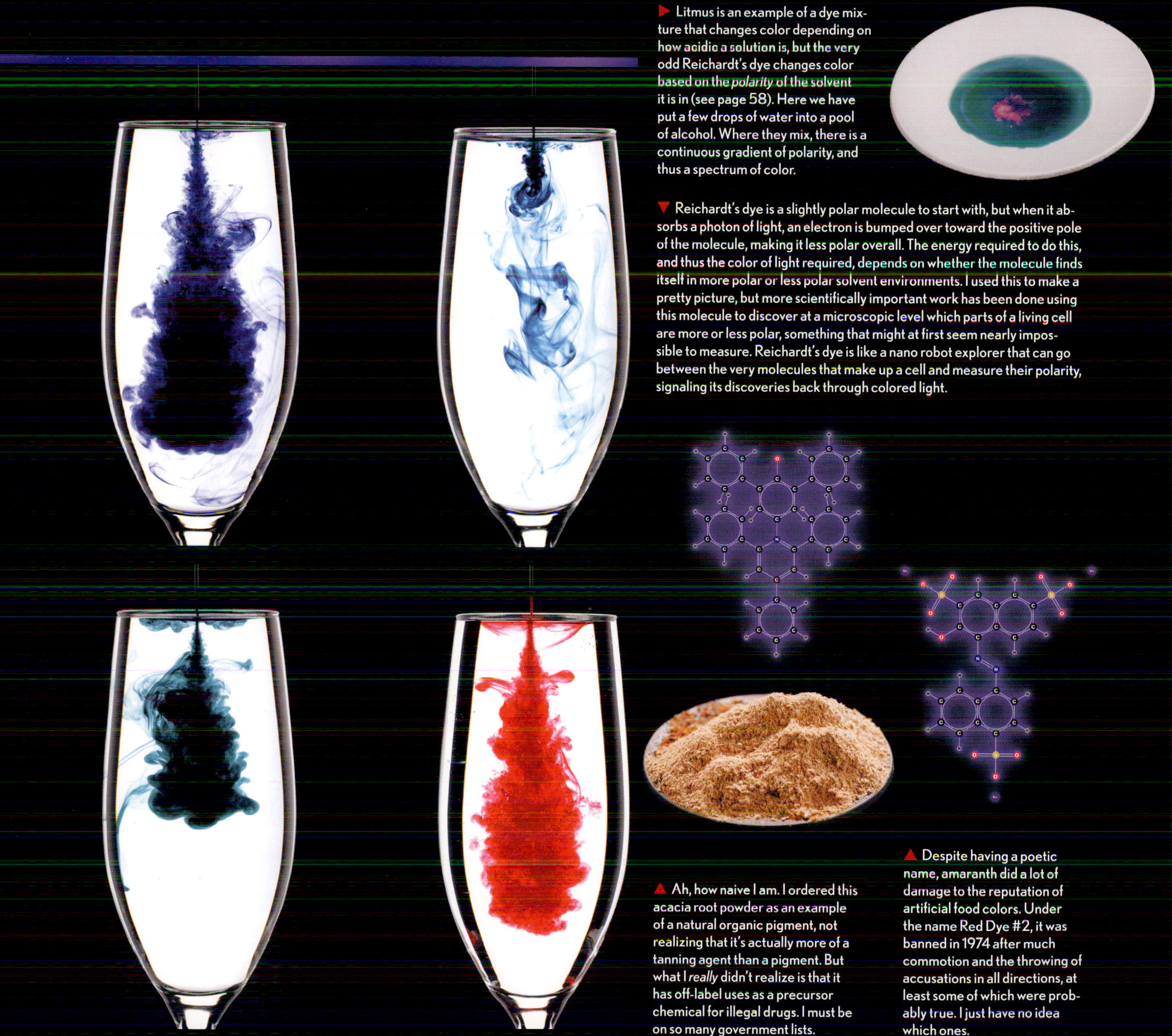

▶ Litmus is an example of a dye mixture that changes color depending on how acidic a solution is, but the very odd Reichardt's dye changes color based on the *polarity* of the solvent it is in (see page 58). Here we have put a few drops of water into a pool of alcohol. Where they mix, there is a continuous gradient of polarity, and thus a spectrum of color.

▼ Reichardt's dye is a slightly polar molecule to start with, but when it absorbs a photon of light, an electron is bumped over toward the positive pole of the molecule, making it less polar overall. The energy required to do this, and thus the color of light required, depends on whether the molecule finds itself in more polar or less polar solvent environments. I used this to make a pretty picture, but more scientifically important work has been done using this molecule to discover at a microscopic level which parts of a living cell are more or less polar, something that might at first seem nearly impossible to measure. Reichardt's dye is like a nano robot explorer that can go between the very molecules that make up a cell and measure their polarity, signaling its discoveries back through colored light.

▲ Ah, how naive I am. I ordered this acacia root powder as an example of a natural organic pigment, not realizing that it's actually more of a tanning agent than a pigment. But what I *really* didn't realize is that it has off-label uses as a precursor chemical for illegal drugs. I must be on so many government lists.

▲ Despite having a poetic name, amaranth did a lot of damage to the reputation of artificial food colors. Under the name Red Dye #2, it was banned in 1974 after much commotion and the throwing of accusations in all directions, at least some of which were probably true. I just have no idea which ones.

Colors Good Enough To Eat

FOOD COLORING has a bad reputation because it seems like a frivolous injection of potentially harmful chemicals into our food. Incidents such as the banning (in the United States) of Red Dye #2 in 1976 didn't help. But the reality is that many "food colors" are quite literally that: colors derived from food, just transplanted into other food. They may be harmful, but they'd be naturally so, and just as harmful in the original food they came from, where people expect them and don't complain.

Other food colors are synthetic or mineral in nature, but even there the scope for concern is relatively limited. Because food colors usually shouldn't materially affect the taste of the food, and because taste is a very sensitive sense, only compounds that are very intensely colored are typically candidates to be food colors. When you're adding only a few parts per million of a substance, it *may* cause harm, but it's not what you'd normally expect. Chances are, there are many other things in the food, present in much larger quantities, that are more likely to cause harm, be they of synthetic or natural origin.

But it's reasonable to put special emphasis on testing the safety of things that you put in your mouth or paint on your skin (where the requirements for safety are just a bit less strict than in food).

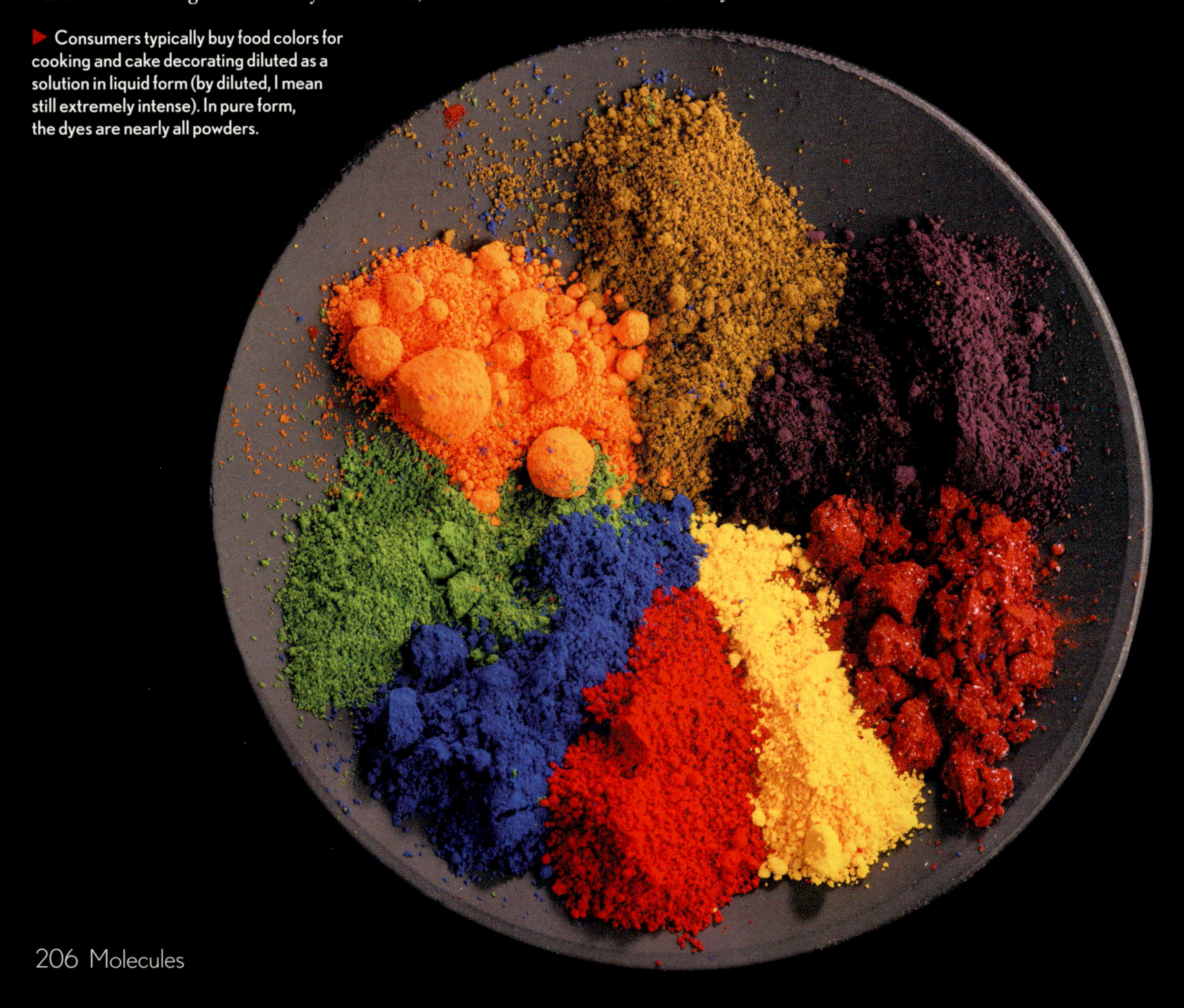

▶ Consumers typically buy food colors for cooking and cake decorating diluted as a solution in liquid form (by diluted, I mean still extremely intense). In pure form, the dyes are nearly all powders.

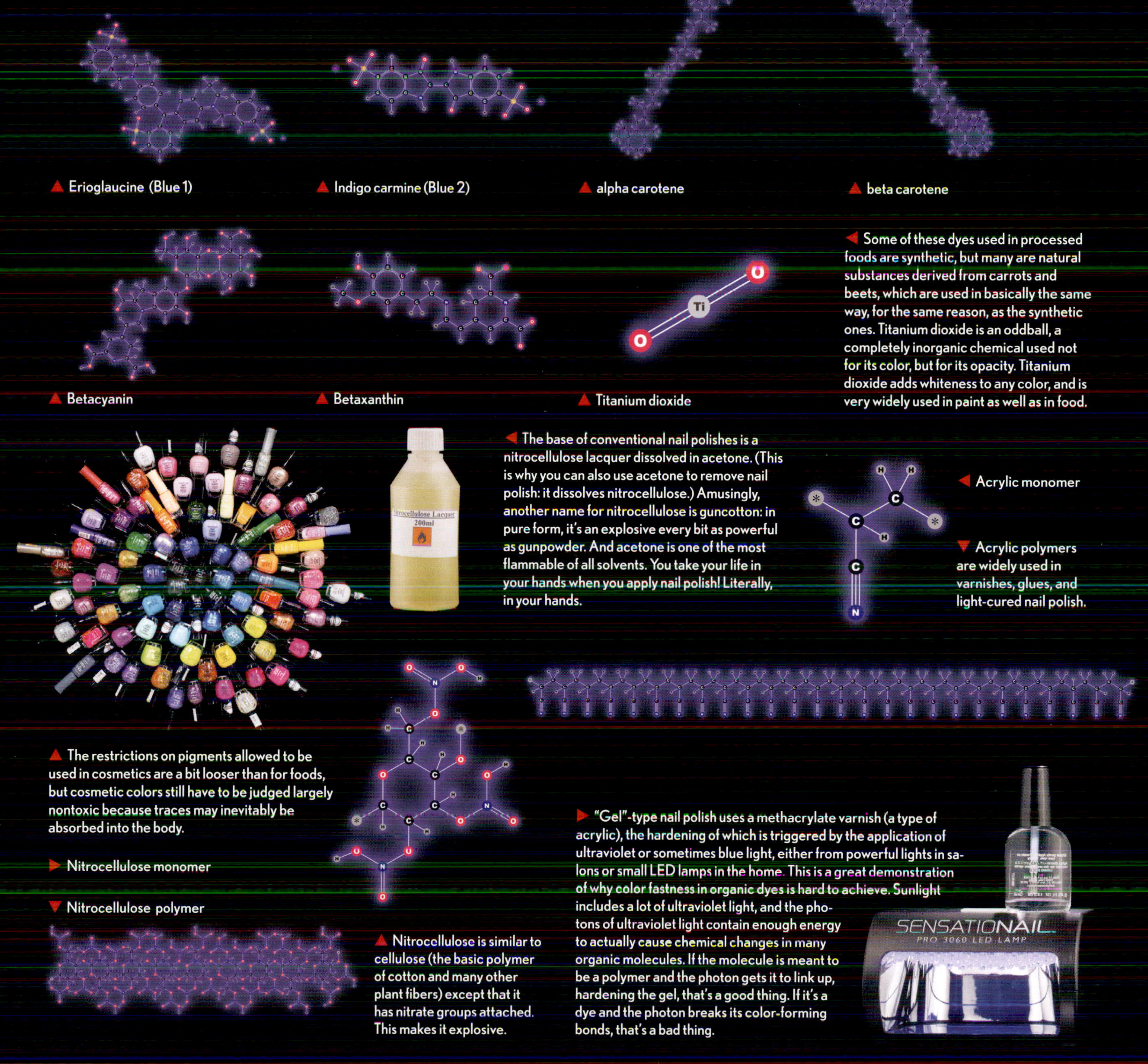

▲ Erioglaucine (Blue 1)

▲ Indigo carmine (Blue 2)

▲ alpha carotene

▲ beta carotene

◀ Some of these dyes used in processed foods are synthetic, but many are natural substances derived from carrots and beets, which are used in basically the same way, for the same reason, as the synthetic ones. Titanium dioxide is an oddball, a completely inorganic chemical used not for its color, but for its opacity. Titanium dioxide adds whiteness to any color, and is very widely used in paint as well as in food.

▲ Betacyanin

▲ Betaxanthin

▲ Titanium dioxide

◀ The base of conventional nail polishes is a nitrocellulose lacquer dissolved in acetone. (This is why you can also use acetone to remove nail polish: it dissolves nitrocellulose.) Amusingly, another name for nitrocellulose is guncotton: in pure form, it's an explosive every bit as powerful as gunpowder. And acetone is one of the most flammable of all solvents. You take your life in your hands when you apply nail polish! Literally, in your hands.

◀ Acrylic monomer

▼ Acrylic polymers are widely used in varnishes, glues, and light-cured nail polish.

▲ The restrictions on pigments allowed to be used in cosmetics are a bit looser than for foods, but cosmetic colors still have to be judged largely nontoxic because traces may inevitably be absorbed into the body.

▶ Nitrocellulose monomer

▼ Nitrocellulose polymer

▲ Nitrocellulose is similar to cellulose (the basic polymer of cotton and many other plant fibers) except that it has nitrate groups attached. This makes it explosive.

▶ "Gel"-type nail polish uses a methacrylate varnish (a type of acrylic), the hardening of which is triggered by the application of ultraviolet or sometimes blue light, either from powerful lights in salons or small LED lamps in the home. This is a great demonstration of why color fastness in organic dyes is hard to achieve. Sunlight includes a lot of ultraviolet light, and the photons of ultraviolet light contain enough energy to actually cause chemical changes in many organic molecules. If the molecule is meant to be a polymer and the photon gets it to link up, hardening the gel, that's a good thing. If it's a dye and the photon breaks its color-forming bonds, that's a bad thing.

▲ Nature uses all kinds of food colors. An array of brightly colored fruits and vegetables covers very nearly the whole spectrum: bright green from chlorophyl, brightly saturated poison red from cyanidins, blues from delphinidin and pelargonidin glucosides, and a great many hues in between. About the only color that doesn't seem to appear in fruit is a proper non-purplish blue. (As a side note, it is a testament to the power of modern transportation that *all* of these fruits and vegetables could be purchased at local stores, for reasonable prices, in the dead of winter in central Illinois.)

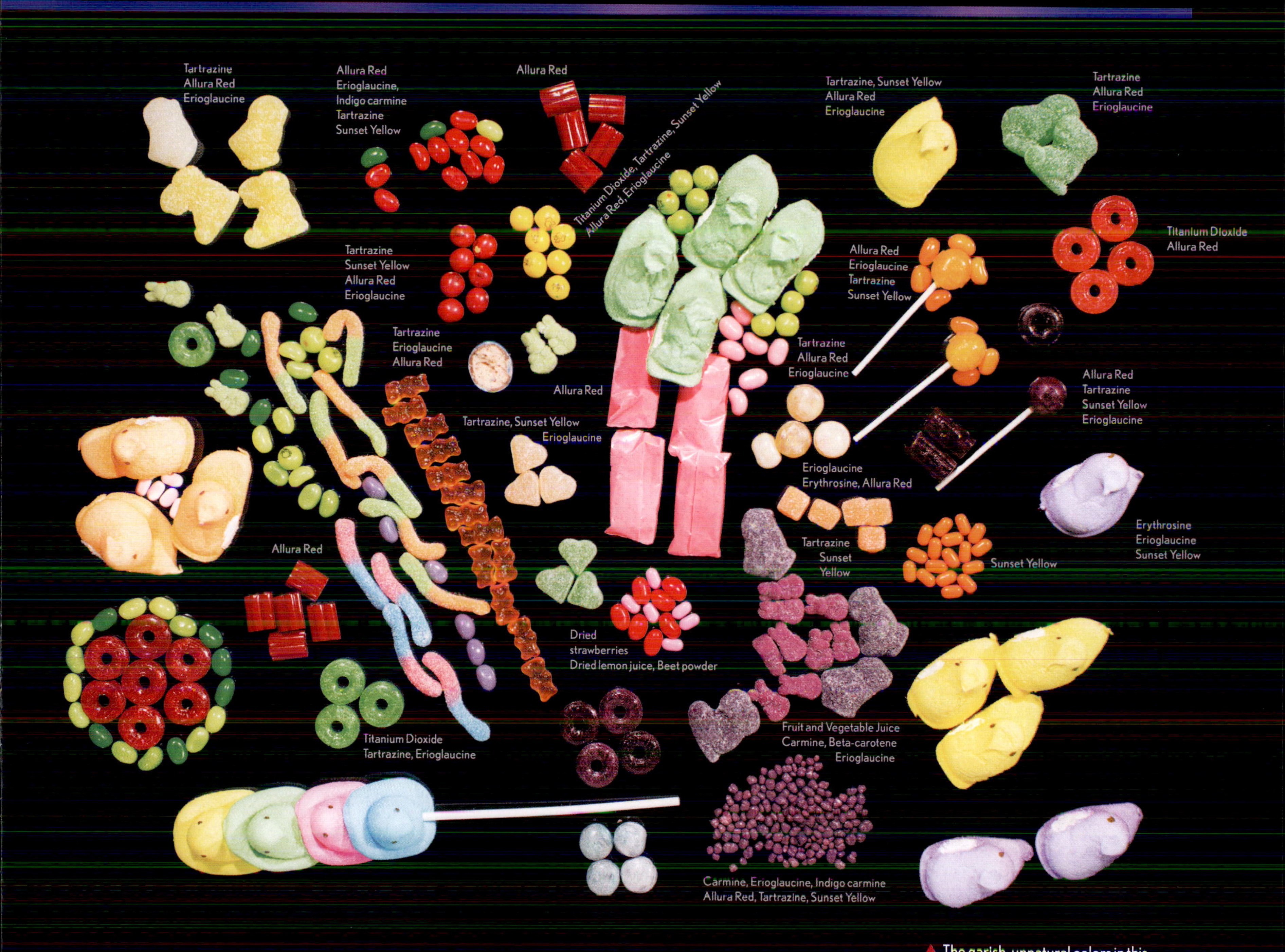

Tartrazine
Allura Red
Erioglaucine
Allura Red
Erioglaucine,
Indigo carmine
Tartrazine
Sunset Yellow
Allura Red
Titanium Dioxide, Tartrazine, Sunset Yellow
Allura Red, Erioglaucine
Tartrazine, Sunset Yellow
Allura Red
Erioglaucine
Tartrazine
Allura Red
Erioglaucine
Titanium Dioxide
Allura Red
Tartrazine
Sunset Yellow
Allura Red
Erioglaucine
Allura Red
Erioglaucine
Tartrazine
Sunset Yellow
Tartrazine
Erioglaucine
Allura Red
Allura Red
Tartrazine
Allura Red
Erioglaucine
Allura Red
Tartrazine
Sunset Yellow
Erioglaucine
Tartrazine, Sunset Yellow
Erioglaucine
Erioglaucine
Erythrosine, Allura Red
Allura Red
Tartrazine
Sunset
Yellow
Erythrosine
Erioglaucine
Sunset Yellow
Sunset Yellow
Dried
strawberries
Dried lemon juice, Beet powder
Titanium Dioxide
Tartrazine, Erioglaucine
Fruit and Vegetable Juice
Carmine, Beta-carotene
Erioglaucine
Carmine, Erioglaucine, Indigo carmine
Allura Red, Tartrazine, Sunset Yellow

Colors Good Enough To Eat

▶ The food colors in natural foods are generally larger molecules than you see in synthetic food colors. Some serve an important purpose other than being colorful (for example, creating chemical energy from light in the notable case of chlorophyll). Several are actually good for you, including beta-carotene, which the body converts into vitamin A. Others, such as betanin (the red of beets) are potentially toxic in high doses.

▶ Alpha-carotene

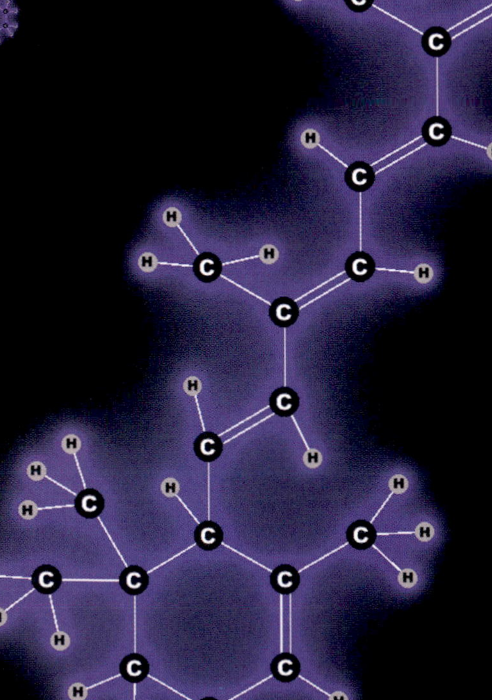

▲ Erythrosine (Red 3)

▲ Beta-carotene

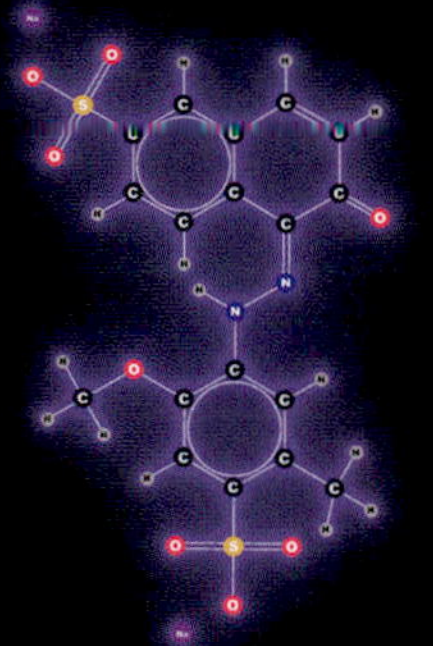

▲ Allura red (Red 40)

▲ Tartrazine (Yellow 5)

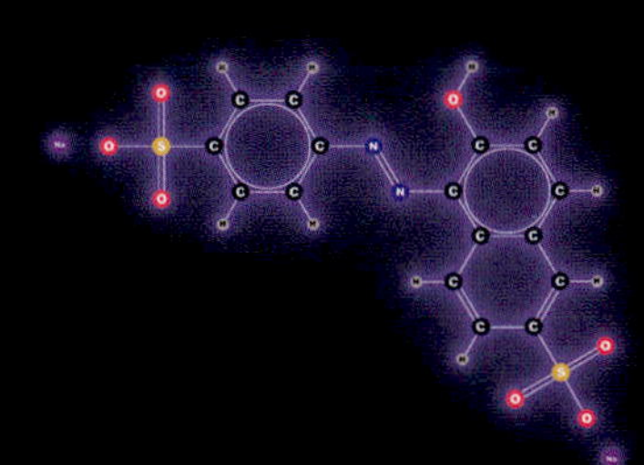

▲ Sunset yellow (Yellow 6)

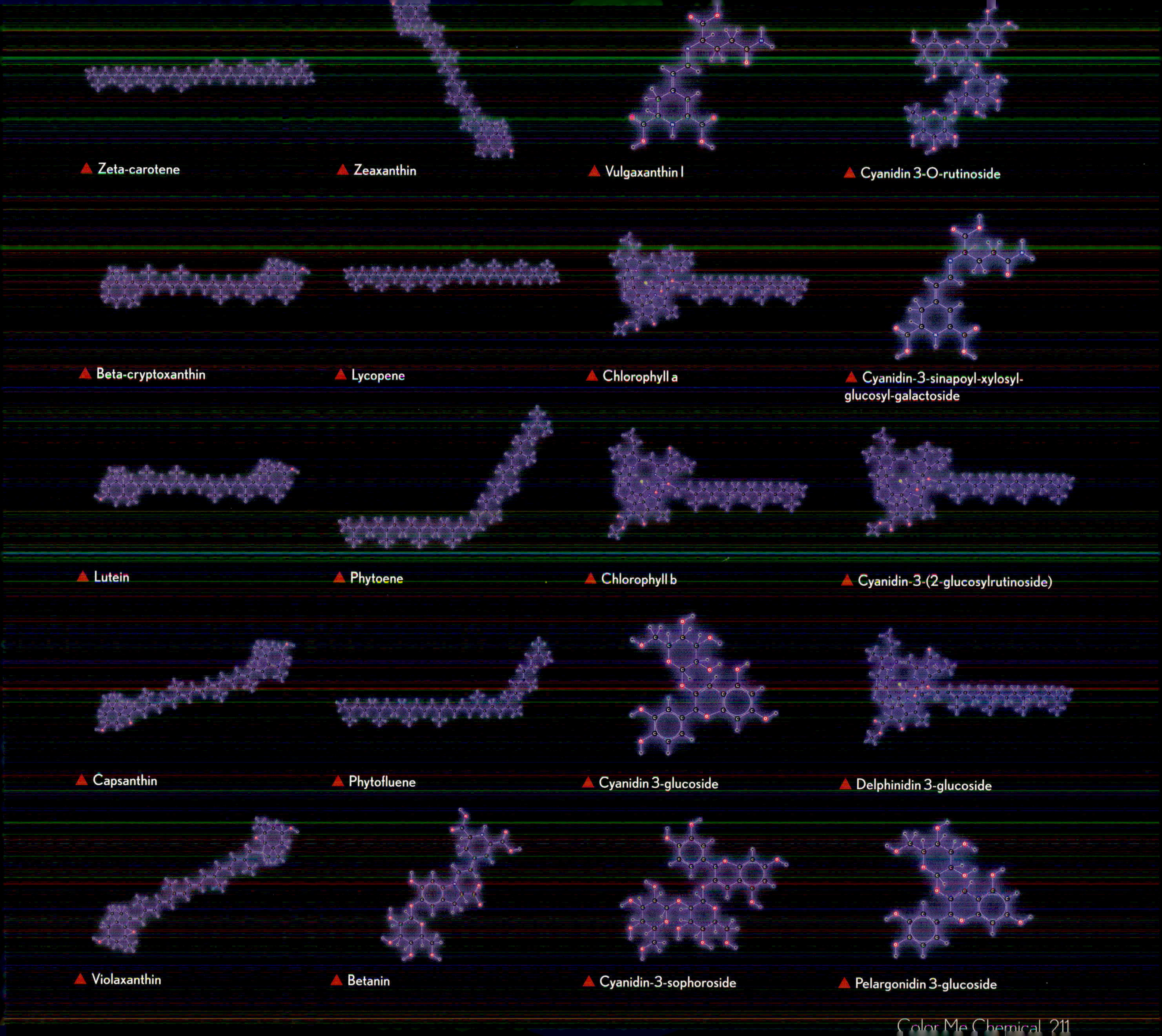
Zeta-carotene
Zeaxanthin
Vulgaxanthin I
Cyanidin 3-O-rutinoside
Beta-cryptoxanthin
Lycopene
Chlorophyll a
Cyanidin-3-sinapoyl-xylosyl-glucosyl-galactoside
Lutein
Phytoene
Chlorophyll b
Cyanidin-3-(2-glucosylrutinoside)
Capsanthin
Phytofluene
Cyanidin 3-glucoside
Delphinidin 3-glucoside
Violaxanthin
Betanin
Cyanidin-3-sophoroside
Pelargonidin 3-glucoside

Art for the Ages

A PERSISTENT PROBLEM with many organic dyes is that they are not lightfast. They fade over time because the simple fact that they absorb the energy of visible light, rather than reflect, deflect, or entirely ignore it, means that they are vulnerable to damage by that very light. Their color originates from delicate structures and goes away if those structures are broken.

But there's another way in which selective absorption can be achieved: through the energy levels generated by the crystal structures of inorganic compounds. These are almost completely immune from damage by light because the crystal only goes together one way. Even if the light dislodges atoms from where they belong, they can't go far. The logic of the crystal dictates that they fall right back into the spot from which they came.

The classic pigments used by artists to create lasting works of art—oil paintings, frescoes, and the like—are often simple, inorganic compounds whose colors will never change and indeed cannot fade as long as the elemental compositions of the compounds remain fixed.

The problem is that only a limited range of colors can be created from inorganic pigments, and there is particularly a shortage of bright, highly saturated colors. Many of those that do exist come from grinding up brightly colored stones. Another name for brightly colored stones is "gems" or at least "semiprecious stones." That sort of pigment, lapis lazuli for example, can be very expensive, which explains why some colors were available only to the wealthy until the invention of a full spectrum of synthetic organic dyes.

▶ Some of the oldest pigments, dating back to cave paintings from the beginnings of human history, are oxides of iron and magnesium. They cover the whole spectrum from earth tone to slightly different earth tone. Basically, we are looking at variations on rust. The lightest, ochre, is almost entirely iron oxide. Sienna adds about 5 percent magnesium oxide, and umber can have up to about 20 percent magnesium oxide. Both sienna and umber can be "burnt," which means literally heated to convert some of the iron oxide into its hematite form, which is darker.

▶ Yellow ochre (hydrated iron oxide)

▼ Raw sienna (iron oxide with 5% magnesium oxide)

◀ Burnt sienna (roasted raw sienna)

▶ Burnt umber (roasted raw umber)

◀ Raw umber (iron oxide with 5-20% magnesium oxide)

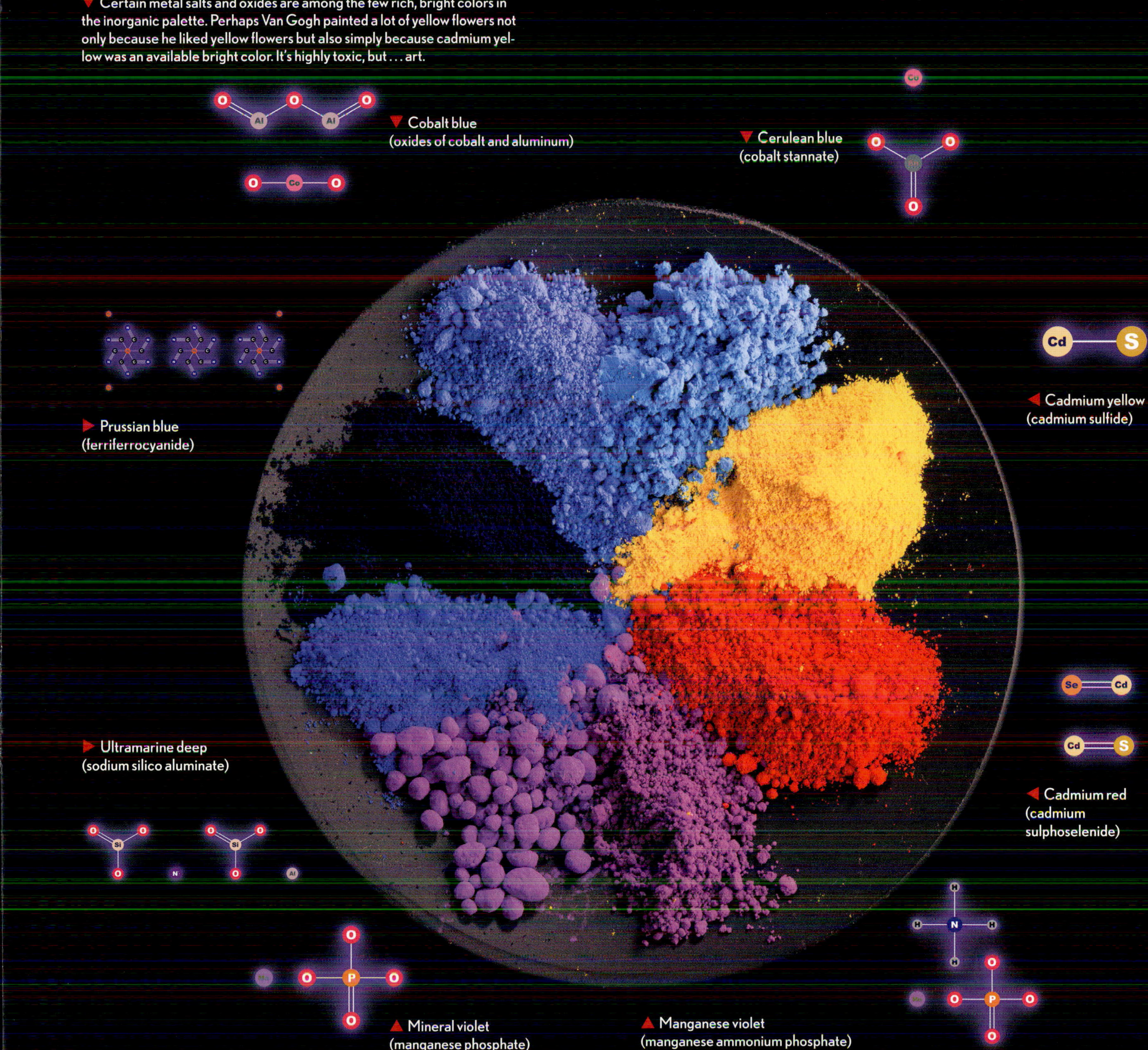

Certain metal salts and oxides are among the few rich, bright colors in the inorganic palette. Perhaps Van Gogh painted a lot of yellow flowers not only because he liked yellow flowers but also simply because cadmium yellow was an available bright color. It's highly toxic, but . . . art.

Art for the Ages

Semiprecious stones were once a prime source of bright colors. Some, such as lapis, were so expensive that they were used only for the most important figures in a painting. Others, such as galena, cinnabar, and realgar, contribute to the spectrum of toxicity in paints as well as the spectrum of color, being made of, respectively, lead, mercury, and arsenic salts.

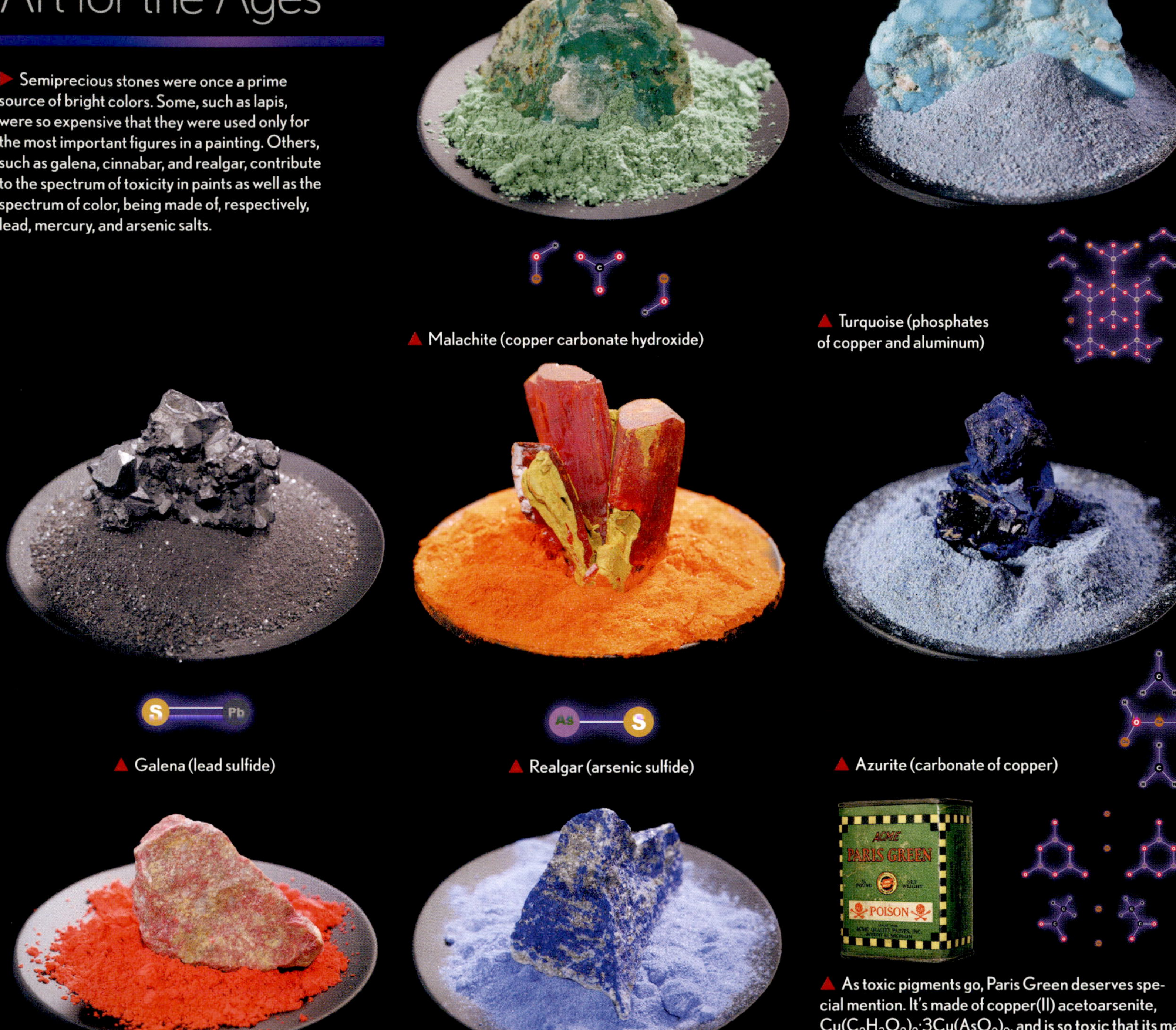

▲ Malachite (copper carbonate hydroxide)

▲ Turquoise (phosphates of copper and aluminum)

▲ Galena (lead sulfide)

▲ Realgar (arsenic sulfide)

▲ Azurite (carbonate of copper)

▲ As toxic pigments go, Paris Green deserves special mention. It's made of copper(II) acetoarsenite, $Cu(C_2H_3O_2)_2 \cdot 3Cu(AsO_2)_2$, and is so toxic that its main use outside the art field was for killing insects and rodents. (Two grams is a lethal dose for an average human.) It, and the related poison Scheele's Green, were used to dye wallpaper of the Victorian era, making many people sick or dead from their fashionable green walls when decay caused by damp weather released the poison. The cure? Move to a dry climate away from your poison walls.

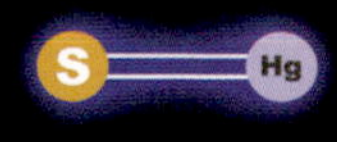

▲ Cinnabar (mercury sulfide), called vermilion as a pigment.

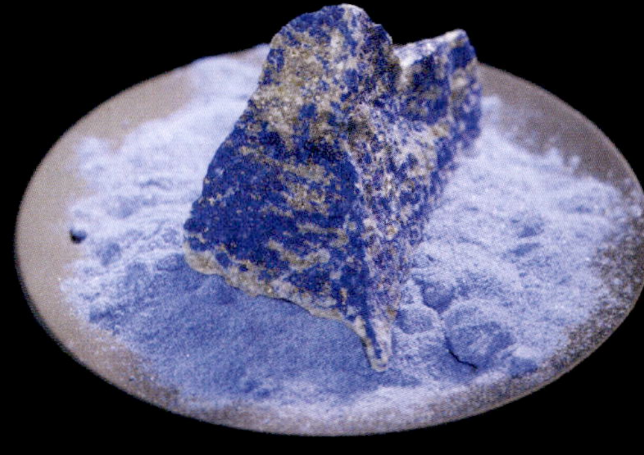

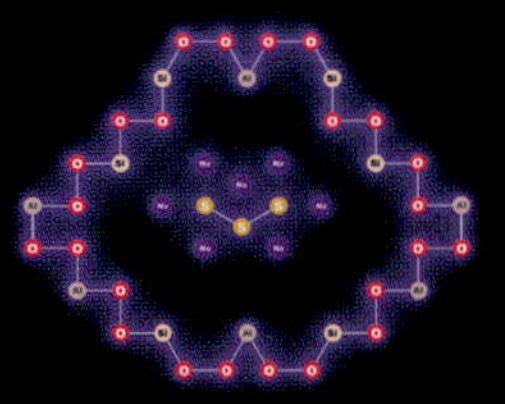

▲ Lapis lazuli (a mixture of minerals including lazurite), called ultramarine as a pigment.

▶ Ebony & ivory may be opposites on piano keyboards, but in the world of pigments, "ivory" is black as well. Ivory black was, and very rarely still is, made of actual ivory, which is heated until it chars into a deep, pure black. Today, the pigment is almost always made by charring bones, and not even elephant bones. A very similar pigment can be made of graphite or soot, both of which are nearly pure carbon. (Bone and ivory black also contain phosphates.) There are several choices for white, the most common by far being titanium dioxide. This material is used in a great many paints, not because it's white, but because it's good at making paint opaque. If a house paint has good "coverage ability," chances are it has a lot of titanium dioxide in it, whatever its visible color.

▼ Titanium dioxide

▼ Ivory black (carbon) Graphite

▲ Zinc oxide

▲ Calcium carbonate

▼ Traditional Chinese watercolor paints are made from a spectrum of natural organic and inorganic pigments that are combined with a binder to hold them to rough paper. The proper pigments are, by definition, of ancient origin, given that this style of artistic expression has an unbroken line of tradition dating more than two thousand years. That makes the wide and vivid palette particularly impressive.

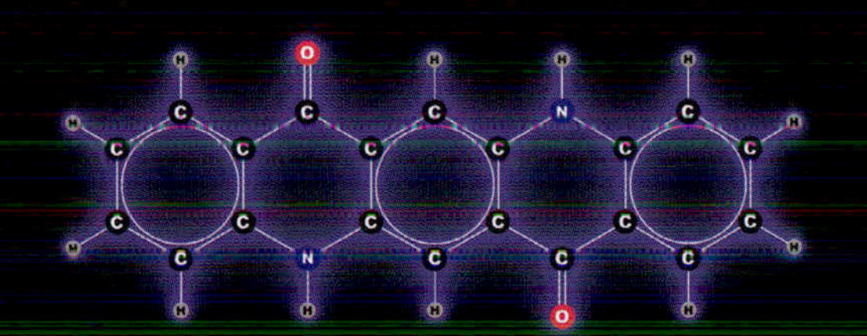

▲ Quinacridone *looks* stable, and it is stable. These ring structures are all solid citizens of the molecular world. Their bonds are strong, they are not particularly eager to react with anything, and they do not care to interact with light, even ultraviolet light. Yet this molecule is a pigment.

▶ Older organic dyes suffer from not being entirely lightfast, but high-tech organic pigments exist whose molecular structures are robust enough to survive bombardment even by the high-intensity ultraviolet rays of direct sunlight. Quinacridone red, made of five strong, linked rings, is used for outdoor signs and automobile paints, which are among the harshest environments for any pigment to survive. It is able to create color despite having such a strong set of bonds (which ordinarily would not absorb visible light) because its crystal structure places its molecules in alignment with each other in a way that allows electrons to transition between them, giving rise to color as a solid-state phenomenon more so than as a molecular phenomenon as in most organic dyes. It's still not as stable as, say, sienna, which has been the color of the Earth for as long as there has been an Earth and will stay that way until there is no longer an eye to see it nor a soul to hear its name.

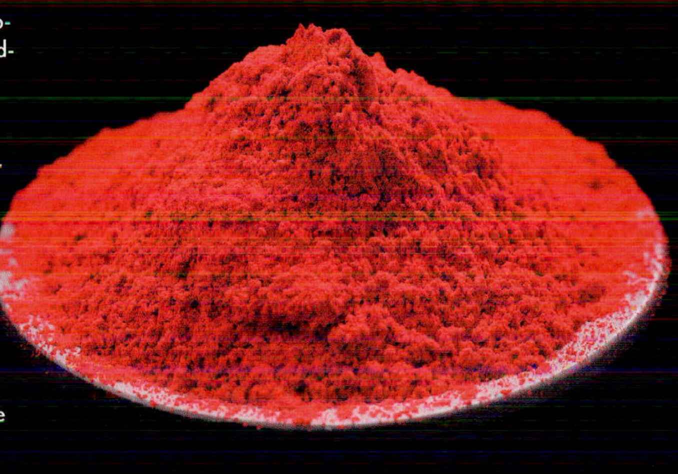

Chapter 13

I Hate That Molecule

IN THIS CHAPTER, I am going to tell you about some chemical compounds that make people very, very angry. I don't mean compounds that people hate because they are clearly harmful. I mean compounds that get caught up in a whirlwind of politics, leaving a residue of stupid that can last for generations, or molecules that showcase the worst of human nature, the greed and shortsightedness that leads to misery and injustice.

The poster child for abused molecules in the early part of the twenty-first century is thimerosal, a molecule used as an antiseptic and antifungal in some vaccine formulations. The trouble began with a study published in 1998 that purported to find a link between childhood vaccines and autism. The study was suspect and widely criticized from the start. Twelve years too late, it was withdrawn completely by the journal—but not before an anti-childhood-vaccination movement took hold. It's very hard to measure how many children may have died as a result of this movement, but the number may well have been in the hundreds and quite possibly over a thousand.

The number of children who developed autism as a result of being given vaccines containing thimerosal is much easier to determine: it is known to be exactly zero.

◀ CO_2 rises from dry ice, like the fog of war that clouds the debate about global warming.

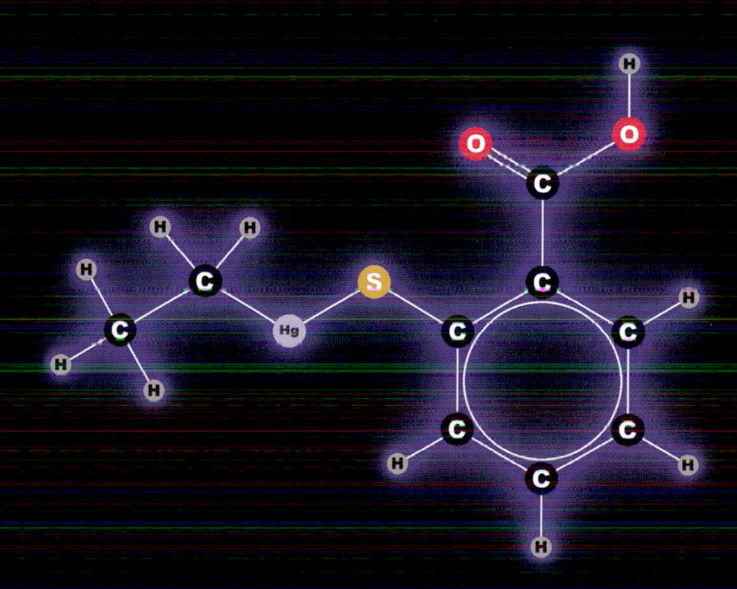

▶ At one point, while looking for an explanation for rising rates of childhood autism, attention focused on thimerosal, which actually looks quite scary as a molecule. Notice the Hg atom in the middle? That's mercury, and worse, it's mercury bound to some organic groups. The thing on the right is thiosalicylic acid, which is essentially aspirin with a sulfur atom. The thing on the left is the really scary part: That's an ethyl group (two carbons). If you split the mercury–sulfur bond, you're left with an ethylmercury ion. If that doesn't cause you to feel just a bit of panic, you don't fully understand the situation.

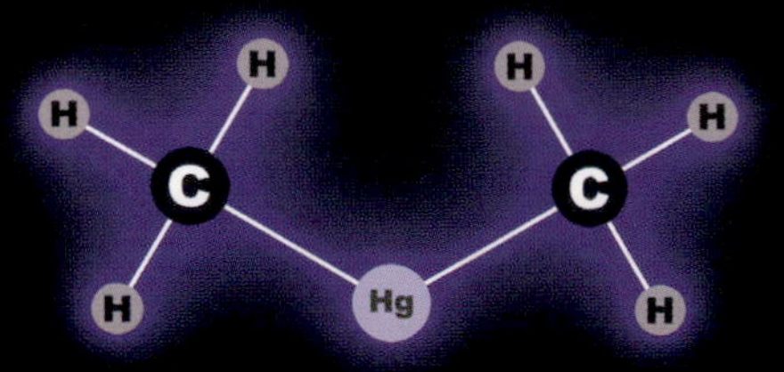

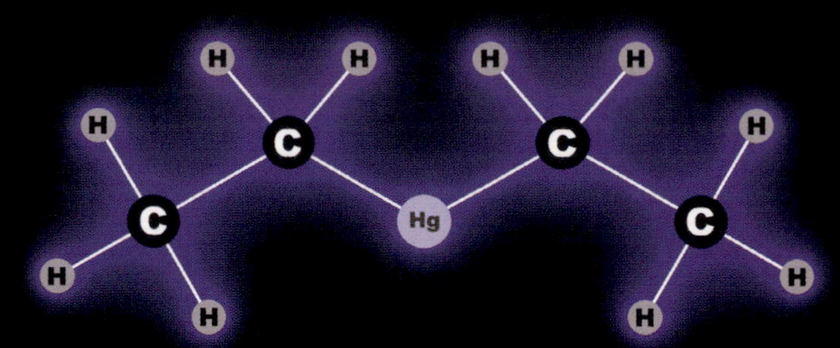

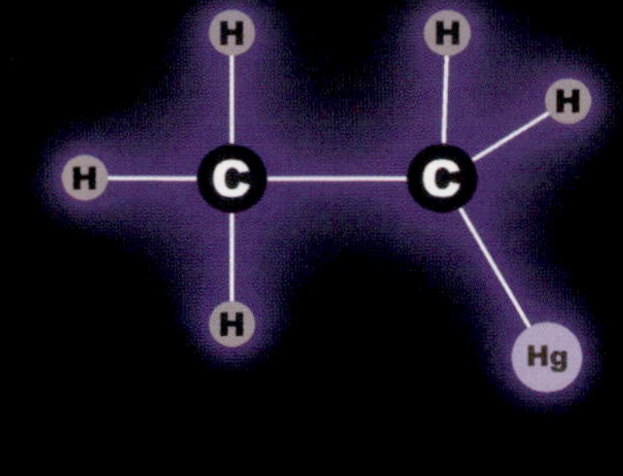

▲ Dimethylmercury and diethylmercury accumulate in the brain and cause serious nervous system damage even in low concentrations. They are among the most neurotoxic substances known. They're retained in the body for a long time and tend to add up, so any exposure is a problem. They also accumulate in the fatty tissue of animals. Some of the mercury emitted by, for example, coal-fired power plants, eventually gets transformed into these sorts of compounds and then works its way back to us in the form of, for example, tuna fish. For this and other reasons, great efforts should be expended to limit the release of mercury into the environment. But neither of these compounds come from thimerosal. It just looks like they *might*.

▲ When thimerosal is broken down in the body, one product is the ethylmercury ion. That's terrifying. Were it to be transformed in some way into diethylmercury or any other similar organomercury compound, it could be a real problem. But all indications are that this just doesn't happen. The situation has been investigated in tremendous detail, and it seems quite definitely the case that this ion is cleared from the body in a period of a few weeks. It doesn't have time to go through the sorts of transformation that happen to mercury left out in the environment for years. You wouldn't want to take in large quantities of ethylmercury ions for an extended period—that would be tempting fate because *some* amount of it could get stuck long enough to cause harm. But half a dozen tiny doses over a lifetime? No, that's not a problem.

▲ Even though it seems to be entirely harmless, why is thimerosal present in vaccines in the first place? Why not just solve the problem by taking it out? In 1928, twelve of twenty-one children given a diphtheria vaccine (without thimerosal) died of bacterial infection. That's the kind of thing that gets people's attention. Thimerosal is and was the only substance known to preserve vaccine potency in multidose vials while preventing dangerous infectious organism contamination. If you want to be able to give a large number of vaccinations on a worldwide scale in a cost-effective way, you have two choices: use thimerosal or don't. And if you choose the latter, you will watch children die from infections that could have been prevented. The antivaccine crowd recklessly suggested that we just stop giving vaccines altogether. But we didn't start administering vaccines without reason! It's impossible to count how many people have died of now-preventable diseases, such as diphtheria, before vaccines were available, but in the modern age alone it's in the hundreds of millions.

▲ There is a simple way to avoid using thimerosal: give all vaccinations from single-dose vials, effectively eliminating the possibility of contamination. Sounds great, right? It is, but only if you're rich. In fact, in an unnecessary response to pressure from the antivaccination movement, thimerosal basically isn't used anymore in childhood vaccines in countries wealthy enough to be able to administer single-dose vials. But in poor countries, where thousands of children die every year from preventable diseases, multidose vials that contain thimerosal, which won't harm them anyway, are the only sensible option.

▶ Thimerosal is still used in a few vaccines and in some other specialized applications. This is an old Boy Scout snakebite kit preserved with a 0.1 percent thimerosal solution.

Destroying the Atmosphere for Fun and Profit

THE HISTORY OF thimerosal is rage-inducing because the substance helped a lot of people, and it deserves better than the shabby treatment it got. The compounds we're going to look at next make the blood boil for the opposite reason: they have actually caused a lot of harm, and were aided and abetted in this by people who not only should have known better but *did* know better and chose to turn a blind eye or even break the law to protect their own interests.

◀ Putting leaded gas into a car designed for unleaded gas is harmful to the public: it ruins the catalytic converter and thus results in greatly increased pollution output from the car, not to mention lead emissions. So all cars designed for unleaded gas have gas tank fill openings that are too small to fit the standard nozzle used for leaded gas. The smaller nozzles for unleaded gas of course fit both kinds of tanks. If you use unleaded gas in an engine designed for leaded, you risk knocking and damage to the engine, but neither of these are harmful to the environment. So that's your private problem, not a matter of public concern.

Getting the Lead Out

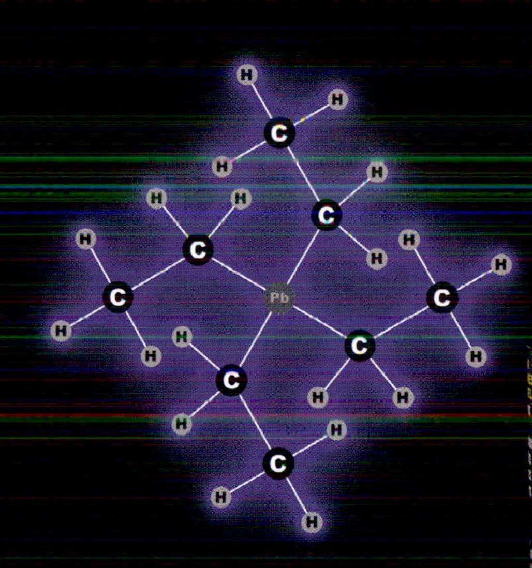

▲ For decades, tetraethyllead was added to gasoline for use in cars. It's an "antiknock" compound that makes some kinds of engines run better (see page 74). Why use this particular compound? Because it's cheap and it works. Why not use it? Because lead in almost any form is an insidious neurotoxin. There does not appear to be *any* level of lead exposure that does not cause harm to the brain. Lead, and tetraethyl lead in particular, were known to be toxic long before leaded gasoline was invented and warnings were issued that leaded gasoline was a bad idea. Dozens of workers died in the plants that produced it, and it's quite clear that companies behaved very badly in their attempt to preserve their ability to use this cheap and effective chemical despite the incontrovertible fact that it was causing tremendous harm. As long as it was only workers being killed, their cover ups worked, but by the 1970s, it became clear that the public at large was also being poisoned. Leaded gasoline has now been banned on all public roads in nearly every country around the world.

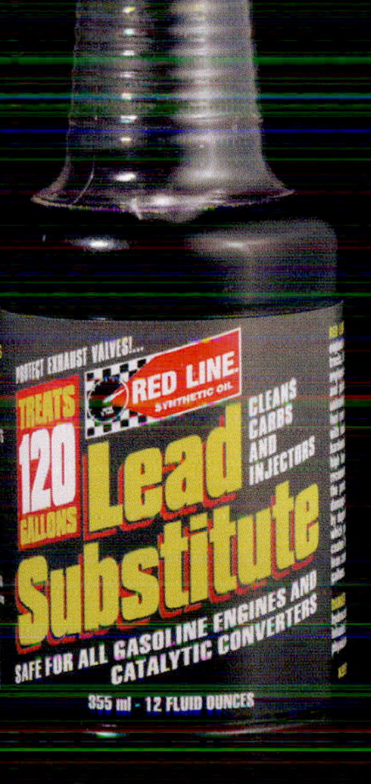

▼ Owners of antique cars that cannot run without leaded fuels were outraged by the ban. Partly as a favor to them and partly because certain kinds of tractor and aircraft engines also can't run without it, leaded fuel is still available. Additives like this are available to add lead back into unleaded fuel to make it compatible with older or specialized engines. Just don't get caught using it on the road, where it is illegal.

▲ Newer fuels use different additives to achieve a high octane rating, including ethanol and actual isooctane, but additives like this are available to push the octane number even higher for use in very high-compression, high-performance racing engines and the like. This one does so with sodium sulfonate, nonane (which is just like octane but with nine carbons instead of eight), and a proprietary mix of other hydrocarbons.

Save the Ozone!

▼ Chlorofluorocarbons (CFCs) are amazing! They're nonflammable, completely nontoxic, able to liquefy at very convenient pressures, have a high heat of vaporization (which means they make great refrigerant gases), and so on. So it's a real shame that they turn out to destroy the Earth's ozone layer as efficiently as they do. Under pressure from lobbyists, governments around the world dragged their feet for decades after it became clear that these substances really needed to be removed from circulation. It was in these battles that an entire industry dedicated to spreading misinformation about climate change sharpened its claws and honed its tactics for the greater battle to come: the battle over CO_2.

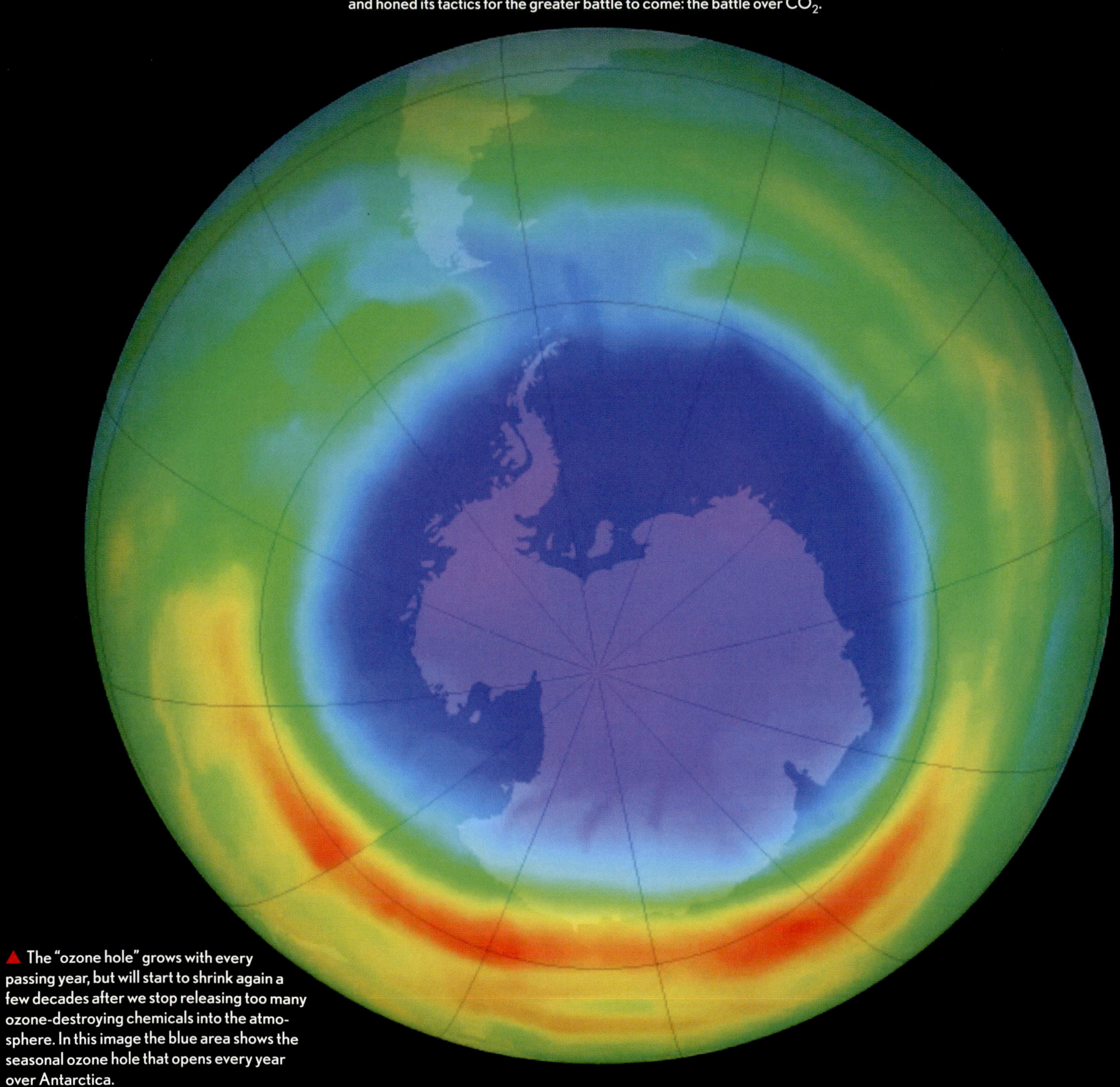

▲ The "ozone hole" grows with every passing year, but will start to shrink again a few decades after we stop releasing too many ozone-destroying chemicals into the atmosphere. In this image the blue area shows the seasonal ozone hole that opens every year over Antarctica.

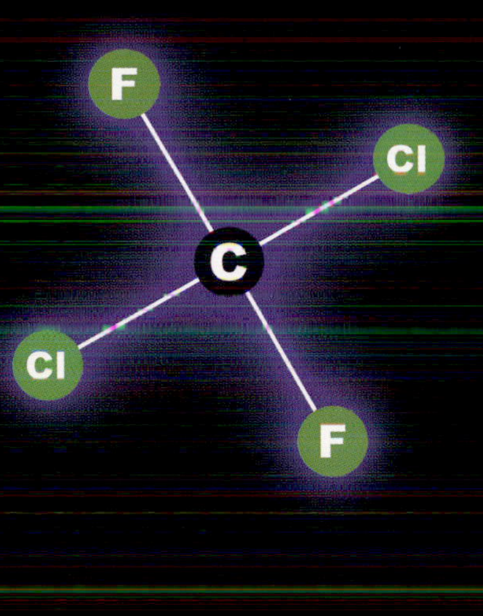

◀ R-22a sounds like a replacement for R-22, right? But if you look at the molecular diagram, to the right, showing what it's made of, you'll notice there is no chlorine and no fluorine, only hydrogen. R-22a is straight up propane—the same thing you burn in your grill. In other words, it's *insane* to put it in a refrigerator. The potential for disaster is obvious to say the least.

◀ CFCs used to be the nearly universal spray-can propellant (the compressed gas that drives the contents out). Putting CFCs in spray cans, where their only function is to be released into the atmosphere, was sensibly the first use that was banned.

▼ Now that CFCs are forbidden, aerosol cans have gotten a lot more exciting! Although CFCs are nonflammable, a common alternative used today is propane, the gas used for grilling. Propane, like CFCs, liquefies at a fairly low pressure, allowing plenty of propellant gas to be stored in the can without making the pressure inside unreasonably high. This can of hairspray doesn't use propane, but it does use dimethyl ether, a similarly exciting propellant.

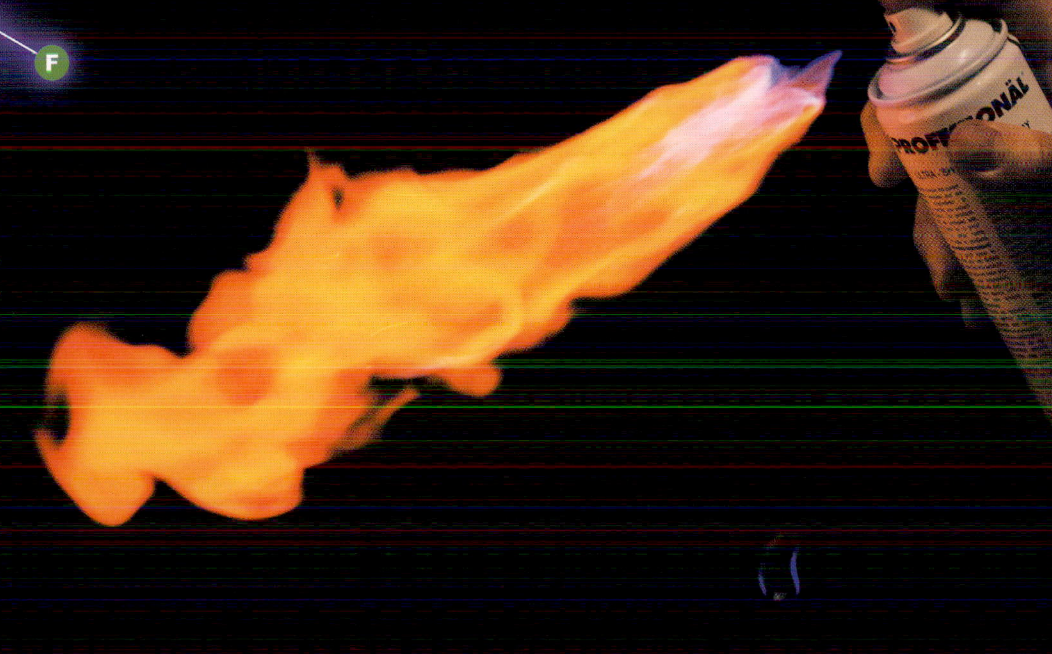

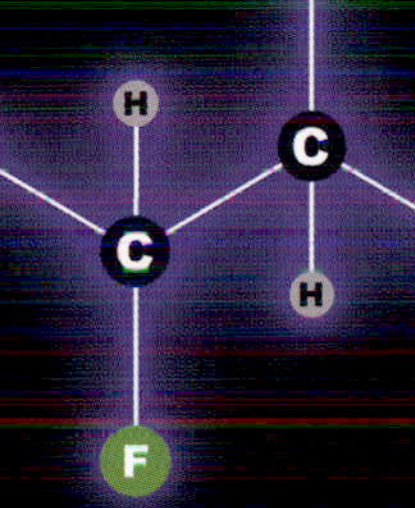

▲ CFCs have been banned for use in most types of air conditioning and refrigeration equipment. As a result, there's a lively gray market in the stuff at prices that would have thrilled the original manufacturers: This ten-pound tank of one of the worst offenders, R-22 (chlorodifluoromethane), cost me almost two hundred dollars. The ban has caused a considerable amount of grumbling from people who say the gas could be recovered or isn't that harmful in the first place. But it seems quite clear that it *is* that harmful, and realistically, if it's out there in millions of car air conditioners, it's going to get out sooner or later.

▲ R134a is a fluorocarbon, as opposed to a chlorofluorocarbon, meaning it has only fluorine atoms substituting for hydrogen, not both chlorine and fluorine. This makes it far less harmful. However, it only works in refrigeration systems that have been specifically designed for it, not ones designed to use R-22.

Save the Planet!

The fights over leaded gasoline and CFCs were mere foothills to the towering mother of all atmospheric chemistry fights: carbon dioxide. Both lead and CFCs are, in a sense, peripheral to the activities they are involved in. No one really cares what additive is used in gasoline, as long as the octane rating is good enough. No one really cares what is propelling the hairspray, as long as they look good on a Saturday night. But carbon dioxide is different. It comes inevitably from the core component, the bulk substance, of the fuels we burn for transportation, electricity, and heat. It's emitted in amounts vastly greater than any other chemical (other than water) being released into the atmosphere by human activity. And the only way to stop that is to completely reengineer the entire energy economy on a planet-wide scale, replacing fossil fuels with something, anything, else. There are going to be some massive winners and some massive losers in that shift. The losers know who they are.

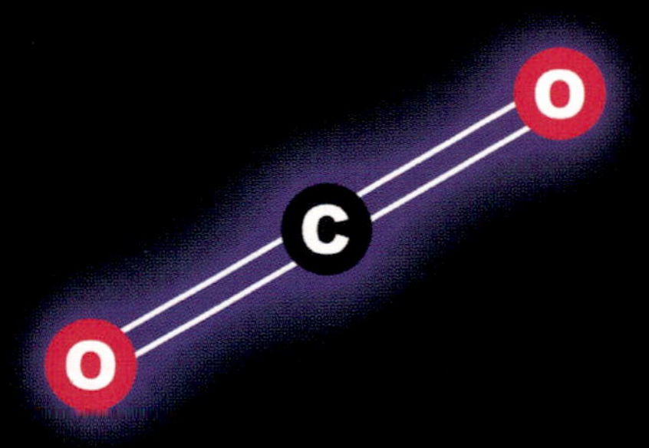

Dry ice is pure carbon dioxide in frozen form. In a few generations, it will be clear to everyone that it would have been smarter not to emit so much. Our children will work together to fix the massive problem we created for them. They will say people should have known better, and they will be talking not about themselves but about *us*. There are people, well-paid people, fighting right now to confuse the issue in the eyes of the public, to deny that there is a problem, and to buy their employers a few more years of unfettered profits. They are, as I write this, starting to have to face up to the fact that the question is no longer whether they will be on the winning or losing side of the debate. The only question now is which side of history they want to be on.

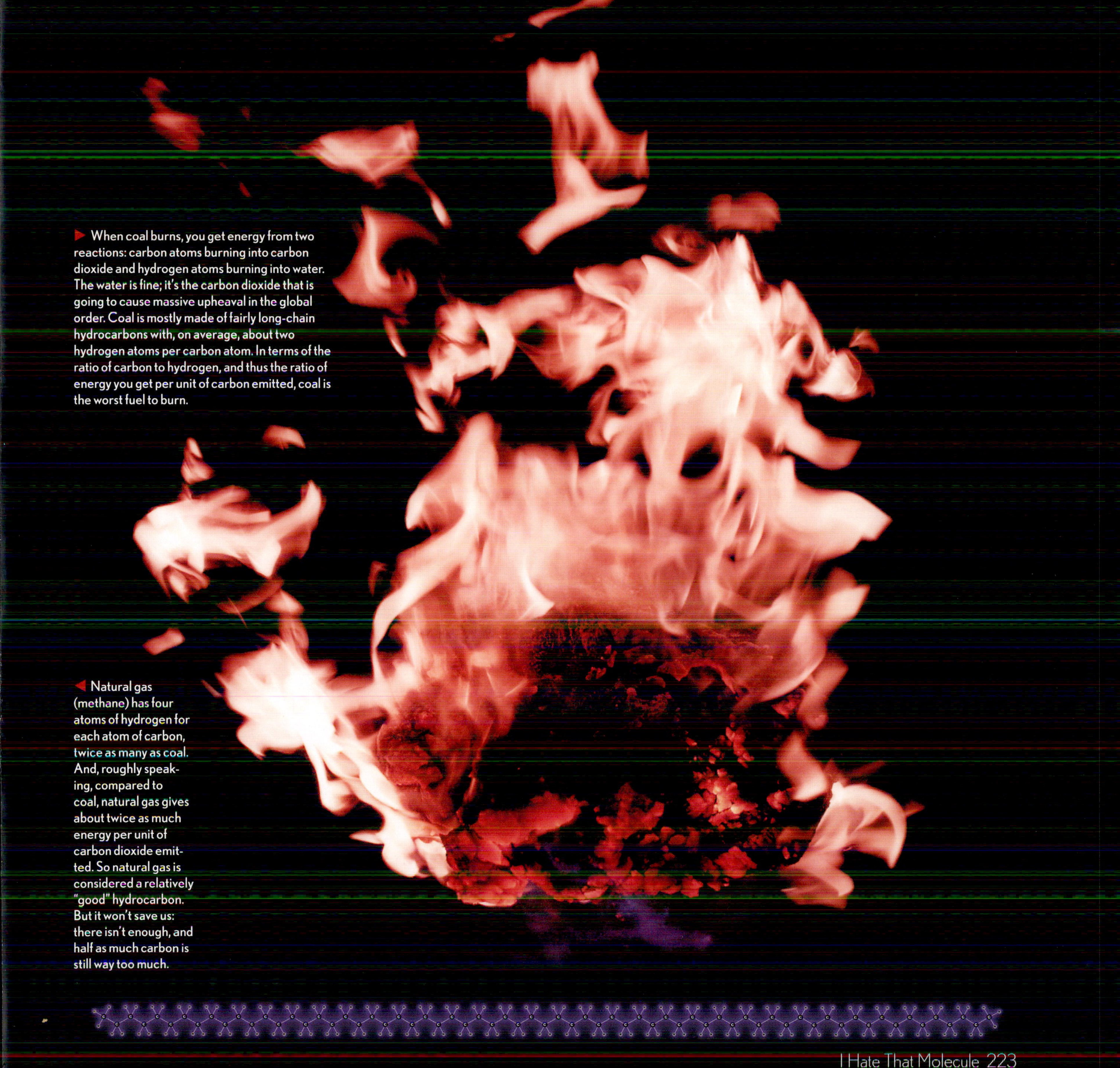

▶ When coal burns, you get energy from two reactions: carbon atoms burning into carbon dioxide and hydrogen atoms burning into water. The water is fine; it's the carbon dioxide that is going to cause massive upheaval in the global order. Coal is mostly made of fairly long-chain hydrocarbons with, on average, about two hydrogen atoms per carbon atom. In terms of the ratio of carbon to hydrogen, and thus the ratio of energy you get per unit of carbon emitted, coal is the worst fuel to burn.

◀ Natural gas (methane) has four atoms of hydrogen for each atom of carbon, twice as many as coal. And, roughly speaking, compared to coal, natural gas gives about twice as much energy per unit of carbon dioxide emitted. So natural gas is considered a relatively "good" hydrocarbon. But it won't save us: there isn't enough, and half as much carbon is still way too much.

A Chemical Also Used To Make Rubber Shoes

▲ Many harsh, dangerous chemicals are used to make, pure, natural, healthy products. For example, sodium hydroxide, known traditionally as caustic lye or just lye, is used to make all-natural, organic soaps as well as to make lye bread, pretzels, hominy, and a number of other humble and wholesome traditional foods. Most small-scale soap makers use it in the form of commercially manufactured food-grade lye. (It is possible to make soap using just the lye and related compounds washed from wood ashes, but this is very rare even in the handmade soaps community. And it's still the same chemical in the wood ash, just mixed with other substances.)

NEXT, LET'S LOOK AT a compound that enrages me not because it's good or because it's bad—I actually don't know whether this particular compound is harmful. What is so outrageous about it is the ignorance on display in the *way* people talk about it.

A national restaurant chain recently announced that it would stop using azodicarbonamide in its bread after a campaign against its use was launched. The headlines related to this drive mostly highlighted the fact that this chemical is also used in making rubber shoes and yoga mats. The harmful effects listed in the petition drive included the fact that when a truck full of it overturned, this was treated as a toxic chemical spill. Would you want such a horrible substance in your food?

Azodicarbonamide may actually be suspect as a food ingredient. *But not because its used in making shoes or is a toxic hazard in pure form!* Not only are those two facts less important than other possible problems with the substance, but they have *absolutely no bearing* on the discussion.

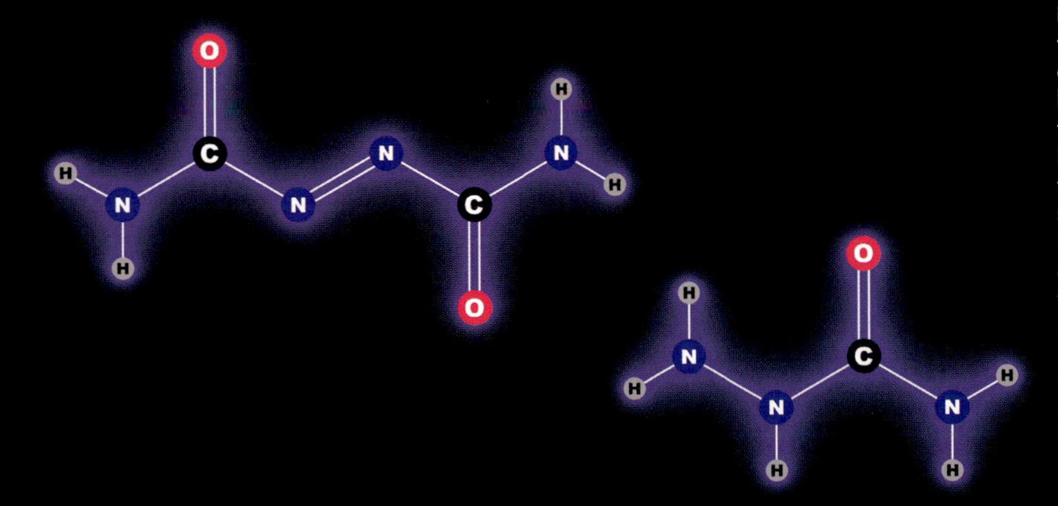

▲ Azodicarbonamide partially breaks down on heating into semicarbazide, which has shown some indications of causing cancer (in animals at high doses). Is it harmful in foods? That's an interesting and important question. Reasonable people differ on the subject. It should be looked into. But the fact that azodicarbonamide is also, separately, independently, and for other reasons, used to make rubber shoes simply has no connection to the question. It's like saying that you shouldn't drink water because it is a powerful solvent used in the chemical industry to dilute acids.

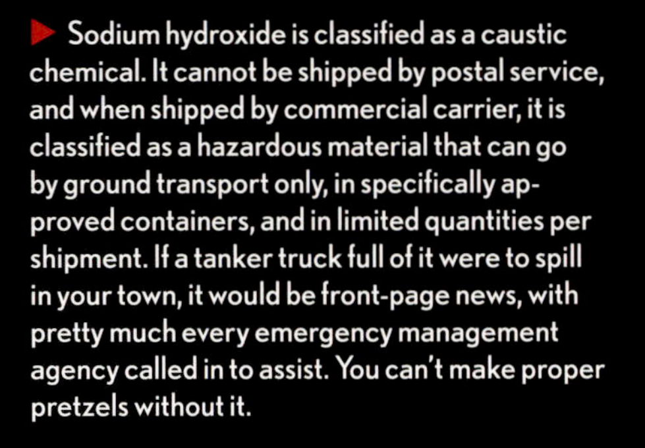

▶ Sodium hydroxide is classified as a caustic chemical. It cannot be shipped by postal service, and when shipped by commercial carrier, it is classified as a hazardous material that can go by ground transport only, in specifically approved containers, and in limited quantities per shipment. If a tanker truck full of it were to spill in your town, it would be front-page news, with pretty much every emergency management agency called in to assist. You can't make proper pretzels without it.

▶ This is the earliest, most archetypical soap. It's made from animal or vegetable fat and lye. Nothing else is required to make an effective soap. The toxic, corrosive lye remains in the soap in the form of its sodium ion bound to the fatty acids and its hydroxide ion combined with the acid hydrogen to form water (some of which may be driven off before the soap is finished). The reaction between fat and lye is not unlike the reaction on the opposite page between lye and the fat, skin, and muscle in a chicken foot. It's a terribly important fact about chemicals that they are able to transform themselves utterly, leaving no trace of what they used to be. So if someone tells you not to use a product because of a precursor chemical used in its manufacture, ask them if they prefer natural soap or artificial detergent. I guarantee they will fall for the trap.

▶ Some of my fondest memories of childhood involve lye bread rolls from the bakery down at the end of Sonnenbergstrasse. Such beautiful trees. A lifetime ago. If anyone ever mounts a successful campaign to have lye bread banned because lye is also one of the most intensely caustic chemicals in the world, I am going to be seriously miffed.

The Most Horrible Very Bad Inorganic Compound Ever

AND FINALLY we arrive at a compound that is bad, that everyone agrees is bad, and that has a public discourse about it that is rational and well informed most of the time. Yet it *still* manages to create massive anger in people who understand what's going on.

Asbestos was once universally hailed as a wonder-substance—the best thermal insulator anyone could hope for. It is stable, resistant to chemical attack, heat-proof, strong, cheap, and useful. But for a generation now, it has been, year after year, one of the world's number one causes of lawsuits. This started out for a good reason: despite being a very useful substance, asbestos also undeniably causes lung cancer. Workers in asbestos factories died from cancers clearly caused by the asbestos they were working with. Some companies actively suppressed damning evidence. They didn't just look the other way; they proactively covered up the facts.

If ever there was a noble cause for a personal injury lawyer to take up, this was it! And for years, that's what these lawyers did: sought just compensation for people willfully harmed by despicable corporations.

Then they started to run out of real victims. Companies cleaned up their act. Asbestos was removed from daily life worldwide. People who had been exposed to asbestos through malice grew old and died. But the lawsuits continued, taking on a life of their own.

Lawyers recruited people suffering from terrible cancers and showed them rooms full of consumer products, asking them to try to remember if they'd ever used or even seen any of those products. If they said yes, a lawsuit was promptly filed against the company that made the product. Suits were often filed by lawyers who had no plausible reason to believe there was a connection, and frequently, the companies in question had never before been suspected of any wrongdoing.

Of course we feel sorry for someone suffering and dying from cancer, and of course we want them to be taken care of in their last days and given the money that can bring them some comfort. But doing that by creating a second victim out of an innocent company that did nothing wrong, and whose products harmed no one, is not justice. It is the opposite.

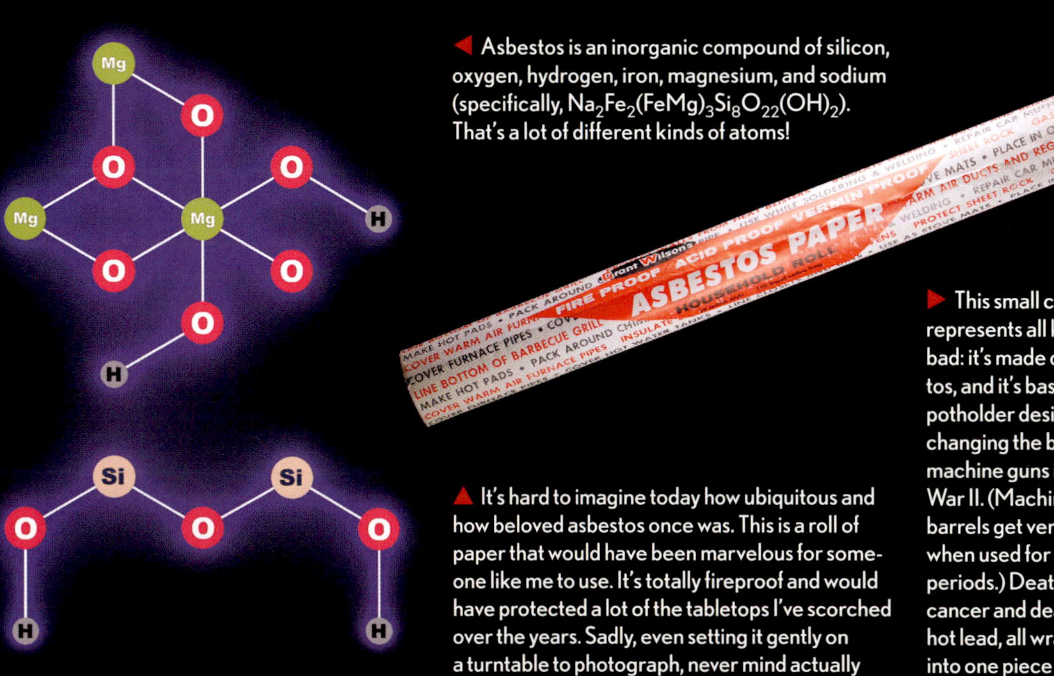

◀ Asbestos is an inorganic compound of silicon, oxygen, hydrogen, iron, magnesium, and sodium (specifically, $Na_2Fe_2(FeMg)_3Si_8O_{22}(OH)_2$). That's a lot of different kinds of atoms!

▲ It's hard to imagine today how ubiquitous and how beloved asbestos once was. This is a roll of paper that would have been marvelous for someone like me to use. It's totally fireproof and would have protected a lot of the tabletops I've scorched over the years. Sadly, even setting it gently on a turntable to photograph, never mind actually unrolling it, made me very nervous, and it's now tightly wrapped in plastic for safekeeping.

▶ The fibers in asbestos are microscopically sharp, able to reach into the very DNA of cells and break it, leading to mutations that may eventually turn into cancer. And because asbestos is so chemically inert, its fibers last virtually forever once inhaled into the lungs, so they can do their dirty work over decades.

▶ This small cloth represents all kinds of bad: it's made of asbestos, and it's basically a potholder designed for changing the barrels of machine guns in World War II. (Machine gun barrels get very hot when used for long periods.) Death by cancer and death by hot lead, all wrapped into one piece of cloth! Only asbestos can pull that one off.

	T	C	A	G	
T	TTT=Phenylalanine(F)	TCT=Serine(S)	TAT=Tyrosine(Y)	TGT=Cysteine(C)	T
	TTC=Phenylalanine(F)	TCC=Serine(S)	TAC=Tyrosine(Y)	TGC=Cysteine(C)	C
	TTA=Leucine(L)	TCA=Serine(S)	TAA=STOP	TGA=STOP	A
	TTG=Leucine(L)	TCG=Serine(S)	TAG=STOP	TGG=Tryptophan(W)	G
C	CTT=Leucine(L)	CCT= Proline(P)	CAT=Histidine(H)	CGT =Arginine(R)	T
	CTC = Leucine(L)	CCC=Proline(P)	CAC=Histidine(H)	CGC=Arginine(R)	C
	CTA =Leucine(L)	CCA=Proline(P)	CAA=Glutamine(Q)	CGA=Arginine(R)	A
	CTG=Leucine(L)	CCG=Proline(P)	CAG=Glutamine(Q)	CGG=Arginine(R)	G
A	ATT=Isoleucine(I)	ACT = Threonine(T)	AAT=Asparagine(N)	AGT=Serine(S)	T
	ATC=Isoleucine(I)	ACC=Threonine(T)	AAC=Asparagine(N)	AGC=Serine(S)	C
	ATA=Isoleucine(I)	ACA=Threonine(T)	AAA =Lysine(K)	AGA=Arginine(R)	A
	ATG=Methionine(M)	ACG=Threonine(T)	AAG=Lysine(K)	AGG= Arginine(R)	G
G	GTT=Valine(V)	GCT=Alanine(A)	GAT=Aspartic acid(D)	GGT=Glycine(G)	T
	GTC=Valine(V)	GCC=Alanine(A)	GAC=Aspartic acid(D)	GGC=Glycine(G)	C
	GTA=Valine(V)	GCA=Alanine(A)	GAA= Glutamic acid(E)	GGA=Glycine(G)	A
	GTG=Valine(V)	GCG=Alanine(A)	GAG=Glutamic acid(E)	GGG= Glycine(G)	G

Chapter 14

Machines of Life

YOU MAY HAVE NOTICED that I haven't said much about one very important class of molecules: the giant ones that run the machinery of life. DNA, RNA, and proteins are all molecules, but they are very different in nature from the others we've talked about so far. They are more like books and robots than they are like other molecules.

All of them are constructed from a small number of simple units assembled into long chains. In this sense, they are like the polymers we discussed on page 103. But those polymers repeat the same units over and over in the same pattern or in a semirandom pattern. There is no significant amount of *information* contained in the order of the units in the polymer. This is not so with the molecules I'm talking about here.

DNA is all about information. It's made of a sequence of *nucleotides* (of which there are four kinds) whose specific order can encode nearly all the information needed to grow, operate, and reproduce an entire living organism. DNA has no real function other than to have the information in it copied and used. The analogy is often made that individual nucleotides are like letters of an alphabet, and a DNA molecule is like a book written in that alphabet.

This is much more than just a useful analogy; it is about as close as you can get to the literal truth. The units are given letters, G, A, T, and C, which stand for the molecules guanine, adenine, thymine, and cytosine, respectively. So a strand of DNA can be described by listing these letters in the order those molecular units occur in the polymer sequence. A typical strand of DNA is many tens of millions of letters long.

The letters are grouped into "words," each of which is exactly three letters long. Those words in turn are grouped into "sentences," each of which contains the information necessary to construct one protein. The words are called codons, and the sentences are called genes. A gene can be anywhere from less than a thousand to over a million letters long.

The complete human genome (the set of DNA needed to build and operate a human) consists of twenty-two books (called chromosomes) written in those sentences. These books contain a total of about three billion letters. (For comparison, the complete set of seven Harry Potter books contains around five million letters.)

Proteins are similarly made of long chains of simple units in a definite order, but rather than encoding information intended to be copied, they are the machines, messengers, and structures that operate the body. Each protein is built from a specific sequence of up to twenty-one different *amino acids*. The sequence of amino acids in a protein determines everything about its shape, and thus its function. And it's this sequence that is written in the words of DNA.

When a cell needs to make a protein, the DNA that says how to make it is copied (by a machine called RNA polymerase, which is itself made of proteins) into a strand of RNA (which is the same idea as DNA, just made of slightly different chemical units). The RNA then goes to another machine (also made of proteins) called a ribosome, which reads the words in order and uses them to assemble the corresponding protein sequence. Each word (three letters) in the DNA corresponds to one particular amino acid in the protein.

◀ This table shows which three-letter words in DNA translate into which amino acids in proteins. For example, CAA and CAG (which stand for the sequences of molecules cytosine–adenine–adenine and cytosine–adenine–guanine, respectively) both translate into the molecule glutamine, represented in protein sequences as the letter Q. It's so computer-like! Three of the sixty-four possible words are "stop codons" that tell the protein synthesis machine to stop working and release the protein it has been building.

t's Not About the Molecules

HAVE INTENTIONALLY not included my ypical diagrams of molecular structures in his chapter because, although DNA, RNA, and proteins are, of course, molecules made of atoms, that's really not the best way to think about them. They are more easily understood n the language of computer science than in he language of chemistry. And, indeed, the ield of "computational biology" is one of the hottest areas of study right now. Hackers, the ort of people who write computer code, are ncreasingly getting interested in hacking the genome and writing in the language of life nstead of the language of silicon.

The table on this page is without a doubt one of the most mind-blowing things you're ikely to encounter, ever. It is *the code*. This table lists which three-letter words in DNA translate into which amino acid in a protein. Using this code, you can read DNA like a book, which is exactly what the protein synthesis mechanism in living cells does. But you can also *write* in the book. The name for that is genetic engineering, and it is every bit as much a field of engineering as computer engineering or mechanical engineering—the same ways of thinking, the same instinct to tinker, adjust, and invent apply here as well. That's scary, and it's exciting, and it's the future.

When we look back on the present age, there is absolutely no doubt that we will see it as the era of DNA, the time in which we took over the very foundations of life, understood them, and turned them to our use—or to our doom. I have a background in computers, which has shown me how a simple idea, an understanding of how to *program machines*, can lead to unimaginable power. A new generation, perhaps including you, will take this paradigm of programming into the world of life. You will build new creatures from the ground up and reprogram existing ones, including us.

Whether we survive the reprogramming of life is very much an open question, just as it remains an open question whether we will survive the invention of nuclear weapons. Let's hope that our better instincts will prevail here, as they have so far, and that the technology of life will be used mainly for good. (I'd like more hair, by the way, in case you feel like working on that particular bit of DNA.)

ATG GCC CGT ACT AAG CAG ACT GCT CGC AAG
TCG ACC GGC GGC AAG GCC CCG AGG AAG CAG
CTG GCC ACC AAG GCG GCC CGC AAG AGC GCG
CCG GCC ACG GGC GGG GTG AAG AAG CCG CAC
CGC TAC CGG CCC GGC ACC GTA GCC CTG CGG
GAG ATC CGG CGC TAC CAG AAG TCC ACG GAG
CTG CTG ATC CGC AAG CTG CCC TTC CAG CGG
CTG GTA CGC GAG ATC GCG CAG GAC TTT AAG
ACG GAC CTG CGC TTC CAG AGC TCG GCC GTG
ATG GCG CTG CAG GAG GCC AGC GAG GCC TAC
CTG GTG GGG CTG TTC GAA GAC ACG AAC CTG
TGC GCC ATC CAC GCC AAG CGC GTG ACC ATT
ATG CCC AAG GAC ATC CAG CTG GCC CGC CGC
ATC CGT GGA GAG CGG GCT TAA

▲ This is the DNA coding sequence for a very small protein, known as histone H3.2 (human variant). It's an unassuming member of the "+" strand of chromosome 1, where it occupies letters 149,824,217 through 149,824,627. Take a minute to wrap your head around the fact that *we actually know this*. I didn't make those numbers up; they come straight from the human genome database, which lists the names, precise locations, and functions of tens of thousands of these sequences. The whole darned thing, the entire human genome, has been sequenced from start to finish. (Though right now we know the precise function of only a small fraction of it.) The map has been made; it's just a question of time before all the blank areas are colored in.

MARTKQTARK
STGGKAPRKQ
LATKAARKSA
PATGGVKKPH
RYRPGTVALR
EIRRYQKSTE
LLIRKLPFQR
LVREIAQDFK
TDLRFQSSAV
MALQEASEAY
LVGLFEDTNL
CAIHAKRVTI
MPKDIQLARR
IRGERA

▲ This sequence looks similar to the one to the left, but notice that it uses different letters and is shorter. It shows the sequence of amino acids in the protein coded for by the longer DNA sequence. Because each amino acid is coded for by three letters in DNA, the sequence is exactly one-third as long. (The letters representing each amino acid are given in the coding table. For example, leucine is represented by the letter **L**.)

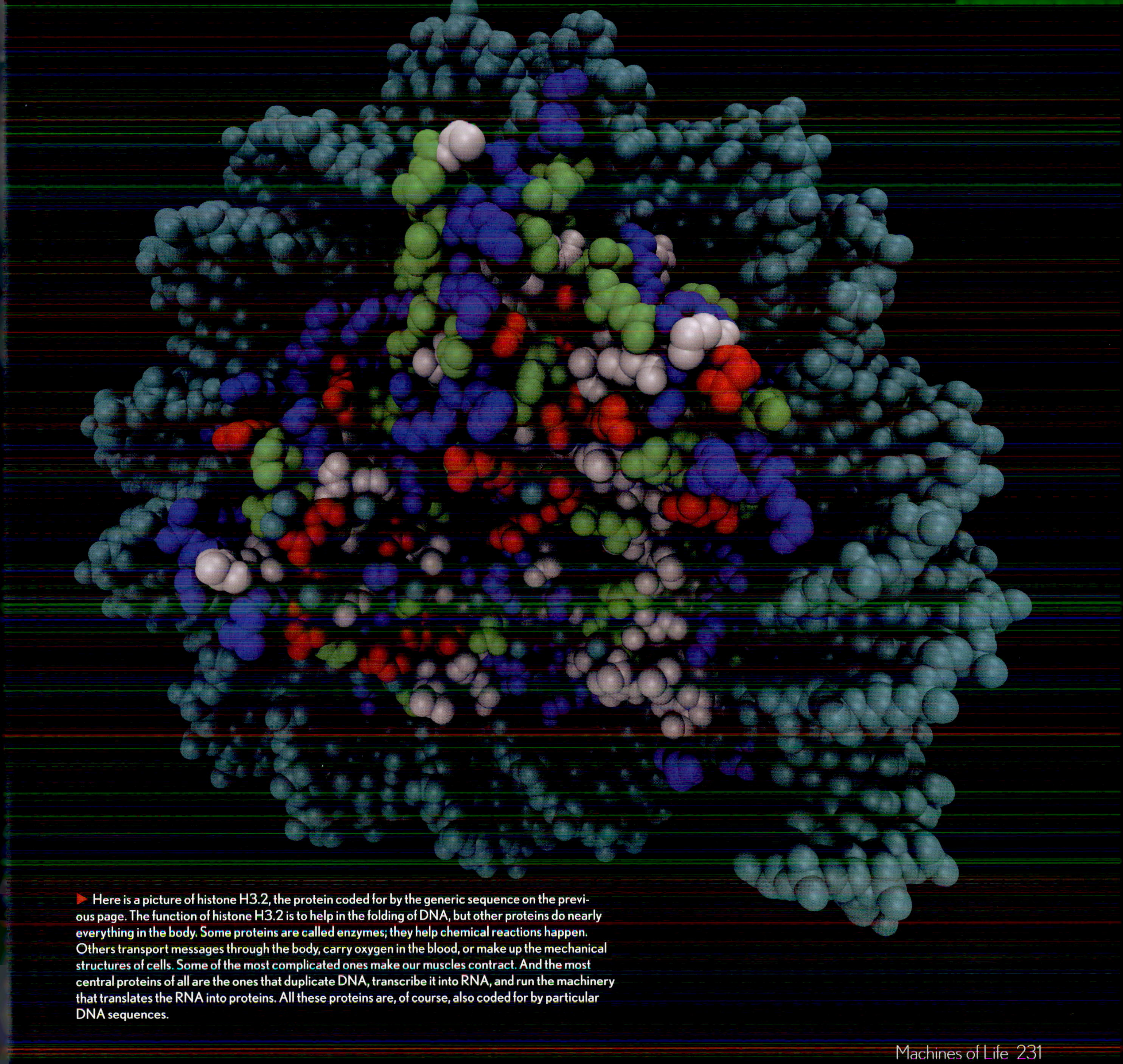

▶ Here is a picture of histone H3.2, the protein coded for by the generic sequence on the previous page. The function of histone H3.2 is to help in the folding of DNA, but other proteins do nearly everything in the body. Some proteins are called enzymes; they help chemical reactions happen. Others transport messages through the body, carry oxygen in the blood, or make up the mechanical structures of cells. Some of the most complicated ones make our muscles contract. And the most central proteins of all are the ones that duplicate DNA, transcribe it into RNA, and run the machinery that translates the RNA into proteins. All these proteins are, of course, also coded for by particular DNA sequences.

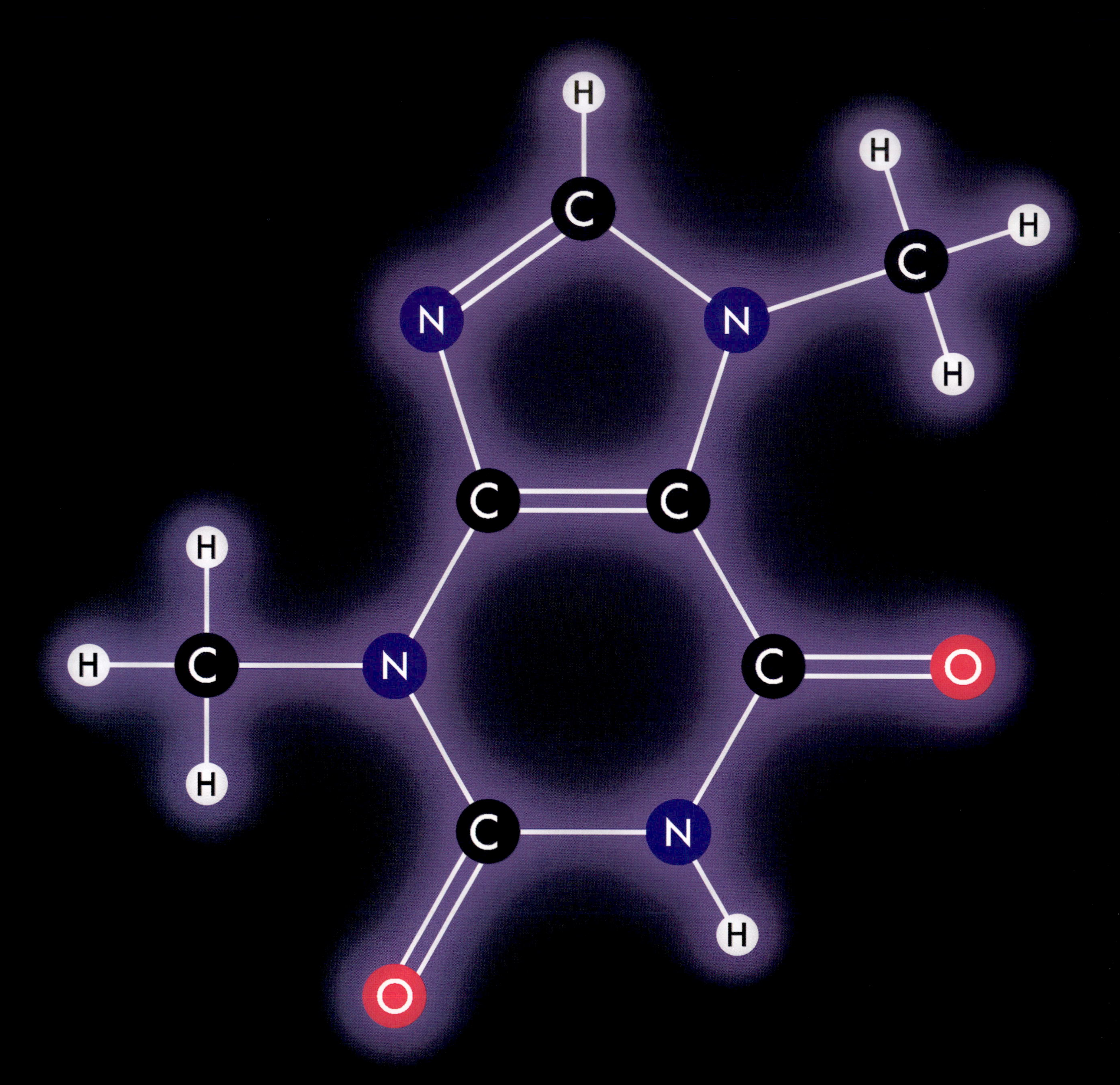
H
H
C
H
C
N
N
H
C
C
H
H
C
N
C
O
H
C
N
H
O

Acknowledgments

AS WITH ANY BOOK, people suffered to make this one possible. First on that list would be my kids and my girlfriend, who I thank for not running away from home and/or breaking up with me more than twice. Next would be my editor Becky Koh, who may have been temped to do the same as the deadline came rushing towards us like a freight train on a track lubricated with the finest synthetic motor oil.

And of course I have to thank my collaborator and photographer Nick Mann for taking nearly all the photographs in this book. He comes after the others only because he didn't suffer: I think he had as much fun shooting the objects in this book as I did collecting them. For a period of months every day was Christmas at the studio as package after package, sometimes a dozen a day, arrived with weird and wonderful things to photograph. We shot over five hundred objects for this project!

Additional photography, encouragement, and support, moral and otherwise, came from my long-time collaborator Max Whitby, a central figure in all our chemistry and elements-related endeavors. Without Max I would have given up on this kind of thing long ago.

Invaluable research assistance, and innumerable molecular structure files, were provided by Deanna Gribb, who also complained about my questionable electron configuration diagrams until they were at least sort of defensible. Thanks to Barry Isralewitz for 3D renderings of some large molecules, and thanks to David Eisenman for editing the whole manuscript and calling me out on some of my more outlandish ideas. Additional research assistance was provided by Koatie Pasley.

I'd like to thank John Farrell Kuhns of the H. M. S. Beagle science shop for producing the most beautiful modern kid's chemistry set ever. It's people like this who keep the dream of science alive for the next generation, and ours. And finally I'd like to thank Rachel for supplying the snake poop, without which this book would not have been possible.

Additional photography credits

Pg. 25: *The Alchemist*, 1937. Newell Convers Wyeth. Used with kind permission of the Chemical Heritage Foundation Collections, Philadelphia, PA

Pg. 35: Copper roof © 2014 Shutterstock

Pg. 38: Waterfall © 2014 Max Whitby. Used with permission.

Pg. 51: Silver cyanate © 2014 Max Whitby. Used with permission.

Pg. 65: Wood pulp foam © 2005 Jocelyn Saurini. Creative Commons Attribution License.

Pg. 72: Ethane balloon explosion © 2014 Max Whitby. Used with permission.

Pg. 91: Blast furnace © 2012 Jamie Cabreza. Used with permission; Aluminum plant © 2014 Street Crane Co. Ltd. Used with permission.

Pg. 142: Opium poppy © 2012 Pierre-Arnaud Chouvy. Used with permission.

Pg. 160: Sugar beets © 2012 Free photos and Art. Creative Commons Attribution License.

Pg. 185: Marigold extract © 2014 Max Whitby. Used with permission.

Pg. 220: Ozone hole data visualization © 2012 NASA. Used with permission.

2D ball and stick renderings of molecular structures were created by the author using a combination of Wolfram Chemical Data sources and structure files from chemspider.com and other sources.

The purple luminous glow was computed with *Mathematica* using an artificial electrostatic field model. It represents the field strength were there to be a point charge on each atom and a line of charge along each bond line. (This does not represent anything physically meaningful, but it provides a subjective impression of the fuzzy nature of atoms, and it looks pretty.)

Molecules that required manual touchup to their structure were edited with Marvin 6.2.2, 2014, ChemAxon (http://www.chemaxon.com). Thanks to Deanna Gribb for wrangling mol files.

Some molecules are so complex that they can only be visualized effectively as 3D structures. These were rendered with VMD molecular visualization software, ©2014 University of Illinois. Humphrey, W., Dalke, A. and Schulten, K., "VMD - Visual Molecular Dynamics," J. Molec. Graphics, 1996, vol. 14, pp. 33-38.

Index

Numbers in italics indicate photographs or diagrams

Index

Index

Index